Advanced Computational Methods in Mechanical and Materials Engineering

Computational Intelligence Techniques
Series Editor:

The objective of this series is to provide researchers a platform to present state of the art innovations, research, and design and implement methodological and algorithmic solutions to data processing problems, designing, and analyzing evolving trends in health informatics and computer-aided diagnosis. This series provides support and aid to researchers involved in designing decision support systems that will permit societal acceptance of ambient intelligence. The overall goal of this series is to present the latest snapshot of ongoing research as well as to shed further light on future directions in this space. The series presents novel technical studies as well as position and vision papers comprising hypothetical/speculative scenarios. The book series seeks to compile all aspects of computational intelligence techniques from fundamental principles to current advanced concepts. For this series, we invite researchers, academicians, and professionals to contribute, expressing their ideas and research in the application of intelligent techniques to the field of engineering in handbook, reference, or monograph volumes.

Computational Intelligence Techniques and Their Applications to Software Engineering Problems
Ankita Bansal, Abha Jain, Sarika Jain, Vishal Jain, Ankur Choudhary

Smart Computational Intelligence in Biomedical and Health Informatics
Amit Kumar Manocha, Mandeep Singh, Shruti Jain, Vishal Jain

Advanced Computational Methods in Mechanical and Materials Engineering
Ashwani Kumar, Yatika Gori, Nitesh Dutt, Yogesh Kumar Singla, and Ambrish Maurya

For more information about this series, please visit: https://www.routledge.com/Computational-Intelligence-Techniques/book-series/CIT

Advanced Computational Methods in Mechanical and Materials Engineering

Edited by

Ashwani Kumar

Yatika Gori

Nitesh Dutt

Yogesh Kumar Singla

Ambrish Maurya

CRC Press
Taylor & Francis Group
Boca Raton London New York

CRC Press is an imprint of the
Taylor & Francis Group, an **Informa** business

First edition published 2022
by CRC Press
6000 Broken Sound Parkway NW, Suite 300, Boca Raton, FL 33487-2742

and by CRC Press
2 Park Square, Milton Park, Abingdon, Oxon, OX14 4RN

© 2022 Taylor & Francis Group, LLC

CRC Press is an imprint of Taylor & Francis Group, LLC

ISBN: 978-1-032-05291-5 (hbk)
ISBN: 978-1-032-06421-5 (pbk)
ISBN: 978-1-003-20223-3 (ebk)

DOI: 10.1201/9781003202233

Typeset in Times
by SPi Technologies India Pvt Ltd (Straive)

Dedication

This book is dedicated to all budding researchers...

Contents

SECTION A Manufacturing Engineering

SECTION B Mechanical Design Engineering

SECTION C   Materials Engineering

Introduction

The modern era of product design and analysis, complex geometry analysis, and nonlinear analysis has shifted from traditional solving methods to advanced computational techniques. This provides us with the opportunity to fill this gap by introducing *Advanced Computational Methods in Mechanical and Materials Engineering*, which provides in-depth knowledge about the field's methods, simulation, and experiments.

Advanced Computational Methods in Mechanical and Materials Engineering provides easier, fast, efficient, and reliable computational methods for solving engineering, geometrical, mathematical, scientific, and statistical problems with the help of computers to all researchers. These techniques provide additional insights into various engineering problems. The book will bring to light the application of computational techniques in mechanical engineering fields with 15 chapters (*manufacturing engineering (Section A: 6 chapters), mechanical design engineering (Section B: 5 chapters), and materials engineering (Section C: 4 chapters)*). Each chapter deals with separate engineering problems and its solving procedures are self-explanatory. This book's teaching goes beyond being a purely theoretical text by providing real examples of numerical methods in real engineering practices, charts, and techniques to aid learning. The authors have covered numerous topics in an informative and engaging way, and provide research objectives for future research work.

Moreover, the applicability of the computational techniques presented in the book covers the automotive industry, aerospace industry, manufacturing engineering, nuclear and mechanical engineering.

Preface

Advanced Computational Methods in Mechanical and Materials Engineering is a pivotal research work of mechanical engineering. The book has 15 chapters divided in three sections (**Section A: Manufacturing Engineering, 6 chapters; Section B: Mechanical Design Engineering, 5 chapters; and Section C: Materials Engineering, 4 chapters**).

The literature review, problem definition, methodology, experimental setup, results, validation, and future scope provided in each chapter has in-depth analysis of the topic covered in each chapter. The chapters are well organized and easy to follow. The above help to ensure the completeness of the book and to satisfy the needs of the researcher in different areas related to computational techniques.

The chapters presented in book highlights impactful use of different computational methods in engineering fields. **Chapter 1** highlights **industrial revolution 4.0, sustainability** and states that Industry 4.0 Technologies are the keys of future industries' sustainability. The chapter provides the comprehensive analysis of Industry 4.0 technologies and implementation techniques to adopt the sustainable manufacturing. It describes the strategic knowledge about innovative technologies and the methods that used to get sustainability in production. As science and technology progresses, it enhances industries' main focus on developing these technologies with Industry 4.0 in mind, so as to strengthen economy and society on an environmental level. To continue the flow of manufacturing, section A, **Chapter 2** deals with **smart manufacturing**. Manufacturing units are leaning toward the smart manufacturing (SM) systems proving to be the recent and a new global trend to achieve the desired objectives in terms of quality, costs, process, precision, and safety. Cobots are designed to share the workspace with humans, making automation easier than ever for all smaller as well as bigger size industries with its ease of design, easy handling, simple control, low cost, and its associated programmability features, making this a robust collaborative system comparing to the robotic system. In this chapter, some important aspects of the collaborative systems namely, the safety measures, on the basis of ISO standards that reveal what extent these collaborative systems needed to be used, as well as the assembly line requirements of the cobotic system based on cobot programming control or some other means of control like control through vision, voice and gesture command, have been addressed.

Chapter 3 incorporates *machine learning techniques in welding*. Machine Learning is defined as the domain of computer science which overcomes the disadvantages of the conventional computing method lacking behind to provide the automating solutions to complex engineering problems. In this chapter, an attempt is made to highlight the various application of machine learning algorithms in the Friction Stir Welding process ranging from the determination of the defects to the determination of various mechanical and microstructure properties. It has been observed that mainly supervised learning-based machine learning algorithms such as artificial neural networks (ANN), regression models, support vector machines (SVM), and decision trees are widely being used in the friction stir welding process.

Chapter 4 highlights *intelligent modeling in steelmaking*. ANN is increasingly used in the process control of steelmaking. Application area includes flow control, flow optimization, parametric optimization etc. **Chapter 5** deals with the *continuous casting* process. Continuous casting increased productivity, higher yield, and reduced costs, directly affecting any steelmaking operation's bottom line. Thus, continuous casting is a dominant process for the production of finished and semi-finished steel. Mathematical model for fluid flow, solidification, and inclusion removal with electromagnetic forces is described in this chapter. The main study objective is inclusion behavior analysis. **Chapter 6** deals with inclusion motion in the *flash welding* process. Due to its high efficiency and good welding accuracy, flash butt welding (FBW) is usually used in the automotive industry, pipeline construction, sheet welding, hot rolled coils, rail welding, ship structure welding, and ship mooring equipment chains. In this chapter author studied the oxide inclusion distribution in FBW joints under various welding parameters to ensure good efficiency.

The section on mechanical engineering design (section B) starts from Chapter 7. **Chapter 7** highlights different methods for *roundness evaluation*. A simple yet efficient hybrid approach, discussed in this chapter, combines the LS and probabilistic global search Lausanne (PGSL) methods to accurately evaluate the circularity error. This approach possesses both global and local search features. The proposed approach has been tested on different sizes of benchmark datasets and it was found to give accurate results, which demonstrates its robustness. The results conform to the MZ criterion specified by ISO. In continuation, **Chapter 8** deals with the magnetorheological fluid assisted finishing (MFAF) process, which is one of the advanced surface finishing techniques. Authors have studied different process parameters of MFAF for better finishing.

Chapter 9 is part of *nuclear engineering* where author main objective of the study is to model the effect of a fuel bundle or fuel simulator on the deflection of PT. In essence, they will explore the applicability of numerical techniques with a 3D model of the experimental channel and predict the deflection of the channel. **Chapter 10** highlights new computational method for calculating the *crack driving force* in presence of an Interface at various orientations. With recent application of the concept of configurational forces, a true crack driving force in terms of J-integral can be calculated which is independent of the materials' constitutive relations and type of loading. The concept can be used therefore, for growing cracks, cracks in cyclic loading, creep loading as well as viscoelastic and viscoplastic conditions. In **Chapter 11**, authors have used the *FEM technique* for thermal contact conductance calculation. In this work two method of rough surface generation has been adopted and a contact has been created between two models of rough surface. The analysis has been performed on the FE model with varying loading condition of different surface roughness and different materials to get the real contact area and thus thermal contact conductance (TCC). Effect of varying parameters on real contact area has been obtained. Increase in Roughness (10–40µm) results in increase in real contact area for Al and MS.

Section C materials engineering starts with **Chapter 12**. In this chapter, different types of *viscoelastic composites* are addressed for the damping materials in the UCLD/CLD treatment of structural vibration. Among these different viscoelastic composites, the damping capabilities of the most advanced ones, namely 0–3 VEC and virtual evolved packet core (VEPC), are focused by evaluating their relative damping performance in the CLD treatment for passive control of plate vibration. **Chapter 13** highlights the ***thermal behaviors of the CNT strengthen composite laminated plate***. It is evaluated liable to uniform temperature fluctuation. The mathematical equation pattern of selected physical issue is done by using higher order shear deformation theory and von- Kármán's nonlinearity. The Chebyshev polynomial applied for the spatial discretization of the field variables is found to be efficient in resolving the boundary value problem. **Chapter 14** includes the Mori-Tanaka and FEM scheme's mathematical background for the elastic and thermo-elastic analysis of polymer-CNT composites. Aligned and random non-coated and coated CNTs composites have been analyzed for the practical evaluation of the elastic and thermal properties using the schemes. The effect of volume fraction, aspect ratio, orientation, and coating are the study's main variables. In **Chapter 15** the presented study is aimed to develop a magnetic abrasive finish technique to remove the scratch and burrs from the internal surface of the rare Earth-based Al-6061 composite housing by optimizing process parameters using Box–Behnken design approach. This approach aspires to produce parts within very close tolerance limits and ultimately helps in reducing the cost.

The variety of models presented in the book is up-to-date and potentiality valuable to researchers in mechanical engineering disciplines and covers a wide range of industries.

Dr. Ashwani Kumar
Yatika Gori
Dr. Nitesh Dutt
Dr. Yogesh Kumar Singla
Dr. Ambrish Maurya

Editors

Dr. Ashwani Kumar received his Ph.D. (Mechanical Engineering) in the area of Mechanical Vibration and Design. He is currently working as Senior Lecturer, Mechanical Engineering (Gazetted Officer Class II) at Technical Education Department, Uttar Pradesh (Government of Uttar Pradesh) India. He has more than 11 years of research and academic experience in mechanical engineering. He is Series Editor of book series "Advances in Manufacturing, Design and Computational Intelligence Techniques" published by CRC Press (Taylor & Francis) USA. He is Associate Editor for *International Journal of Mathematical, Engineering and Management Sciences (IJMEMS)*, indexed in ESCI/Scopus and DOAJ. He is editorial board member of four international journals and acts as review board member of 20 prestigious (Indexed in SCI/SCIE/Scopus) international journals with high impact factor. In addition, he has published 85 research articles in journals, book chapters, and conferences. He has authored/co-authored and edited 13 books on mechanical and materials engineering. He is associated with International Conferences as Invited Speaker/ Advisory Board/ Review Board member. He has been awarded as *Best Teacher* for excellence in academic and research. He has successfully guided four M.E./M.Tech theses and currently guiding one Ph.D. and 1 M.Tech thesis. In administration, he is working as coordinator for AICTE, E.O.A., Nodal officer for PMKVY-TI Scheme (Government of India) and internal coordinator for the CDTP scheme (Government of Uttar Pradesh). He is currently involved in the research areas of Machine Learning, Advanced Materials, Machining, and Manufacturing Techniques, Biodegradable Composites, Heavy Vehicle Dynamics and Coriolis Mass Flow Sensor.

Yatika Gori is pursuing her Ph.D. in Tribology and working as Assistant Professor in Department of Mechanical Engineering Graphic Era University Dehradun Uttarakhand India. She has more than 8 years of research and academic experience. She is working in area of FEA, Crack Propagation, Welding and Tribology. She has published 16 research articles in International Journals. Her current article in Tribology International has 4.127 Impact Factor. She is a core team member of NAAC and NBA accreditation work in Mechanical Engineering Department. She is an editor of two international books on mechanical and materials engineering published by CRC press, Taylor & Francis.

Dr. Nitesh Dutt is working as Assistant Professor in Department of Mechanical Engineering, College of Engineering Roorkee, Uttarakhand India. He has over seven years of teaching and research experience. He has completed his bachelor's, master's and Ph.D. from IIT Roorkee. He has published 11 research article of high impact factor in SCI journals. His main area of research is nuclear engineering and design, heat and mass transfer, thermodynamics, fluid mechanics, refrigeration and air conditioning thermodynamics, computational fluid dynamics (CFD).

Dr. Yogesh Kumar Singla is associated with Case Western Reserve University, USA, and did his Ph.D. at IIT Roorkee. He has more than 6 years of research and teaching experience. He is working in the area of manufacturing, welding, surface engineering, tribology, materials characterization, welding metallurgy and mechanical behavior of metals. He has 10 SCI publications and 5 Scopus Indexed publications. Apart from this, 4 SCI papers are in process. In his career, he has successfully guided one Ph.D. and 4 M.E. theses. He is a reviewer of many SCI journals of Elsevier and Springer having impact factor ranging from 0.7-5.289. He is about to submit one patent on solar dryer. He has given many expert talks at State and Central Level Institutes/Universities. He also has the experience of conducting a 10-day Faculty Development Program. In addition to this, he is an Editor of two international books entitled *Advanced Computational Methods in Mechanical and Materials Engineering* and *Advanced Materials for Bio-Mechanical Applications* to be published by CRC Press (Taylor & Francis Group https://taylorandfrancis.com/books/) in 2021.

Dr. Ambrish Maurya is working as an Assistant Professor in the Department of Mechanical Engineering at National Institute of Technology Patna, Bihar, India. He has a teaching experience of more than 5 years and during his academic career, he has guided several Master's Thesis and B. Tech. projects and presently he is also guiding five Ph.D. scholars. Dr. Maurya has completed his Master of Technology in Production and Industrial Engineering from the Indian Institute of Technology Roorkee in 2010. He carried out his Ph.D. research at the Indian Institute of Technology Roorkee in the area of physical and mathematical modeling of continuous casting process of steelmaking, during 2012–2017. He also has research expertise in the area of processing and fabrication of metal matrix composite, polymer composites, advanced casting technologies, computational fluid dynamics, and solar energy. Dr. Maurya has published more than 20 research articles and book chapters in international journals and conferences. He has also reviewed for several journals such as *Journal of Alloys and Compounds*, *Journal of Material Processing*, among others.

Acknowledgment

We express our heartfelt gratitude to **CRC Press (Taylor & Francis Group)** and the editorial team for their guidance and support during completion of this book. We are sincerely grateful to reviewers for their suggestions and illuminating views for each book chapter presented here in *Advanced Computational Methods in Mechanical and Materials Engineering*.

Contributors

Gaurav Arora
Composite Design and Manufacturing
  Research Group
School of Engineering, Indian Institute
  of Technology Mandi
Mandi, Himachal Pradesh, India

Masood Ashraf
Department of Mechanical Engineering
Prince Sattam Bin Abdulaziz University
Al-Kharj, Saudi Arabia

Fisal Asiri
Department of Mathematics
Taibah University
Madinah, Saudi Arabia

Manas Das
Department of Mechanical Engineering
Indian Institute of Technology Guwahati
Assam, India

Nitesh Dutt
College of Engineering Roorkee,
Roorkee
Uttarakhand, India

Fehim Findik
Metallurgy and Materials Engineering
  Department
Sakarya Applied Sciences University
Sakarya, Turkey

Yatika Gori
Department of Mechanical Engineering
Graphic Era University
Dehradun Uttarakhand, India

Janardhan Gorti
National Institute of Technology
Rourkela, India

Abhay Gupta
Department of Mechanical Engineering
Indian Institute of Technology
  Guwahati
Guwahati, India

Sajan Kapil
Department of Mechanical Engineering
Indian Institute of Technology Guwahati
Assam, India

Vipul Kumar Gupta
Department of Mechanical & Industrial
  Engineering
Indian Institute of Technology, Roorkee
Uttarakhand, India

Pramod Kumar Jain
Department of Mechanical & Industrial
  Engineering
Indian Institute of Technology,
  Roorkee
Uttarakhand, India

Pradeep Kumar Jha
Department of Mechanical and
  Industrial Engineering
Indian Institute of Technology Roorkee
Uttarakhand, India

Ravinder Singh Joshi
Department of Mechanical
  Engineering
Thapar Institute of Engineering and
  Technology Patiala
India

Kaushik Kumar
Department of Mechanical Engineering
Birla Institute of Technology,
Mesra, Ranchi, India

Ambrish Maurya
Department of Mechanical Engineering
National Institute of Technology Patna
Bihar India

Deepak Mehra
Jodhpur Institute of Engineering &
    Technology
Rajasthan, India

Akshansh Mishra
Department of Mechanical Engineering
Politecnico Di Milano
Italy

Himanshu Pathak
Composite Design and Manufacturing
    Research Group
School of Engineering, Indian Institute
    of Technology Mandi
Mandi, Himachal Pradesh, India

J. Paulo Davim
Department of Mechanical Engineering
University of Aveiro, Portugal

Ashwani Kumar
Technical Education Department Uttar
    Pradesh
Kanpur India

Rajneesh Kumar
Department of Mechanical and
    Industrial Engineering
Indian Institute of Technology Roorkee
Uttarakhand, India

Ranjan Kumar
Department of Mechanical Engineering
Birla Institute of Technology,
Mesra, Ranchi, India

Vinod Kumar
Department of Mechanical Engineering
Thapar Institute of Engineering and
    Technology Patiala
India

Satyajit Panda
Department of Mechanical
    Engineering,
Indian Institute of Technology Guwahati
Guwahati, India

Atul Singh Rajput
Department of Mechanical
    Engineering
Indian Institute of Technology Guwahati
Assam, India

Sachin Rana
Department of Mechanical Engineering
ABES Institute of Technology
    Ghaziabad
U.P. India

Rajidi Shashidhar Reddy
Department of Mechanical
    Engineering
Indian Institute of Technology
    Guwahati, Guwahati, India

Pradeep K. Sahoo
Botswana International University of
    Science and Technology (BIUST)
Botswana

Saumya Shah
Indian Institute of Technology (BHU)
Varanasi, India
Department of Mechanical Engineering
Meerut Institute of Engineering and
    Technology
Meerut, India

Mayur Sharma
Daikin Air-Conditioning India Private
    Limited,
Gurugram Haryana, India

Neelesh Kumar Sharma
Department of Mechanical
    Engineering
Indian Institute of Technology Patna
Bihar India

Vipin Kumar Sharma
Department of Mechanical Engineering
Meerut Institute of Engineering &
   Technology
India

K.K. Shukla
Department of Applied Mechanics
MNNIT
Allahabad, Prayagraj, India

Md Irfanul Haque Siddiqui
Department of Mechanical Engineering
King Saud University
Riyadh, Saudi Arabia

Ankit R. Singh
Indian Institute of Technology Bombay
Mumbai, India

Yogesh Kumar Singla
Department of Materials Science &
   Engineering
Case Western Reserve University, USA

D. S. Srinivasu
Indian Institute of Technology Madras
Chennai, India

Virendra Suthar
Jodhpur Institute of Engineering &
   Technology
Rajasthan, India

Shivraman Thapliyal
Mechanical Engineering Department
National Institute of Technology
Warangal, India

Abhishek Tiwari
Department of Metallurgical and
   Materials Engineering
Indian Institute of Technology Ropar
Rupnagar, Punjab, India

N. Venkaiah
Indian Institute of Technology
   Tirupati
Tirupati, India

Section A

Manufacturing Engineering

1 Integration of Technologies to Foster Sustainable Manufacturing

Virendra Suthar and Deepak Mehra
Jodhpur Institute of Engineering and Technology Rajasthan, Rajasthan, India

J. Paulo Davim
University of Aveiro, Portugal

CONTENTS

1.1 INTRODUCTION

The coming of the Industrial Revolution 4.0 has created the demand for all industries to adopt sustainability. The I4.0 describes a collective form of smart technologies within intelligent manufacturing systems. It improves the product life cycle, minimizes negative impact on the environment, reduces the generation of waste, etc. Basically, the new industrial revolution presents the model to gain high sustainable products. It implies the structured network of all technologies to gain a high quality of manufacture. Product quality is improved with the virtual network of machines like cyber-physical systems (CPS) and Internet of Things (IoT) [1]. By using this

DOI: 10.1201/9781003202233-2

organized structure, any growing industry may now raise its profit margin with the advantage of sustainability. Produce innovation is highly recommended to achieve the smart product criteria. Efficient and intelligent manufacturing concepts require feasible knowledge of the new innovative technologies. It offers new opportunities in terms of energy, waste material removal, etc., for the manufacturing industry. The value creation modules contribute to all dimensions of society, environment, and economy. The sustainability approach considers the product's proper design and remanufacturing to attain recommended efficiency. The use of manufacturing resources and efficient use of assets are demanding criteria in an industrial sustainability approach. From the perspective of new technologies, a machine's life cycle requires proper functioning and structure to get smart products. Various sensor-based devices are achieving a new level of simulation techniques, which is highly beneficial from an industry perspective [2].

All the virtual data are properly secured and linked, achieving a high level of usability for industry. Machine-to-machine communication has become the best result-oriented solution for achieving effectiveness in manufacturing. The sustainable modules define how a product is improved with creating the new life cycle and made the smart industries. The main focus is on establishing the sustainable manufacturing field so that current and future generations can access innovative industrialist products with alternative technologies. Thus, the new Industrial Revolution 4.0 has now opened the gates of global sustainability. Many countries are already achieving sustainability with high-grade systems, which is an important consideration for society, the environment, and the economy. A better understanding, skills, and knowledge of Industry 4.0 key technologies will open up opportunities for sustainable manufacturing in industries.

1.2 THE KEY TECHNOLOGIES OF INDUSTRIES 4.0

Many digital manufacturing technologies come under the umbrella of Industry 4.0 production scenarios, which significantly incorporates all the best possible solutions to establish the vision and mission of Industry 4.0. The advanced revolution totally changed the design techniques of products, which is why we are moving to an era of smart manufacturing. These technologies are constitutes of optimization techniques, which lead to the efficient output of an industry.

1. The Industrial IoT (IIOT) – The industrialized world is now interconnected via networks of computers. Human–human communication has since changed to human–computer with IIOT. It has increased with the new advances in mobile devices. It is defined as a unique tool to connect things and people by incorporation with each other in a real environment [3]. Through this virtual technology, the systems are digitalized. The organization and supply chain management are integrated with the virtual decision-making elements, which establish effective analytical real-time product services. The role of this technology increases due to advancements in radio frequency identification (RFID) technology, various wireless sensors, and cloud computing technologies. The connections of industrial products with

machines create a virtual pattern. Thus, with digitalized information, it is the connection of industrial products.

2. Cloud Computing – This virtual technology is the alternate platform for IT resources. The criterion of this prominent tool is to reduce the costs, direct and indirect, of IT sectors. The main aim of this technology is to collect and process the data in an integrated manner so that the whole data collections are available for organization design [4]. The solutions are analyzed with high processing software, which improves data storage processing requirements.

3. Big Data – The massive amount of data generated from different objects are combined in one unit, which is termed Big Data. The technology is an important consideration regarding the time and cost of any manufacturing industry [5]. Systematic data collection is a helpful operation for decision-making and to solve the complicated production operation. Thus, this technology is characterized as a value-added opportunity in the product life cycle.

4. Simulation – It helps to analyze the operations and decision-making product cycle to understand the most effective manufacturing process for complex systems by using mathematical models. This technology enhances the capabilities of solving real-life problems in an experimental manner. It also allows us to find insights into the complex operational problems, thus helping to gather the best possible realistic information of the system. It helps to gain an effective approach to products.

5. Augmented Reality – The novel technology increases human–machine performance by artificial information to simulate assistance and connect the environment with real-world objects. The necessary part of this technology is to connect electronic devices with the virtual world by technical components. It allows the visualization of the product model in the real world. The planning and evaluation techniques are comprehended with the virtual information to develop the specific model in the industry. Thus, it digitally combines the real and virtual products to make 3D and real-time environments [6].

6. Additive Manufacturing – This technology is defined as creating 3D objects with the deposition of material layers to under layers in a computer-structured manner. The complexity of the product is effectively translated into physical design parameters to obtain a valuable product life cycle. It creates virtual prototypes for simulation and analysis. It is mainly used for innovative product prototyping in production. This technology involves layer–digital manufacturing with a computer-integrated system which makes the process highly feasible with the environmental characteristics. Thus it is an effective tool for making prototypes and gaining the best possible time reduction in the manufacturing industries [7].

7. Autonomous Robots – The technology's manufacturing efficiency is highly increased because of its high durability and the flexibility of robots in production. The communication and control between robots are constantly providing the desired products and services in now–a–days. The robots are programmed with artificial intelligence methods to create solutions with

little human interaction. The position of robots with components is integrated with a computer, so that quality and productivity of industry are to be grown. Material handling, picking, and the capability to do repetitive tasks more efficiently is the one specialization of this robotics technology [8].

8. Cyber Security – This is the information security tool of the manufacturing process for protecting and detecting data in the most authentic manner. The privacy of virtual data is highly secured with protection software. It develops the optimum product design security techniques, which are highly recommended in the industries. All sensitive information is prevented from unauthorized elements, which is more important for industry from a privacy perspective. The innovative creative data must be secured for the transparency of Industry 4.0 technologies in industry and cybersecurity proves to be an effective solution for this problem [9]. Thus, by this technology, all the valuable information allows the creation of safety in a virtual environment.

9. Artificial Intelligence (AI) – This performs specific tasks in a logical sequence with perception from the environment and combining the ability of machines with humans for risk management within the industries. AI now fully changed the inventory management system, and production efficiency is highly improved with the various functionalities of this technology.

10. Autonomous Vehicles – These vehicles are imported with high intelligent motion adaptability without guiding or control. The response rate to make a specific action in an unexpected situation is much better with the evolution of Industry 4.0. It transformed the supply chain management into smart management with the potential impacts of these vehicles. The shipping costs and risks of an accident are lower after adopting these technologies in production. It also reduces greenhouse gas emissions, which is beneficial for sustainable manufacturing [10]. The loading and unloading of the component with large weights is not feasible for humans, and these tasks are performed with higher accuracy and with little or no human interaction.

11. Big Data Analysis – In any industry, there is the challenge of storing large amounts of data in computer systems. This technology eliminated these storage problems and built a digital system for all the processes to store and secure for supply chain management [11]. The analyses improve marketing, connecting secured digital networks to generate many new opportunities with effective solutions.

12. Machine Learning – The new industrial revolution provides the digital opportunity for machine learning to be used for inside extractions of data processing using computer algorithms. This advanced technology promotes better visualization techniques with machine communication. The processes of data selection, processing, and validation are effectively transmitted in machine language to build whole processes with a machine-to-machine interface.

13. Horizontal and Vertical Integration – The two networks are essential requirements of Industry 4.0 implementation because these connect the production chain logically with higher flexibility and an automated approach. It is the foundation of an interconnected ecosystem for creating effective industrial

networks. All production activities are closely bonded with one another, so that process transparency and distribution are performed without any failure. Any failure in these activities automatically affects the schedule, with delay ensuing in the production period. So the production floors consist of closed value chains to support automation in production processes. In Figure 1.1, the closed loop of horizontal and vertical integration is illustrated. The smart supply chain, smart services, and cyber-physical production systems in smart manufacturing are closely connected with all production activities [12]. The main aim of this technological advancement is to reduce production time to perform activities in a logical sequence, so that the reaction time of industry is faster and industrial growth is improved. The energy consumption of all the processes also improved with this integration. Cost-effective products and the elimination of uneven delays in processes are major advantages of this technology. Therefore, this is the necessary step to adopt the I4.0 automation to promote sustainable manufacturing.

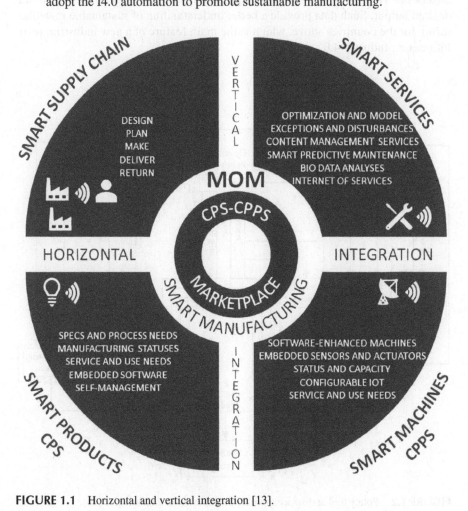

FIGURE 1.1 Horizontal and vertical integration [13].

1.3 POLICY TOOLS AND SMART APPROACH FOR INDUSTRIAL INNOVATION FOR SUSTAINABLE MANUFACTURING

As science and technology advances, countries like China, Germany, and the USA adopted innovative policy frameworks for sustainable manufacturing. Figure 1.2 illustrates the policy tool and smart approach for innovative technologies. The 12 policy tools are enlisted in Government to connect the industrial innovation terminology [14]. This innovative technology not only focuses on developing new ideas but also aims for the commercialization of applications in a systematic way. All 12 tools can be categorized into three divisions of innovation chain: supply mode, demand mode, and environmental mode. In these modes, all the logistic activities play an important role in developing innovative approaches in the industries. These policy analyses will require the revitalization of industries to understand Industry 4.0 automation. The fundamental policy tool covers the vital areas to fulfill the requirement in industry with the maximum desired output. Such data provide a better understanding of sustainable manufacturing for the countries above, which is the main feature of a new industrial revolution, i.e., Industry 4.0.

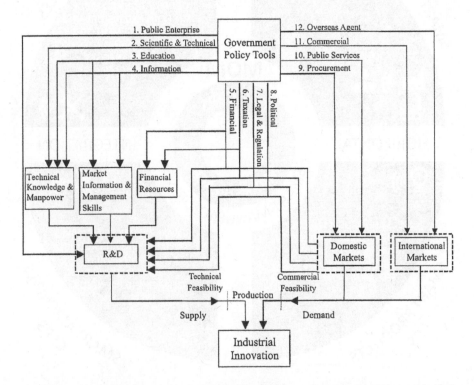

FIGURE 1.2   Policy tool and smart approach for innovative technologies [15].

1.4 CHINA'S INDUSTRY 4.0 APPROACH

Looking at today's industry scenario, Chinese Industries decided to transform their production system in the green transformation approach. The high investment, high material resources, pollution etc. are raised the concern to improve the whole economic development. With the calculations of total energy, GDP, and environmental policies, Chinese industries adopted the new revolution technologies, i.e., Industry 4.0, the green transformation of industries. The main vision and mission are to gain sustainability in the manufacturing process to reduce the negative effect on the environment, material wastage, and improves the product life cycle and total energy consumption [16]. The upgraded innovative industries will promote the best optimization method for all manufacturing processes, which leads to energy-efficient, lightweight, minimum power consumption resources in the product life cycle. China aims to strengthen innovative technologies to accelerate energy-efficient production, which leads to the green development of industrial fields. Figure 1.3 illustrates the green manufacturing development model of China. China builds the green manufacturing model to support the environmental safety industries as well as the way to establish the key technologies. It is essential to promote green products concerning advanced green manufacturing technologies, which help improve the country's whole structure.

The new engine of China's industrial field, i.e., green manufacturing development, will change its whole economic policy for the better over the next 10–20 years. Since Germany also established the green manufacturing industries in the interests of long-term efficient growth, it also focuses on the idea that traditional manufacturing methods can be updated to new revolution technologies in line with sustainable manufacturing. This can be an optimistic reference for China to develop, alongside Germany's strategies.

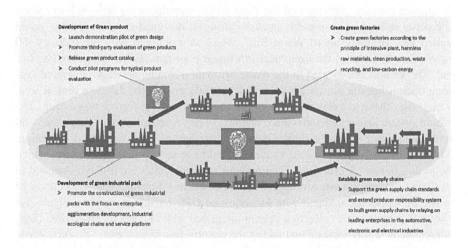

FIGURE 1.3 China green manufacturing development models [17].

1.5 GERMANY'S INDUSTRY 4.0 APPROACH

Germany is now involved in this consideration of sustainability requirements. While dealing with increasing economic demand, it revolutionized its industrial field. Thus, Germany established the biological transformation of manufacturing industries regarding Industry 4.0 [18]. Biological transformation in industries means to develop the scientific principles processes of a biological field with a technical component in industries to achieve sustainable manufacturing. Due to the huge opportunities generated by scientific knowledge, the systematically integrated concept of biological principles leads to value creation models in industry. The strategic initiative of the country is sustainable manufacturing, so they adopt the communication between biological and technical components with the Industry 4.0 technologies. The key technologies of these initiatives are the following:

1. Biosensors, Pneuro – robotics with digital twin
2. Biological Additive Manufacturing
3. Bio-based Energy Carriers
4. Bioleaching and Photobioreactors

Germany's main pillars of industries are innovative technologies to maintain a stable economy. Knowledge of 3D printing, AI, machine learning, fiber composite technology, etc., significantly helps reduce CO_2 emissions, reduces the wastage of resources, improves compatibility and standardization, increasing product quality, generating value creation models, increasing energy-efficient methods, etc. The latest flexible production technologies are making the Standards of Germany powerful and durable. The novel biological value creation technologies have become a solid foundation in the country, with a sound sustainable manufacturing rating.

1.6 SINGAPORE: CASE STUDY

Singapore country implemented the Zero Waste manufacturing approach to reduce the waste to support sustainable manufacturing. It designed the Zero Waste system called the "Bigbelly Smart Waste Bin", which is monitored with the Industry 4.0 revolution. It includes the Graphical information system (GIS), sensors, CPS, IoT etc. The sensors are located in the waste bin with a remote network-integrated system. It identifies the storage bin rubbish level and transfers the data to a central area, which uses these to know when to empty the bin or enable further processing. The whole system is updated with GPS technology [19]. A sensor also identifies the solid and liquid waste so that it further helps to differentiate them for further recycling. The bins use solar photovoltaic electricity and have an internal wi-fi system, which helps as a hotspot for the area. Data systems and algorithms are highly integrated to optimize the daily waste storage to send the waste efficiently. The collections of waste by trucks are updated with location and GPS navigation system. Figure 1.4 shows the smart waste collection system integrated with I4.0 key technologies. Thus, Singapore is highly focused on making the country free from waste, i.e., Zero Waste ecosystems, with the implementation of automation in the industrial fields.

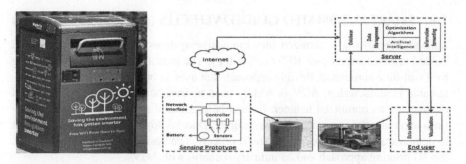

FIGURE 1.4 Big belly smart waste bin collection system [20].

1.7 GERMANY'S CASE STUDY

As the requirement for sustainable manufacturing increases, Germany has decided to make I4.0 technologies a key strategy. The case study we discuss focuses on a wheel hub engine for a solar car [21]. Basically, it is designed to analyze lightweight components with the help of the Life Cycle Assessment tool. This tool examines the environmental awareness policy and impacts of products for the whole life cycle. Thus, it generates opportunities for industries to manufacture the product with the minimum impact on the environment. The study shows how we eco-design the vehicle components for a lightweight strategy in the automobile sector. In this, the two types of engine – C type and D type – are analyzed. Figure 1.5 illustrates the C type and D type engines for the solar car. Firstly, the weight and material of both engines are decided, and then modeling and simulation are done on software. The main aim of this study is to get a key strategy for which types of design have the best effect on the environment, rather than what is the best type of engine. Thus, the study shows that D type has a lower environmental effect than C type engine in 5 to 6 categories.

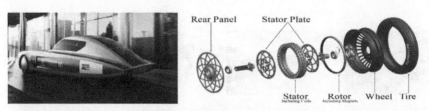

Component Name	D-Type		C-Type	
	Weight [kg]	Material	Weight [kg]	Material
Wheel	5.3	S420/DP600	1.6	AlCu4PbMgMn
Tire	1.6	Rubber	1.6	Rubber
Rotor	0.5	280 30 AP	0.8	280 30 AP
Magnets	0.4	NdFeB	0.5	NdFeB
Stator	1.9	280 30 AP	2.3	280 30 AP
Coils	2.4	Copper	1.9	Copper
Stator Plate	0.6	AW 7075	0.5	AW 7075
Wheel Bearing	1.6	AW 7075/100Cr6	0.7	AlCu4PbMgMn
Rear Panel	1	AW 7075	4.8	AW 7075
Sealing Wall	0.2	GFRP	-	-
Electronics	4	Copper, Plastics	3.7	Copper, Plastics
Various	0.3	Various	0.2	Various
Total	19.8		18.6	

FIGURE 1.5 C types and D type engine for solar car [22].

1.8 SMART AUTOMATED GUIDED VEHICLES (AGV) – INDUSTRY 4.0

Industry scenario now changed into innovative systems by the use of automated guided vehicles which uses RFID technology. The industries become more flexible by a real-time automated, flexible approach that uses an optimization method in sustainable manufacturing. AGV is widely used as transport materials and industrial elements in an optimized manner. It must be capable of sustaining different operations to require the smart decision-making RFID technology. This technology is used to demonstrate motion control for smart objects. The design of the AGV setup uses the 3D printing approach and an industry scenario with AGVs.

Figure 1.6 illustrates real-time automated guided vehicles, which are demonstrated as a virtual industrial scenery model. The location and movement of these vehicles are virtually tested with the help of GPS and cyber-physical systems. Actual production will be optimized with higher success rates, and failure modes will also be resolved to achieve efficiency.

1.9 INDONESIA'S CASE STUDY

With the aim of becoming a sustainable country, Indonesia designed the comprehensive sustainable smart waste management system [24]. It uses the information and communication technology, IoT, and the automated subsystems of Industry 4.0. The integration of these technologies not only removes landfill waste but also generates many opportunities like electricity from the waste, which is more reliable for attaining sustainable development goals. The technology first collects and receives the waste data, which are transferred securely via a network of computers. After that, a low-cost waste management system is adopted to recover the problem waste, which contributes to the economy. A high level of skills is required to effectively operate a digitized waste management system [25]. Figure 1.7 shows Indonesia's sustainable waste management model, which is integral to governance, economy, environment, and society. The whole process of waste collection, waste sorting, technical material

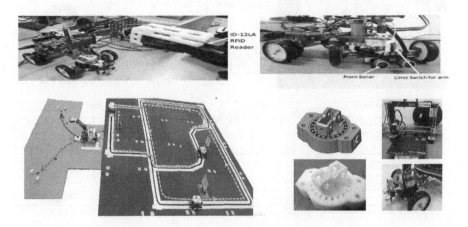

FIGURE 1.6   Smart, automated guided vehicles [23].

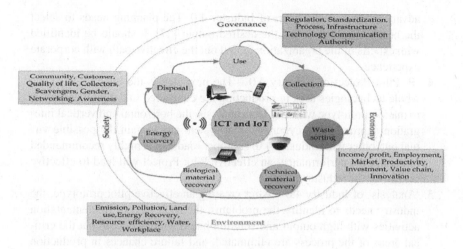

FIGURE 1.7 Sustainable smart waste management system [26].

recovery, biological material recovery, energy recovery, disposal, and use are connected with the digital twin technology. Thus, these are effective efforts that have made a significant contribution to sustainable manufacturing.

1.10 NEW ZEALAND'S CASE STUDY

The Industrial Revolution 4.0 changes the way of industry by opening lots of opportunities by automation. New Zealand also adopted the Industry 4.0 implementation techniques in their industries. After a survey of various industries, the country proposed the work implementation approach of Industry 4.0. The approach is known as the "ARPPAD" Industry 4.0 model.

1. 'A' Awareness of Industry 4.0 – Revolution is also necessary for manufacturing quality work and the potential to demonstrate the processes. Too little knowledge about automation is not a conducive approach to create a digitalized product. Manufacturers should have good skills, knowledge to gain proper techniques to establish automation in industries. Industry agility is increased if the manufacturers have proper capabilities and competence regarding digitalization in today's scenario. Systematic, structured training is essential to enhance the performance of the whole industrial field.

2. 'R' Readiness about Industry 4.0 – A comprehensive study of the company with the current status of industry demand is the next step to determine the critical areas of weakness to evaluate the digital planning. The actual reading areas about industry structure are really important to corporate the organization with automation. Proper scheduling should be established to identify the performance level on behalf of today's market scenario.

3. 'P' Planning of Industry 4.0 – After defining the structure of an organization, it is essential to develop a proper strategy for technologies. Technological

advancement is at the heart of Industry 4.0. The planning needs to select the best technology to get the desired output [27]. It should be identified with experts of the organization to find out the effective path with corporate experience.

4. 'P' Pilot Project of Industry 4.0 – The next step is the integration of the whole technologies uses in production are connected to a digital computer to make a prototype with Big Data analysis, the horizontal and vertical integration approach, and cyber-physical system etc. to gain the possible virtual outcomes with Industry 4.0. A proper roadmap is highly recommended to predict the performance; an effective Pilot Project will lead to effective results for the industry.

5. 'Analysis' of Industry 4.0 – After creating an effective pilot prototype, the industry needs to identify the problems, the right path of implementation activities with high output and solution-based approaches so that the critical areas of the process are eliminated, and failure chances in production reduces, which is focused areas of industry. This is also called a feedback platform of all technological solutions so that the result-oriented analysis is securely maintained [28].

6. "Digitalization" of Industry 4.0 – This is the final and major step of Industry 4.0 because the whole concept of growth and marketing are dependent on this. The market demands a highly accurate and precise product, so industry should take a step-based approach to generate innovative production ideas, to create a quality-based and economical product. Sustainable manufacturing method should be adopted to digitalize the production field.

For better productivity and performance, the model presented by New Zealand country will significantly contribute to future planning to develop an effective Industry 4.0 platform. It will take research and design capability knowledge about some efficient countries like Germany, China, etc., to better construct these technologies. New Zealand is an emerging country in terms of automation, so currently, it aims to find the potential demand of Industry 4.0 with the reference of developed countries in digitalization. As discussed earlier, the smart manufacturing approach finds the way – the ARPPAD model, which concerns the future planning of all industrial sectors in this country. After analysis of this model, the whole production should be compared with the standard and quality of high rating products so that industry shall find out which technologies are beneficial for a particular production or not. That step is also known as feedback of all production systems [29]. This will lead to the discovery of the strengths and weaknesses of whole industries. After all, the production system should be compared with the sustainable manufacturing parameters because the main aim of Industry 4.0 is adopting sustainable manufacturing, which offers another level of production system. Thus, this approach and solution will definitely resolve the traditional manufacturing problems to transform digital manufacturing from the perspective of sustainable manufacturing. Remarkable opportunities will emerge in this country to improve economic growth at a production level [30].

1.11 CONCLUSION AND FUTURE SCOPE

The study provides a comprehensive analysis of Industry 4.0 technologies and implementation techniques to adopt the sustainable manufacturing. It describes the strategic knowledge about innovative technologies and the methods used to get sustainability in production. As the science and technology development enhances the industries' main focus on developing these technologies with Industry 4.0 policy, so the industry, as well as the whole country, become strong pillars of economy, society, and environmental welfare. The production scenario has totally changed after adopting automation like CPS and IoT, which connects the whole process, generating the smart product with an updated version of Industry 4.0. The product life cycle and quality are increasing with digital twin technology. In terms of the future, material and resources must be used effectively with the aim of not compromising future generations. As the rating of the ozone layer diminishes, the Industry 4.0 technologies reduce the CO_2 emissions with the aim of environmental security. Cost-effective products with good performance characteristics are in high demand; digitalization has also fulfilled this requirement. The main problem of today's industries is the large amount of waste generation, and sustainable smart waste management systems have solved the problem. As we have seen in Singapore's case study, the country is significantly developing the Zero Waste management system using key automated technologies. Indonesia also stabilized automation in industries to prevent waste generation and waste utilization with the best perspective to attain sustainable manufacturing in the whole society. The revolution in industries should redesign the whole manufacturing process, as we have seen in New Zealand's case study results which contributes to establishing the proper structure to implement this automation with high flexibility and agility. The role of Smart Automated Guided Vehicles is important to create the smart factory scenario, which is a necessary condition of Industry 4.0 adoption. China's green transformation and Germany's biological transformation of industries both aim to create sustainable manufacturing because these will change countries' economic growth. Therefore, Industry 4.0 Technologies are key to future sustainability for industry, which will develop in the world as technological revolutions continue.

REFERENCES

1. Jusco, S., Strop, A., Hollinger, T., Chowan, T., & Ebony, J. (2020). Development of manufacturing execution systems in accordance with Industry 4.0 requirements: A review of standard-and ontology-based methodologies and tools. *Computers in Industry*, *123*, 103300.
2. Mie He, R., Bauernhansl, T., Beckett, M., Bracer, C., Dimmer, A., Dorsal, W. G., ... & Wolperdinger, M. (2020). The biological transformation of industrial manufacturing–Technologies, status and scenarios for a sustainable future of the German manufacturing industry. *Journal of Manufacturing Systems*, *54*, 50–61.
3. da Silva, E. R., Shinohara, A. C., Nielsen, C. P., de Lima, E. P., & Angelis, J. (2020). Operating Digital Manufacturing in Industry 4.0: the role of advanced manufacturing technologies. *Procedia CIRP*, *93*, 174–179.

4. Alcácer, V., & Cruz-Machado, V. (2019). Scanning the industry 4.0: A literature review on technologies for manufacturing systems. *Engineering Science and Technology, An International Journal, 22*(3), 899–919.

5. Carvalho, N., Chain, O., Czarina, E., & Gerolamo, M. (2018). Manufacturing in the fourth industrial revolution: A positive prospect in sustainable manufacturing. *Procedia Manufacturing, 21*, 671–678.

6. Mehami, J., Nawi, M., & Zhong, R. Y. (2018). Smart automated guided vehicles for manufacturing in the context of Industry 4.0. *Procedia Manufacturing, 26*, 1077–1086.

7. Mehra, D., Mahapatra, M. M., & Harsha, S. P. (2018). Processing of RZ5-10wt% TiC in-situ magnesium matrix composite. *Journal of Magnesium and Alloys, 6*(1), 100–105.

8. Li, F., Zhang, T., Sha, Q., Pei, X., Song, Y., & Li, C. (2020). Green Reformation of Chinese Traditional Manufacturing Industry: Approach and Potential for Cooperation. *Procedia Manufacturing, 43*, 285–292.

9. Hopkins, J. L. (2021). An investigation into emerging industry 4.0 technologies as drivers of supply chain innovation in Australia. *Computers in Industry, 125*, 103323.

10. Ebrahimi, M., Baboli, A., & Rother, E. (2019). The evolution of world class manufacturing toward Industry 4.0: A case study in the automotive industry. *Ifac-Papersonline, 52*(10), 188–194.

11. Fatimah, Y. A., Govindan, K., Murniningsih, R., & Setiawan, A. (2020). Industry 4.0 based sustainable circular economy approach for smart waste management system to achieve sustainable development goals: A case study of Indonesia. *Journal of Cleaner Production, 269*, 122263.

12. Mehra, D., Mahapatra, M. M., & Harsha, S. P. (2018). Effect of wear parameters on dry abrasive wear of RZ5-TiC in situ composite. *Industrial Lubrication and Tribology, 70*, 256–263.

13. Sharma, P. K., Sood, A., Setia, R. K., Tur, N. S., Mehra, D., & Singh, H. (2008). Mapping of macronutrients in soils of Amritsar district (Punjab) A GIS approach. *J. Indian Soc. Soil Sci, 56*(1), 34–41.

14. Leng, J., Ruan, G., Jiang, P., Xu, K., Liu, Q., Zhou, X., & Liu, C. (2020). Blockchain-empowered sustainable manufacturing and product lifecycle management in industry 4.0: A survey. *Renewable and Sustainable Energy Reviews, 132*, 110112.

15. Hamzeh, R., Zhong, R., & Xu, X. W. (2018). A survey study on Industry 4.0 for New Zealand manufacturing. *Procedia Manufacturing, 26*, 49–57.

16. Kerdlap, P., Low, J. S. C., & Ramakrishna, S. (2019). Zero waste manufacturing: A framework and review of technology, research, and implementation barriers for enabling a circular economy transition in Singapore. *Resources, Conservation and Recycling, 151*, 104438.

17. Mehra, D., Mahapatra, M. M., & Harsha, S. P. (2018). Optimizations of RZ5-TiC magnesium matrix composite wear parameters using Taguchi approach. *Industrial Lubrication and Tribology, 70*, 907–914 ISSN: 0036–8792.

18. Kuo, C. C., Shyu, J. Z., & Ding, K. (2019). Industrial revitalization via industry 4.0—A comparative policy analysis among China, Germany and the USA. *Global Transitions, 1*, 3–14.

19. Li, L. (2018). China's manufacturing locus in 2025: With a comparison of "Made-in-China 2025" and "Industry 4.0". *Technological Forecasting and Social Change, 135*, 66–74.

20. Nara, E. O. B., da Costa, M. B., Baierle, I. C., Schaefer, J. L., Benitez, G. B., do Santos, L. M. A. L., & Benitez, L. B. (2021). Expected impact of industry 4.0 technologies on sustainable development: A study in the context of Brazil's plastic industry. *Sustainable Production and Consumption, 25*, 102–122.
21. Reinhardt, R., Pautzke, F., Schröter, M., & Wiemers, M. (2017, September). *A case study of sustainable manufacturing strategy: comparative LCA of wheel hub engine for solar car application.* In *2017 International Conference on Research and Education in Mechatronics (REM)* (pp. 1–6). IEEE.
22. Sung, T. K. (2018). Industry 4.0: A Korea perspective. *Technological forecasting and social change, 132*, 40–45.
23. Sahu, S., Mehra, D., & Agarwal, R. D. (2012). Characterization and thermal analysis of hydroxyapatite bioceramic powder synthesized by Sol-Gel technique. *Biomaterials for Spinal Surgery, 3*, 281–289.
24. Li, X., Hui, E. C. M., Lang, W., Zheng, S., & Qin, X. (2020). Transition from factor-driven to innovation-driven urbanization in China: A study of manufacturing industry auçtomation in Dongguan City. *China Economic Review, 59*, 101382.
25. Shahrubudin, N., Koshy, P., Alipal, J., Kadir, M. H. A., & Lee, T. C. (2020). Challenges of 3D printing technology for manufacturing biomedical products: A case study of Malaysian manufacturing firms. *Heliyon, 6*(4), e03734.
26. Schuh, G., Zeller, V., Hicking, J., & Bernardy, A. (2019). Introducing a methodology for smartification of products in manufacturing industry. *Procedia CIRP, 81*, 228–233.
27. Mogos, M. F., Eleftheriadis, R. J., & Myklebust, O. (2019). Enablers and inhibitors of Industry 4.0: Results from a survey of industrial companies in Norway. *Procedia CIRP, 81*, 624–629.
28. Díaz-Chao, Á., Ficapal-Cusí, P., & Torrent-Sellens, J. (2021). Environmental assets, industry 4.0 technologies and firm performance in Spain: A dynamic capabilities path to reward sustainability. *Journal of Cleaner Production, 281*, 125264.
29. Haapala, K. R., Zhao, F., Camelio, J., Sutherland, J. W., Skerlos, S. J., Dornfeld, D. A., ... & Rickli, J. L. (2013). A review of engineering research in sustainable manufacturing. *Journal of Manufacturing Science and Engineering, 135*(4), 10–13.
30. Seliger, G., Kim, H. J., Kernbaum, S., & Zettl, M. (2008). Approaches to sustainable manufacturing. *International Journal of Sustainable Manufacturing, 1*(1–2), 58–77.

2 Intelligence-Assisted Cobots in Smart Manufacturing

Ranjan Kumar and Kaushik Kumar
Birla Institute of Technology Mesra Ranchi, India

CONTENTS

2.1 INTRODUCTION

Cobots stands for the "collaborative robots", first introduced by Colgate [1], which is now very popular and has a wide range of industrial applications instead of robotic systems concerning design ergonomics, productivity, quality, and safety. The autonomous mobile service robots are termed collaborative robots or cobots, are meant to have direct human–machine interactions with sharing payloads. This can be marked as a new separated automatic collaborative system, which is much safer and has been isolated for human safety reasons from the autonomous industrial robots in many ways. Due to the continuous growth in technology, today the manufacturing sectors are realizing global competition, higher complexity, automation, fast product

DOI: 10.1201/9781003202233-3

delivery, as well as the high degree in product variations and demands the intelligent systems to enhance the productivity within the bounded time. In order to guarantee the requested product quality and costs with reduced downtime, the manufacturing units are leaning toward the smart manufacturing (S.M.) systems proving to be the recent and a new global trend to achieve the desired objectives in terms of quality, costs, process, precision, and safety [2]. The product demands are continuously vary-ing according to customers' requirements. Variations in product demands and cus-tomization of products with reduced life cycle are demanding the manufacturers establish a "variant-oriented" production system [3] using the advanced intelligent methodologies and cognitive technologies to enhance the rapid productivity and fast and real-time dynamic responses toward the quality of products comprising the "self-optimization" of production processes.

Cobots are designed to share the workspace with humans, making the automation easier than ever for all smaller as well as bigger size industries with its ease of design, easy handling, simple control, low cost, and associated programmability features, making this a robust collaborative system comparing to the robotic system. This provides flexible automation for manufacturing all-sizes products, processes, and operations from "assembly to painting, palletizing to screw-driving, polishing to packaging", and all other tasks that you can think of. Cobots are meant to interact with humans by producing a software-defined "virtual surface" [1], which provides continuous guidance and constraints toward the motion of shared payloads with very little or no power. The collaborative system of providing the human-computer inter-actions through multiple sensing techniques results in benefits, such as productivity, ergonomics, quality, costs, and safety. Cobots are now established as an alternative emerging approach toward the class of material handling equipment termed "intelli-gent-assist devices (IADs)". In 1995, the collaborations of the Northwestern University, the University of California and the General Motors Corporation took a new initiative toward the development of IADs [5]. The IADs are a kind of hybrid device, designed particularly for assembly-line operators to provide the direct physi-cal interactions between the operator and machine in the assembly-line and reduce the "ergonomic concerns" that arise due to having the live physical and cognitive loading/unloading. It also helps in improving productivity and safety concerns. An example of a typical assembly line can be transformed completely by having the emergence of robots, but the sense of having harsh and unsafe dangerous environ-ments is always associated with these robotic systems [6].

Many surveys have reported accidents during manufacturing. Workers' illness, along with reactions and exertions, illness due to continuous and repeated trauma disorders, and repetitive motion or overexertions make workers' lives difficult and very painful. This impacts the manufacturing sectors' productivity and creates higher costs to companies due to these problems and demands for an IAD. Hence, IADs are considered an alternative approach and promising solution for making workers' lives less stressful by addressing these said problems. In this context, the motor firm Ford, along with Northwestern University and Fanuc Robotics, has put parallel efforts into developing IADs. Deeter et al. [7] have reported the new generations to assist devices for munitions handling developed by Oak Ridge National Laboratory. The capabili-ties of material handling technologies by incorporating the cobotic systems in use have proven its new robust design have it possible to use new alternatives. The design of the IADs is important due to its association with humans, products, and processes. Since, in the last few decades, robotic systems are very useful in manufacturing processes and general assembly (G.A.) areas. From the technological point of view, we conclude that we are still far from the "geometric dimensional variability" and are needed to achieve the reliability levels during mass production. Further, the programming complexities also increase exponentially due to increased customer offerings, resulting in floor space requirements and impacting the overall production costs and processes.

The requirement to meet the desired objectives in this global era of competitiveness by introducing the S.M. system, which includes the core technologies like "cognitive systems, agent technology, cognitive agent, swarm agent etc., with continuous rapid advances in information and communication technologies", is of much importance. The included flexibility and responsive production process with available facilities and wide networks have completely changed the experience of production systems. In this era of industrial globalization, the large volume of available literature has proved the existence and continuous experience of flexibility and changeability taking place as the key elements in the present global market [8–10]. Initially, we were not aware of the use of artificial and intelligent resources in manufacturing or was not much in practice for the mainstream production system. In fact, there existed some limitations in using the robotic system in automation such as: (1) robots were thought of to work as an assistant for the human and generally both maintained a suitable distance for the sake of safety reasons, (2) the robots were given tedious and repetitive kinds of work while the humans were associated with rest of the automated works in production-environment. The agility and problem-solving skills of humans, along with the combination of robotic systems with applying forces, strength with precision, produce a suitable working condition [11] for workers to accomplish their tasks with high efficiency and quality during manufacturing.

Information and communication technologies have made the manufacturing and assembly system much sophisticated in the last few years. Also, smart computing devices such as Bluetooth, wi-fi local area network (LAN), and other sensory equipment have drastically increased production systems' potential and process observability. During the manufacturing processes, the plant workers are supported by these technologies depending upon the "degree of assistance, multiple actuators and sensory devices or data-processing". A typical diagram of influence factors can be seen in Figure 2.1.

Monostori [12] has reported that the application of smart sensory devices has developed such a multi-directional collaborative environment of "cyber-physical production system" (CPPS) by having the interactions between virtual and physical entities where the integration of human and machine can contribute their complementary strengths and potential in a very productive manner. "The potential of Cyber Physical System (CPS) to change every aspect of life is enormous. Concepts such as autonomous cars, robotic surgery, intelligent buildings, smart electric grid, smart manufacturing, and implanted medical devices are just some of the practical examples that have already emerged". Since the use of robotic systems in industrial applications is still very popular but limited due to their exhibition of performing precise and

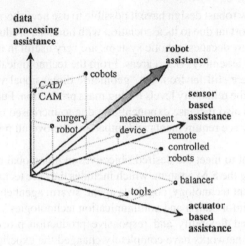

FIGURE 2.1 A typical schematic diagram showing the influence factors [39].

repetitive complex tasks even in unpredictable, harsh, and dangerous environments facing unexpected situations without any failure. "Humans tackle unexpected situations better, are aware of a much larger part of the environment than formally declared and show more dexterity in complex or sensitive tasks. Humans, however, are more prone to error, stress or fatigue [13], and their employment underlies strict health and safety regulations". Technological growth has made it possible to have a workable and collaborative environment by bridging human skills and operational characteristics gaps. Now one can rely on the collaborative systems, i.e., cobots, as their working partner without any fear of facing the "potentially hazardous tools" [14].

This paper is organized as, starting from the introduction about the cobotic systems and their industrial importance along with its utilization and applications in various sectors have well explained in Section 2.1. Further, in Section 2.2, the collaborative systems such IADs are briefly explained with their working and benefits in utilization. In stepping ahead, the human-robot collaboration (HRC) systems, their classification and the deployment framework in assembly lines have been briefly discussed in the subsequent sections in Section 2.3, Section 2.4 and Section 2.5. Further, Section 2.6 deals with the "cobot selection parameters" and the associated safety measures discussed in subsequent Sections 2.7 and 2.8. Finally, this work has been concluded with results, conclusion, and future scopes along with the associated challenges incorporated with the collaborative systems in the last few sections.

2.2 INTELLIGENT-ASSIST DEVICES (IADS)

Collaborative robots, i.e., cobots are IADs was developed by a cobotic company in 2003 based on the fundamental research provided by Colgate and Peskins [1], which are generally used to manipulate the objects by applying direct human–machine collaboration and further the amplified form of this research called "power amplifying assist devices" (IPAD) was done by Fraunhofer IPK [15]. The most important characteristic of this cobotic system is having its integration of human workers with

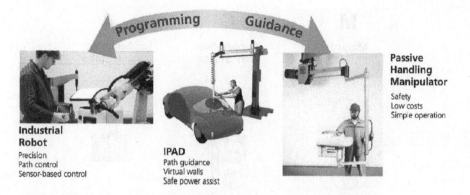

FIGURE 2.2 Four different operative modes of a collaborative system [78].

intelligent devices providing the shared payloads, constraints, and guidance to workers' motion by following the virtual surfaces. The results of implementing the IADs are mostly associated with the concerned ergonomic benefits, enhanced productivity, and improved safety [16, 17]. Kruger [18] described the cobots as a mechanical device that integrates the human–machine to perform the required tasks during manufacturing and assembly lines by directly contacting the human workers in a very friendly and safe environment. The increased productivity demands a higher degree of automation. This relates to productivity in terms of mass production, production flexibility, automation, and a variety of products can be summarized as a result of collaborative systems. The four basic configurations of a collaborative robotic system in four different operating systems can be seen in Figure 2.2.

In this context, Heilala [19] has proposed the relationships between "flexibility, volume, variants and batch-production". The final assembly system must be manual because the assembly process requires variability due to having the multivariant products at the end. Hence, humans are incomparable and can be considered a flexible part of the system that can adapt to the changes in production systems as per requirements by their skills and intelligence [20].

In the last few decades, the concept of "Lean automation" has gained much popularity for enhancing productivity by minimizing wastes and adding some valued activities [21]. Dulchlnos [22] proposed the principle of lean automation. They defined it as: *"the lean automation is a technique which applies the right amount of automation to a given task. It stresses robust, reliable components and minimizes overly complicated solutions"*. So, we can say that the collaborative system is a combination of intelligence, knowledge, flexibility as well as the best use of human skills along with the integration of power, precision, and accuracy. The application of this collaborative system makes the unit automated and capable of producing highly complex products at a very reduced cost with the required level of accuracy [23]. IADs can influence productivity to a higher degree with improvised safety [24] concerns as well as quality products. Let's look at giant industries such as Honda, Toyota, Nissan, GM, etc. We have come to know that they have installed "the multiple production system to handle the instrument panels, engine blocks, struts, transmissions

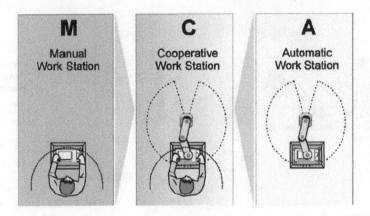

FIGURE 2.3   A typical human–machine interaction in the workplace [39].

and many other parts". IADs are aimed to combine the capabilities of both humans and machines and to utilize these in a super-efficient manner to enhance productivity. A typical human–machine interaction in a workplace can be seen in Figure 2.3.

Prior to the innovative design of cobotic systems, the manipulation of tough jobs was performed either by humans or robots; there was a lack of collaboration, which results in safety issues and sometimes dangerous accidents. In the automobile sector, the assembly of the parts is very complex. It often deals with "poor working ergonomics, high-degree of error generation as well as the product mutilation rate", resulting in poor productivity. IADs are liable to understand these many issues associated with the assembly-line production system. They are designed to provide the best and easy way to enhance productivity by collaborating between humans and machines [25]. The collaborative robots, i.e., cobots [26] are considered to be single or multi-axis devices that are capable of being computer-controlled, providing the benefits of the HRC system for easy adaptation by human workers during the manufacturing process. The main important functionalities associated with IADs include:

- Extension of strength
- Inertia management (i.e., ensures the motions in all directions by providing the control of starting, stopping, or turning forces)
- Virtual path generation for guided and constrained motion
- Integration of auxiliary sensors for some specific purpose such as weighing the parts or tracking the parts in the assembly lines
- Interfacing the plant information system for analyzing errors

The IADs can also be classified as per their axes [24], such as: X and Y axes show the lateral motions, Z shows the vertical motion, roll, pitch and yow. It is possible to have the IADs with single-axis or IADs with multi-axes to work independently or sometimes in a coordinated manner. A typical 3-axes IAD is shown in the following Figure 2.4.

It is evident from the intensive literature survey that many investigations have been provided on the efficient applications of cobots technologies and HRC

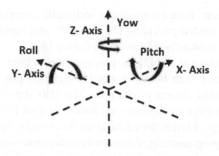

FIGURE 2.4 A schematic 3-axes directions for IADs movement.

applications [27], methodologies on safety for utilizing the HRC system [28] and some more related cobot-specific topics such as LFD [29, 30], augmented reality [31], gesture recognition [32], computer vision [33–35] etc. Baucer et al. [36] have reviewed the cobots' technologies and summarized them in a very lucid way, which enables us to understand the said concepts of HRC. Its working is deeply related to machine learning, action-planning, artificial intelligence, and intention-estimation.

2.3 HUMAN–ROBOT COLLABORATION (HRC) AND ITS CLASSIFICATION

In the last few decades, the HRC has fascinated researchers. The continuous ongoing research has on HRC has paved the way to new dimensions and has opened up multi-dimensional possibilities. The working of intelligent robotic systems in very unpredictable environments has been a long-term subject of interest. The industrial production process has been using intelligent devices for the last few decades. However, it still has some issues related to safety measures and health regulations, limiting this HRC system's practice.

2.3.1 HRC IN ASSEMBLY LINE

Humans habitually use intelligent robotic systems as their assistants for doing repetitive, complex, and dangerous jobs for their convenience. This helps develop automated, fast, safe, and collaborative environments for easy work processes and enhanced productivity [18]. Helms et al. [37] have defined the "robot assistant as a direct interacting, flexible devices that provides sensor based, actuator based and data processing assistance". The efficient interaction between humans and the "artificial mechatronic intelligent assistant" is still very much a challenge for the coming decades. This human–machine interaction is not only possible to have the dialogue but also act as an element of "autonomous or partial autonomous practice of forces by machine on the common workspace or on humans" [38]. The workspace sharing between the robot and machine is important in multiple ways [39], such as holding and handling the object and standing aside, or retrieving the objects whenever required. This makes the human–machine joint performance fairly effective and efficient in material handling and assembly processes [40]. The human–machine or the human–robot interactions have established a well-defined taxonomy, and identified the feasible solutions toward providing the assembly and handling of the bulk and heavy materials with absolute precision and accuracy [41].

In this context, Morioka et al. [42] has proposed a new kind of production assembly system incorporation with human–robot interactions, just to enhance the efficiency in assembly lines during production. The HRC-based "cooperative parts feeding station" and control mechanisms are established on the basis of information support. Further, Krüger [43] developed the framework and proposed the cooperative system along with its programming and control mechanisms which address the HRC capabilities such as "rapid commissioning, compliance control of bimanual interactions in all assembly processes as well as the intuitive planning and programming" for both online and offline programming mode.

2.3.2 CLASSIFICATION OF THE HRC SYSTEM

The classification of the HRC system is important in terms of the classified sets of tasks that are needed to be accomplished. This seems to be very helpful during the planning and design of a process or a system. So, we can say that the tasks must be an integral part of any system's classification for a particular design process [44].

- *Sequential and spatial relationships* of an HRC system are nowadays likely to provide the viable need and responses for automation. Today's the cobots are widely included in a wide variety of applications such as "tele-medicine, tele-operation, assisted vehicle steering" etc., where the tele-operation is more successful but their reach is limited due to its heavier costs [45]. As it is known that the collaborative systems share the same workspace with humans, and their respective activities do not overlap [39]. Also, the collaboration enables the handling of larger workpieces easily and enhances productivity in terms of process, quality, safety, and environment.
- *Collaborative agents* possess a variety of applications that are indeed necessary for industrial need. On the basis of literature survey, one may distinguish the collaborative systems on a single, multiple, and team basis. The group of robotic systems interacts with each other and with environments in a very specified manner and utilizes the resources that provide specific services.
- *Agent relationships* and their collaboration express the robot's (or agent's) action, which is directly determined by human workers. It is important to decide which of the agents (i.e., human or robot) is liable to take the decision and lead. This leads to the "leader–follower relationships" [46].

2.3.3 HUMAN–ROBOT COLLABORATION (HRC) IN ASSEMBLY LINES

We have already discussed the collaborative HRC needs in industrial applications. In assembly lines, the requirement of symbiotic HRC keeps its own importance based on certain characteristics associated with human–robot interactions. A symbiotic HRC system with a closed-loop assembly/packaging architecture can be seen in Figure 2.5.

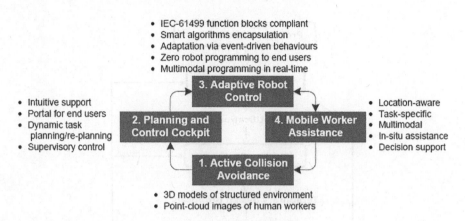

FIGURE 2.5 A symbiotic HRC-based assembly/packaging closed-loop system [10].

2.4 SYMBIOTIC HRC REQUIREMENTS

These collaborative key characteristics play a vital role to distinguish the symbiotic collaborative systems from the conventional HRC as:

- *Instinctive and multivariant programming*: The cobots are equipped with multivariant programming environments. It is easy to perform and does not require any expertise and in-depth knowledge from the workers.
- *No programming skills*: The technological advancements in cobotic systems make workers' lives even simpler because it does not require any coding skills. However, one can program [50] this HRC system easily as per the task requirements. The HRC systems are also equipped with voice command, stereo-visions, as well as gesture-controlled inputs [31, 32, 34], by which the particular system can be efficiently controlled for the particular set of tasks.
- *Intelligent devices collaborations*: Human–machine collaboration is widely engaged with external devices such as "screens, wearable displays and goggles etc.", making life easier in working environments and minimizing errors. In this context, the haptic display [47–49], equipped with cobotic systems, plays an important role in providing safety and stability assurance to the overall systems. In fact, the industrial cobots for commercial purpose comes with a "built-in compliance mode", i.e., the cobot works according to the human's touch and forces exerted on the collaborative system. Also, the researchers are continuously extending their research toward the "enhanced intelligence and friendliness" of the cobotic system [50].

2.5 COBOT DEPLOYMENT FRAMEWORK IN ASSEMBLY LINES

The systematic framework associates the activities in implementing the cobots in the assembly-line processes. The assembly work-cell processes are divided into three phases: "(i) the development phase; (ii) the exploration phase; (iii) the decision phase" [4]. The development phase involves analyzing business needs and the

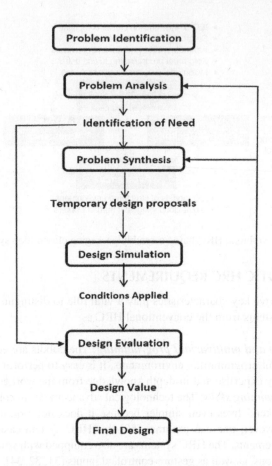

FIGURE 2.6   A schematic diagram showing the engineering design process [51].

corresponding production process for defining the requirements of a "hybrid assembly system". Next, the exploration phase is important to decide the requirements for functional elements such as cobots and workers. Finally, the decision phase involves concluding the design processes and developing the "virtual prototype to achieve the final design". The schematic diagram of an engineering design cycle proposed by Roozenburg [51] can be seen in Figure 2.6.

The search for business potential gains for any business is an important and primary step toward establishing a "new production strategy". This comprises the investments, performance, planning, and final design strategy to achieve the target with the help of "flexible hybrid automation" [52].

2.5.1   HYBRID PRODUCTION REQUIREMENTS

The development of a collaborative environment by implementing the HRC system refers to the "analysis of the assembly process", which involves the analysis of both "product design as well as the assembly sequence for which the cobots are liable to perform automation". The most important and basic objective of any production system is to achieve the required product at the required time and required costs with targeted quality and flexibility, as stated by Chryssolouris [53]. This hybrid production involves the product handling, process and production requirements as stated below:

- **Product handling**: This stage involves the product analysis done by cobots. This full product or payload analysis involves analyzing their "shape, size, material, dimensions, tolerances, weight and surface finish etc., and after this the product is handled carefully by cobots during the assembly process".
- **Process requirements**: The process requirements during the assembly involve the investigation of some important factors, such as the assembly process sequence, material alignments, rotation angle, methods of fastening etc. [54].
- **Production requirements**: This involves the "required production volume, cycle-time, variety in product(s) and the associated complexities of the process".

2.5.2 Development Phase Analysis

The above-mentioned hybrid production requirements described in terms of product, process, and productions are important to investigate. This helps describe the complexities involved in the assembly process for industrial automation based on the collected data from the above-mentioned requirements. Many researchers are involved in defining the complexities involved in manufacturing, as well as trying to quantify them.

ElMaraghy [55] has done commendable work in defining and quantifying the involved "manufacturing operational complexities, product's assembly complexities [56] and assembly system complexities [57]". However, this is not enough to address the problems involved. Still, the requirement is to develop a more comprehensive model for addressing the involved complexities and comprehending the issues associated with the assembly process more precisely.

2.5.3 Production Synthesis

The meaning of synthesis refers to the idea of mixing or fusion or combination. In terms of the production system, "synthesis" is nothing but "making things, connected as a whole". The manufacturing system designers are involved in designing the "flexible collaborative environments" for the manufacturing units to increase productivity at reduced costs. This environment may be rife with safety concerns. Wiendahl [58] described the production system as "a place of adding values by production with the help of production factors".

Nyhuis [59] defines factory objects as "physical and nonphysical factory objects which are designed during the process of factory planning and have ability to change during their life cycle". These factory objects, e.g., means, space and organization, are organized in four levels: station, system, segment, and site-level.

2.6 COBOT SELECTION PARAMETERS

With the continuous technological growth in recent years in collaborative systems, nowadays, the market is flooded with multiple kinds of cobotic system based on their "diverse capabilities, strength, facilities with advanced features and specifications". It is a challenging task to choose the most suitable intelligent system.

a) *Selection parameters*: Cobotic system for one's particular requirements and for "real-time manufacturing environments". It has become increasingly difficult to identify the correct system as per its particular use due to its associated complexities, highly advanced features, and facilities continuously introduced by different manufacturers. Hence, researchers have developed the criterion for smart and intelligent cobots selections based on the "multiple criteria decision-making (MCDM) problem" to overcome such scenarios. In this context, Chatterjee [60] has proposed two important methods such as: (i) selection based on the "compromise ranking method"; and (ii) selection based on an outranking method, to identify and select the best-suited cobotic system at minimum costs and particular specifications according to the use and requirements. Further, "the studies based on the robot's evaluation with listed sets of parameters and evaluation methods have been reviews by Mortensen [61] on the basis of 19 scientific studies". Until now, these selection parameters are only based on the basic cobotic functionality criterion, but it has not regarded the cobots as a "hybrid-automation tool" with some additional parameters for its performance assessment like "ease-of-use, safety measures and social-interactions". The proposed "Domain-theory" presented by Andersen [62] presents a big and diverse picture supplied by the "product design and process". Based on this wide perspective, the selection parameters can be summarized as: (i) functionality; (ii) flexibility; (iii) human–machine interaction or HRC system; and (iv) economy perspectives, as shown in Figure 2.7.

b) *Gripper selection*: The paramount objective for utilizing the building blocks of human–robot collaborations is to establish a perfect balance between "productivity and flexibility". This refers to the material holding system which requires a high level of flexibility as per the required purpose [39].

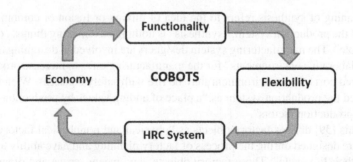

FIGURE 2.7 Cobot selection parameters.

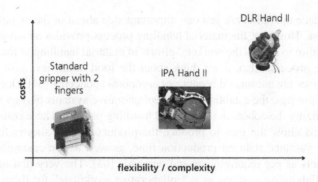

FIGURE 2.8 A spectrum of intelligent gripping systems.

The types of materials to be handled determine the shape and size of the gripper for a robotic system, and which are under continuous development. Kruger [39] has presented the "flexible gripping systems" with their associated challenges at the different gripping stages, as shown in Figure 2.8. While the robotics system has a degree of flexibility as per the required specifications, its performance is still limited in material handling or gripping capabilities. Along with the availability of conventional gripping systems, industries are also focused on developing "intelligent gripping systems" based on the grippers' complexity, flexibility, and costs.

2.7 INDUSTRIAL APPLICATIONS

This section has tried to illustrate some essential applications of implementing a collaborative system, which seems advantageous in productivity, automation, quality, and safety. The collaborative HRC systems can improve the efficiency of some important and selected tasks, which seem to be relevant issues in conventional HRC. There is a natural tendency to use intelligent agents like robots in place of human workers to perform repetitive jobs (pick and place operations), workpiece holding, performing tests, helping workers to handle materials, and other kinds of dull jobs that humans do not like to do. All these functional scenarios highlight that robots are a tool to be used to lessen some human physical burdens and enable some autonomation [63]. However, this type of working environment is still advantageous and humans have relieved themselves from some unenjoyable tasks. Furthermore, collaborative robots are working as the "collaborative workmates endowing the greater extent of autonomy and offers the proactive assistance to human workers". This section summarizes some important industrial applications in collaboration with cobots, such as:

- *Material Handling*: The involved ergonomics concerns have been widely recognized in the current years [1]. The handling process comprises many processing aspects such as transportation, grasping, packaging, palletizing, etc. [64]. The applications such as assembly, product testing, and pick and place operation are just one important step ahead of the manufacturing process. However, the material handling process provides an advantageous algorithm to reduce the workers' efforts in material handling or the material lifting process. Also, if we think about the food industries, most working processes are automated because of ergonomic concerns [64]. Due to some shortcomings, the establishment of collaborative systems plays a vital role in utilizing the cobots in the material handling process. These collaborative systems allow the user to produce the products with a shorter life cycle, fewer variants, reduced production time, as well as the customization of products as per requirements and demands [65]. The very first utilization of collaborative systems as a "collaborative workmate" for these applications was reported in the SIMPLEXITY EU project [65]. This project utilizes the collaborative system or cobots in the "surface finish" application, which requires more sophisticated "human skills and sensitivity" [66]. In this work scenario, the collaborative systems are required to hold the job in accurate required positions and orientations so that the job/workpiece can be exposed to the worker as per need. Also, the "impact in material handling process comprises the repetitive motion, excessive loads, awkward postures and vibrations which all leads us to cumulative trauma disorders such as carpel tunnel syndrome, lower back pain and tennis elbow" [1].

- *Welding operation*: Automatic welding using robotics is one of the most demanding operational areas in manufacturing nowadays. However, some limitations are involved in the effective application of robotics in industrial environments due to some of the associated complexities and available uncertainties in welding [67]. Currently, industries are involved in robotics, but in a very traditional manner "based on lead-through and online-programming", due to having some limitations in terms of complexities, working conditions as per current need, errors in robotic devices, which makes this not suitable to cope-up with humans, the errors occur due to "pre-machining and workpiece fitting" or workpiece distortions due to induced heat [68]. These shortcomings are a problem. Therefore we require other alternatives, such as intelligent robotics technologies with in-built welding equipment with real-time online programming incorporated with guiding and tracking features, including the multiple sensing techniques such as voice, vision, and gestures that provide real-time control over the devices and working conditions. These features are now being developed in collaborative robots to enable these intelligent systems to tackle the associated uncertainties and complexities. The industries are also widely reliant on these collaborative systems to effectively utilize automation for a "walk-through programming approach" for increased productivity [69–71]. Similar approaches have also been proposed using the multi-modal or stereovision-based approaches in [72, 73]. By implementing cobots in manufacturing operations, these intelligent devices get some "autonomy or cognitive capabilities", so that the system can work as an assistant tool.
- *Assembly unit applications*: It's well-known that the utilization of collaborative robots has expanded production capacity. The "hybrid assembly robotic

cells" [39, 64, 74] are the most common type of collaborative robotic systems used widely in industrial applications nowadays. Automated assembly requirements are essential in terms of increased productivity and simplified assembly tasks for the workers, and seem to be particularly beneficial for complex tasks or payloads handling during the assembly process. The sequential assembly operations are well-suited in cooperative assembly environments, where at first a simple task is performed by the robotic system. Then frequently the "complex varied tasks" are performed, giving the products all the features of human assembly [39]. On the other hand, the requirement of "parallel cooperative assembly" is important when multiple parts are needed to be assembled with required precession. In such a process, the timing and coordination between human and robot interactions are critical factors that might severely affect the acceptability and effectiveness of HRC. The analysis of HRC utilization in the assembly cell has been well-described in [39, 74]. This work describes the well-defined "guided assembly operation" using the virtual surface path of the robotic system used for required workpiece lifting and handling the heavier payloads. Furthermore, "the advanced assembly processes can be controlled by means of the robot's compliant motion control that measures the joint torques or contact forces using the force-torque sensor mounted on the robot flange" [39, 75].

2.8 SAFETY MEASURES

According to the "Robot Safety Regulation" [76], it is important to keep the separation between human and machine to avoid any unpredicted and inevitable dangerous situations. In most of the regulations, it is clearly mentioned how precautions and separations have to be maintained, such as by fencing and warnings around the agents. The utilization of industrial cobots has completely redefined the working environments and we are now required to re-think the necessary safety measures because of their intelligence and friendliness in collaborations with workers as they share the workspaces without any danger. A draft for the everyday use of cobotic systems has been proposed by the "Robotics Industries Association" working committee to ensure the safety considerations of IADs [77].

The proposed standards describe the clear distinction between the "hands-on-payload mode" and "hands-on-control mode". Sometimes, the IADs are capable of recognizing the worker's intention of applying the motion on the payload by pushing, but not always it can be detected by the IADs [25]. So, it is important for the operators to notice that in which mode the IADs are currently working, and IADs should not enter in "hands-on-payload mode" without any explicit command given by operator. The IADs should be capable of sensing overforced conditions for safety reasons. This means the IADs should recognize the different "signaling intentions" of the forcing applied by the workers, monitoring the payload to avoid the unusual situations.

Apart from the two abovementioned categories of control mode, the standard also describes the third mode, i.e., the "hands-off mode" as the semi-autonomous mode. In this mode the cobots can move without continuous command from the operator.

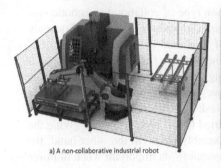

a) A non-collaborative industrial robot   b) A collaborative industrial robot

FIGURE 2.9   A traditional and non-traditional industrial robots [78].

This "hands-on mode" makes the operator's life easier, so that they can concentrate and focus on assembly-line tasks and on the other "line-tracking of the moving assembly lines", or on the "go-fetch" function when the operator requires a new part or workpiece. Hence, the paramount objective is of utilizing the conventional robots or collaborative robots in relation to the safety issues associated with these systems.

Figure 2.9(a) shows the robot's working area surrounded by a sensor-based fence and barrier to ensure human safety. Also, Figure 2.9(b) is of a typical collaborative system, showing the essential safe collaboration with such fence and barriers [78].

Sylla et al. [79] have summarized their work by adopting some scenarios from the ISO 10218-1 [80] report, suggesting the basic collaborative requirements. Some industrial cobots have built-in safety features [26] such as: "MRK SYSTEM-KR 5 S.I. and the COMAU are equipped with tactile sensors to detect contact. The BOSCH APAS has smart capacitive skin that detects the proximity of a human and stops before contact". The most specific safety requirements have been suggested in ISO/TS 15066 [81]. But it is important to note here that the cobots must undergo risk assessment before being implemented in the plant/factory [82–88]. It is not always that the unpredicted dangerous situation occurs due to cobots only, but this may be possible due to the nature of tasks that is to be performed. "For example, even if a cobot moves with a safe speed and force, and stops upon collision, the cobot is dangerous if it is holding sharp tools or parts. Moreover, the cobot cannot discern whether it is colliding with the human's arm, which is uncomfortable but permissible, or with his/her head, which is unacceptable regardless of speed and force. Therefore, a cobot should be ennobled with some additional intelligence and perception abilities to be completely safe. A lot of work has been done on collision avoidance [33, 38, 41, 42], human motion prediction [43, 44], risk assessment through simulation and V.R. [45] and other safety enabling technologies".

2.9   RESULTS

From the assembly point of view, recent trends are essential not only in the manufacturing sector but also in other engineering applications. This is intended to move toward the adaptation of collaborative flexible environments where humans and machines share the workspace and payloads to enhance the ease of design process, quality, productivity, safety, and ergonomic concerns. From the in-depth literature survey, it is evident that the collaboration of human–machine enhances productivity and enables both to be the complementary strengths of each other without applying any limitations to their efficiency. One can use their efficiency as per the task requirement for a particular purpose. These collaborative environments can be efficiently set-up in a very structured manner with having the systematic-analysis and synthesized approach. The manufacturing units are now aiming toward the reduced costs with better quality, and for such manufacturing strategies with implemented automation, for middle and the large volume product with low product variants, is the only alternative in today's competitive market. The IADs are liable to provide the improved production system and also increases the competitions among the industries in terms of increased productivity and quality with essential safety needs. All these scenarios, demands a very flexible, safe, and collaborative environments developed by implementing the symbiotic cobotic system in the workspace. This makes the worker's life very easy and produces the large volumes at reduced cost with required level of precession and required quality without making any harm to the workers working at the same workspace.

The improved technologies of IADs are now involved in manufacturing a family or team of manipulators to share the intermediate working area with the collaborative robots. The cobots are meant to provide the "guidance and control using the servomotors while the worker provides the motive powers for its operation". All such developments and enhancing the productivity led to the same point of being concerned about the safety measures associated with the collaborative systems. This is being critical to consider the safety while having the "human–machine interactions and robotization".

2.10 CONCLUSION

Collaborative robots, i.e., cobots, are much popular nowadays in many industrial applications especially in manufacturing units. On the basis of wide research and in-depth literature survey, we have tried to illustrate the current utilization of cobots over the robotic systems by means of providing the enhanced productivity, ergonomics concerns, safety, and collaborative environments for easy handling of materials. In this paper, some important aspects of the collaborative systems, namely, the safety measures, on the basis of ISO standards that tell us to what extent these collaborative systems needed to be used, as well as the assembly-line requirements of the cobotic system based on cobot programming control or some other means of control like control through vision, voice, and gesture command, have been addressed. The current work has also highlighted HRC classification, their requirements in assembly lines, as well as the framework for HRC deployments. The safety standards for cobots have also been discussed, as well as the wide range of industrial applications along with the utilization of symbiotic collaborative systems. In addition, it is also evident that with many industrial applications, the conventional robotics systems are also in practice at large scale with traditional offline programming. The current work has also introduced many approaches that use multi-modal interactions and augmented reality to this extent. The present work characterizes the core elements and summarizes the typical requirements with highlighting the computational sensing, automatic control along with the communication resources to provide the best possible solutions toward the needs of the ergonomic concerns.

2.11 FUTURE SCOPE

In the past few decades, the manufacturing sectors have grown up drastically by implementing new technological trends such as automation and robotics. Further, the new development in collaborative systems, that is, cobots and their implementation, have taken the manufacturing sectors to a new high. Still, conventional robotics are being used widely. So, future researches are pushing the industries toward the widespread integration of HRC system in manufacturing. The cut-throat competition among industries is forcing the industries to avail the safe, easy collaborative environments for their production process, where the human and machine can work together as their co-workers, and both can utilize their own and other's efficiencies in the productive manner. Other issues related to safety should be addressed for collaborative systems, i.e., in terms of multiple force sensing in different positions and gestures to avoid any of the small risks of danger.

Most of the instinctive developments, the ease of use, easy interface between machine and human in terms of online or offline programming, and their easy control strategies are the part of research and limited to laboratory itself, which have not found any concrete applications till now. These are probably the gaps that are needed to be addressed and to overcome such issues or gaps, the "specific efforts and technology transfer is required" to bring the research in the shopfloor areas. However, the HRC system is not likely to be limited till the bigger firms but should be incorporated in "small and medium sized companies" too. The multiple capabilities and skills of the users play an important role in the design solutions of collaborative systems. The literature survey on the applications of HRC solutions highlights that HRC systems are mostly underused in the industries. These intelligent collaborative robots are mainly considered just an assistant for human workers, simply taking the worker's job responsibility and relieving them from heavy-duty work, saving them from physical fatigue and enhancing their capacity to concentrate on other jobs as well. Future scenarios are needed to endow these collaborative robots equipped with appropriate "cognitive processing skills" that can be used to tackle unpredicted, unwanted, and complex situations at their own enhanced artificial intelligence-based capabilities.

REFERENCES

1. M. Peshkin, and J.E. Colgate, "Cobots", *Industrial Robots: International Journal*, 26, pp. 335–341, 1999.
2. H.-S. Park, "From automation to autonomy – a new trend for smart manufacturing", Chapter 03 in *DAAAM International Scientific Book 2013*, pp. 075–110, Published by DAAAM International, ISBN 978-3-901509-94-0, ISSN 1726-9687, Vienna, Austria, 2013.
3. A.A. Malik, and A. Bilberg, "*Framework to implement collaborative robot in manual assembly: A lean automation approach*", *Proceedings of the 28th DAAAM International Symposium*, pp.1151–1160, 2017. DOI: 10.2507/28th.daaam.proceedings.160

4. L. Peternel, W. Kim, J. Babic, and A. Ajoudani, *"Towards ergonomic control of human-robot co-manipulation and handover"*, *IEEE-RAS 17th International Conference on Humanoid Robotics (Humanoids)*. IEEE, 2017.
5. P. Akella, *"Intelligent assist devices: A new generation of ergonomic tools (workshop)"*, *International Conference on Robotics and Automation*, Detroit, MI, 1999.
6. P. Akella, et al, *"Cobots for the automobile assembly line"*, *Proceedings of International Conference on Robotics and Automation (ICRA)*, IEEE, pp. 728–733, 1999.
7. T.E. Deeter, G.J. Kaury, K.M. Rabeadeau, M.B. Leahy, and T.P. Turner, *"The next generations munitions handler advanced technology demonstrator program"*, *Proceedings of International Conference on Robotics and Automation (ICRA)*, IEEE, 1997. DOI: 10.1109/ROBOT.1997.620061
8. T. Mulc, T. Udiljak, and D. Ciglar, "Structure of reconfigurable manufacturing systems", Chapter 32 in *DAAAM International Scientific Book 2015*, B. Katalinic (Ed.), Published by DAAAM International, ISBN 978–3–902734-05-1, ISSN 1726–9687, Vienna, Austria, pp. 369–390, 2015.
9. J. Fasth-Berglund Stahre, "Cognitive automation strategy for reconfigurable and sustainable assembly systems, *Assembly Automation*, 33 (3), pp. 294–303, 2013. DOI: 10.1108/AA-12-2013-036
10. X.V. Wang Z. Kemény, J. Váncza, and L. Wang, "Human-robot collaborative assembly in cyber-physical production: Classification framework and implementation", *CIRP Annals – Manufacturing Technology*, 2017, DOI: 10.1016/j.cirp.2017.04.101
11. S.A. Elprama, C.I.C. Jewell, A. Jacobs, I.E. Makrini, and B. Vanderborght, *"Attitudes of factory workers towards industrial and collaborative robots"*, *Proceedings of the ACM/IEEE International Conference on Human-Robot Interaction*, pp. 113–114, 2017.
12. L. Monostori, B. Kádár, T. Bauernhansl et al, "Cyber-physical systems in manufacturing", *CIRP Annals – Manufacturing Technology*, 65(2), pp. 621–641, 2016. DOI: 10.1016/j.cirp.2016.06.005
13. T. Arai, R. Kato, and M. Fujita, "Assessment of operator stress induced by robot collaboration in assembly", *CIRP Annals – Manufacturing Technology*, 59(1), pp. 5–8, 2010.
14. T. Fong, C. Thorpe, and C. Baur, "Collaboration, dialogue, human–robot interaction", *Robotics Research, pp.* 255–266, 2003, Springer. DOI: 10.1007/3–540-36460-9_17
15. W. Wannasuphoprasit, P. Akella, M. Peshkin, and J.E. Colgate, *"Cobots: A novel material handling technology"*, *Proceedings of International Mechanical Engineering Congress and Exposition, Anaheim, ASME 98-WA/MH-2*, 1998.
16. J. Kruger, R. Bernhardt, D. Surdilovic, and G. Seliger, "Intelligent assist systems for flexible assembly", *Annals CIRP, 55*, pp. 29–33, 2006.
17. S. Grigorescu, S. Vatau, and A. Dobra, "Dedicated robot-robot cooperation", Chapter 55, *DAAAM International Scientific Book*, B. Katalinic Ed., Published by DAAAM International, ISBN 978–3–901509-74-2, ISSN 1726–9687, Vienna, Austria, pp. 633–644, 2010.
18. J. Krüger, T.K. Lien, and A. Verl, "Cooperation of human and machines in assembly lines", *CIRP Annals – Manufacturing Technology, 58* (2), pp. 628–646, 2009.
19. J. Heilala and P. Voho, "Modular reconfigurable flexible final assembly systems", *Assembly Automation, 21*(1), pp. 20–30, 2001.
20. M. Peschl, J. Roening, and N. Link, *"Human integration in task-driven flexible manufacturing systems"*, *Proceedings of the 23rd International DAAAM Symposium*, Volume 23, No.1, Published by DAAAM International, Vienna, Austria, 2012.
21. D. Kolberg, and D. Zühlke, "Lean automation enabled by industry 4.0 technologies", *IFAC-Papers Online, 48*(3), pp. 1870–1875, 2015.

22. J. Dulchlnos, and P. Massaro, "The time is right for labs to embrace the principles of industrial automation", *Drug World Discovery*, Winter-Issue, *2006*, pp. 25–28, 2005.

23. R. Bernhardt, D. Surdilovic, V. Katschinski, and K. Schröer, "Flexible assembly systems through workplace sharing and time sharing human-machine cooperation", *PISA, IFAC Proceedings*, *40*(3), pp. 247–251, 2007.

24. A. Bicchi, M.A. Peshkin, and J.E. Colgate, "Safety for physical human–robot interaction", *Springer Handbook of Robotics*, B. Siciliano and O. Khatib, Eds. Springer Berlin Heidelberg, pp. 1335–1348, 2008.

25. M. Peshkin, J.E. Colgate, and S.H. Klostermeyer, "*Intelligent assists devices in industrial applications: A review*", *Proceedings 2003 IEEE/RSJ International Conference on Intelligent Robots and Systems (IROS)*, 2003. DOI: 10.1109/IROS.2003.1249248

26. Robotiq, "Cobots EBook", 2018. [Online]. Available: https://blog.robotiq.com/collaborative-robot-ebook

27. S.A. Green, M. Billinghurst, X. Chen, and J.G. Chase, "Human-robot collaboration: A literature review and augmented reality approach in design", *International Journal of Advanced Robotic Systems*, *5*(1), pp. 1–18, 2008.

28. P.A. Lasota, T. Fong, and J.A. Shah et al., "A survey of methods for safe human-robot interaction," *Foundations and Trends® in Robotics*, *5*(4), pp. 261–349, 2017.

29. J. Lee, "A survey of robot learning from demonstrations for human-robot collaboration", *arXiv preprint arXiv:1710.08789*, 2017.

30. Z. Zhu, and H. Hu, "Robot learning from demonstration in robotic assembly: A survey", *Robotics*, *7*(2), p. 17, 2018.

31. S. A. Green, M. Billinghurst, X. Chen, and J.G. Chase, "Human-robot collaboration: A literature review and augmented reality approach in design", *International Journal of Advanced Robotic Systems*, *5*(1), pp. 1–18, 2008.

32. H. Liu, and L. Wang, "Gesture recognition for human-robot collaboration: A review", *International Journal of Industrial Ergonomics*, 68, pp. 355–367, 2018.

33. M.J. Timms, "Letting artificial intelligence in education out of the box: Educational cobots and smart classrooms", *International Journal of Artificial Intelligence in Education*, *26*, pp.701–712, 2016.

34. A. Mohamed, P.F. Culverhouse, et al, "Automating active stereo vision calibration process with cobots", *Control Conferences Africa (CCA 2017)*, *50*(2), pp. 163–168, 2017. DOI: 10.1016/j.ifacol.2017.12.030

35. I. ElMakrini, S.A. Elprama, et al, "Working with walt: How a cobot was developed and inserted on an auto assembly line", *IEEE Robotics and Automation Magazine*, *25*(2), 2018. DOI: 10.1109/MRA.2018.2815947

36. A. Baucher, D. Wollherr, and M. Buss, "Human-robot collaboration: A survey", *International Journal of Humanoid Robotics*, *5*(1), pp. 47–66, 2008.

37. E. Helms, R.D. Schraft, et al, "*rob@work: Robot assistant in industrial environments*", *Proceedings of the 11th IEEE Int. Workshop on Robot and Human Interactive Communication, ROMAN2002*, Berlin, Germany, September 25–27, pp. 399–404, 2002.

38. R.D. Schraft, E. Helms, M. Hans, and S. Thiemermann, "Man–Machine-Interaction and Co-operation for mobile and assisting robots", *Proceedings of the EIS*, *2004*, pp. 67–77, 2004.

39. J. Krüger, T.K. Lien, and A. Verl, "Cooperation of human and machines in assembly lines", *CIRP Annals – Manufacturing Technology 58*(2), pp628–646, 2009.

40. H.A. Yanco, J. Drury, "Classifying human–robot interaction: An updated taxonomy", *IEEE International Conference on Systems, Man and Cybernetics*, *3*, pp. 2841–2846, 2004.

41. G. Michalos, S. Makris, N. Papakostas, D. Mourtzis, and G. Chryssolouris, "Automotive assembly technologies review: challenges and outlook for a flexible and adaptive approach", *CIRP Journal of Manufacturing Science and Technology*, *2*(2), pp. 81–91, 2010.

42. M. Morioka, and S. Sakakibara, "A new cell production assembly system with human–robot cooperation", *CIRP Annals – Manufacturing Technology*, *59*(1), pp. 9–12, 2010.

43. J. Krüger, G. Schreck, and D. Surdilovic "Dual arm robot for flexible and cooperative assembly", *CIRP Annals – Manufacturing Technology*, *60*(1), pp. 5–8, 2011.

44. H.A. Yanco, and J. Drury, *"Classifying human–robot interaction: An updated taxonomy"*, *IEEE International Conference on Systems, Man and Cybernetic*, *3*, pp. 2841–2846, 2004.

45. J.L. Burke, R.R. Murphy, E. Rogers, V.J. Lumelsky, and J. Scholtz, "Final report for the DARPA/NSF interdisciplinary study on human–robot interaction", *IEEE Transactions on Systems, Man, and Cybernetics, Part C*, *34*(2), pp. 103–112, 2004.

46. D.J. Bruemmer, D.A. Few, R.L. Boring, J.L. Marble, M.C. Walton, and C.W. Nielsen, "Shared understanding for collaborative control", *IEEE Transaction on Systems, Man, and Cybernetics-Part A: Systems and Humans*, *35*(4), pp. 494–504, 2005.

47. J.E. Colgate, W. Wannasuphoprasit, and M. Peshkin, *"Cobots: Robots for collaboration with human operators"*. *Proceedings of the International Mechanical Engineering Congress and Exhibition*, Atlanta, GA, 58, pp. 433–439, 1996.

48. J.E. Colgate, W. Wannasuphoprasit, and M. Peshkin, "Nonholonomic haptic display", *IEEE International Conference on Robotics and Automation, Minneapolis*, *1*, pp. 539–544, 1996.

49. J.E. Colgate, and J.M. Brown, *"Factors affecting the Z-width of a haptic display"*, *International Conference on Robotics and Automation, IEEE R&A Society*, San Diego, CA, *4*, pp. 3205–3210, 1994.

50. S. EI Zaatari, M. Marie, and W.D. Li, "Cobot programming for collaborative industrial tasks: An overview", *Robotics and Autonomous Systems*, 2019. DOI: 10.1016/j.robot.2019.03.003.

51. J. Eekels, and N.F. Roozenburg, "A methodological comparison of the structures of scientific research and engineering design: Their similarities and differences", *Design Studies*, *12*(4), pp. 197–203, 1991.

52. E. Francalanza, J. Borg, and C. Constantinescu, "Deriving a systematic approach to changeable manufacturing system design", *Procedia CIRP*, *17*, pp. 166–171, 2014.

53. G. Chryssolouris, D. Mavrikios, and D. Mourtzis, "Manufacturing systems: Skills & competencies for the future", *Procedia CIRP*, *7*, pp. 17–24, 2013.

54. S.J. Hu, J. Ko, L. Weyand, H. ElMaraghy, T. Lien, Y. Koren, et al., "Assembly system design and operations for product variety", *CIRP Annals – Manufacturing Technology*, *60*(2), pp. 715–733, 2011.

55. W. ElMaraghy, and R. Urbanic, "Assessment of manufacturing operational complexity", *CIRP Annals – Manufacturing Technology*, *53*(1), pp. 401–406, 2004.

56. S. Samy, and H. ElMaraghy, "A model for measuring products assembly complexity", *International Journal of Computer Integrated Manufacturing*, *23*(11), pp. 1015–1027, 2010.

57. S. Samy, and H. ElMaraghy, "Complexity mapping of the product and assembly system", *Assembly Automation*, *32*(2), pp. 135–151, 2012.

58. H.-P. Wiendahl, H.A. ElMaraghy, P. Nyhuis, M.F. Zäh, H.-H. Wiendahl, N. Duffie, et al., "Changeable manufacturing – classification, design and operation", *CIRP Annals – Manufacturing Technology*, *56*(2), pp. 783–809, 2007.

59. P. Nyhuis, T. Heinen, and M. Brieke, "Adequate and economic factory transformability and the effects on logistical performance", *International Journal of Flexible Manufacturing Systems*, *19*(3), pp. 286–307. 2007.

60. P. Chatterjee, V.M. Athawale, and S. Chakraborty, "Selection of industrial robots using compromise ranking and outranking methods", *Robotics and Computer-Integrated Manufacturing*, *26*(5), pp. 483–489, 2010.

61. Y.T. Ic, M. Yurdakul, and B. Dengiz, "Development of a decision support system for robot selection", *Robotics and Computer-Integrated Manufacturing*, *29*(4), pp. 142–157, 2013.

62. N.H. Mortensen, and C.T. Hansen, *Structuring as a basis for product modelling*. Crit. Entusiasm, Trondheim: Tapir, pp. 111–128, 1999.

63. C. Heyer, *"Human-robot interaction and future industrial robotics applications"*, *Proceedings of the IEEE/RSJ international conference intelligent robots and systems (IROS)*. IEEE. pp. 4749–4754, 2010.

64. M. Hägele, K. Nilsson, J.N. Pires, and R. Bischoff, "Industrial robotics", in *Springer handbook of robotics*, Siciliano B., Khatib O., Ed. 2nd Springer, pp. 1385–1418, 2016.

65 E. Gambao, M. Hernando Surdilovic, "A new generation of collaborative robots for material handling", *Geron*, *11*(2), p. 368, 2012.

66. *SYMPLEXITY EU* project. http://www.symplexity.eu/

67. A. Wilbert, B. Behrens, O. Dambon, and F. Klocke, *"Robot assisted manufacturing system for high gloss finishing of steel molds"*, *Proceedings of the Intelligent Robotics and Applications. ICIRA 2012*. Lecture Notes in Computer Science 7506. Springer. pp. 673–685, 2012.

68. S. Chen, T. Qiu, T. Lin, L. Wu, J. Tian, W. Lv, et al, *"Intelligent technologies for robotic welding"*, *Lecture Notes in Control and Information Science* 299. Springer. pp. 123–143, 2004.

69. M.H. Ang Jr, W. Lin, and S.-Y. Lim, "A walk-through programmed robot for welding in shipyards", *Industrial Robot: An International Journal*, *26*(5), pp. 377–388, 1999.

70. M.H. Ang, L. Wei, and L.S. Yong, *"An industrial application of control of dynamic behavior of robots – A walk-through programmed welding robot"*, *Proceedings of the IEEE international conference robotics and automation (ICRA)*. IEEE, pp. 2352–2357, 2000.

71. R. Hollmann, A. Rost, M. Hägele, and A.A. Verl, *"HMM-based approach to learning probability models of programming strategies for industrial robots"*, *Proceedings of the IEEE international conference robotics and automation (ICRA)*. IEEE, pp.2965–2970, 2010.

72. G. Du, and P. Zhang, "A markerless human–robot interface using particle filter and Kalman filter for dual robots", *IEEE Transactions on Industrial Electronics*, *62*(4), 2257–2264, 2015.

73. B. Takarics, P.T. Szemes, G. Nemeth, and P. Korondi, *"Welding trajectory reconstruction based on the intelligent space concept"*, *Proceedings of the IEEE conference human system interactions (HSI)*. IEEçE. pp. 791–796, 2008.

74. P. Tsarouchi, A.-S. Matthaiakis, S. Makris, and G. Chryssolouris, "On a human-robot collaboration in an assembly cell", *çInternational Journal of Computer Integrated Manufacturing*, *30*(6), pp. 580–589, 2017.

75. S. Kock, T. Vittor, B. Matthias, H. Jerregard, M. Källman, I. Lundberg, et al., *"Robot concept for scalable, flexible assembly automation: A technology study on a harmless dual-armed robot"*, *Proceedings of the IEEE International symposium assembly and manufacturing (ISAM)*. IEEE. pp. 1–5, 2011.

76. Robot Safety Requirements Standard, ANSI/RIA R15.06–1999.

77. M.A. Peskins, "T15.1 draft standard for trial use for intelligent assist devices – Personnel safety requirements", *Robotic Industries Association*, 2002. Available at, https://peshkin. mech.northwestern.edu/publications/2002_T15.1_DraftStandardForTrialUse_ IntelligentAssistDevicesPersonnelSafetyRequirements.pdf.

78. V. Villani, F. Pini, F. Leali, and C. Secchi, "Survey on human–robot collaboration in industrial settings: Safety, intuitive interfaces and applications", *Mechatronics*, *55*, pp. 248–266, 2018.

79. *Robots and robotic devices: Safety requirements for industrial robots Part 1: Robots*, ISO Standard ISO 10 218–1:2011, 2011.

80. N. Sylla, and S. Mehta, "Implementation of collaborative robot applications: A report from the industrial working group, high speed sustainable manufacturing institute", *Technical Report*, 2017.

81. *Robots and robotic devices – Collaborative robots*, ISO Stanard ISO/TS 15066:2016, 2016.

82. A.R. Vargas, K.C. Arredondo-Soto, et al, "Introduction and configuration of a collaborative robot in an assembly task as a means to decrease occupational risks and increase efficiency in a manufacturing company", *Robotics and Computer-Integrated Manufacturing*, *57*, pp. 315–328, 2019.

83. R. Meziane, M.J.-D. Otis, and H. Ezzaidi, "Human-robot collaboration while sharing production activities in dynamic environment: SPADER system", *Robotics and Computer-Integrated Manufacturing*, *48*, pp. 243–253, 2017.

84. B. Schmidt, and L. Wang, "Depth camera-based collision avoidance via active robot control", *Journal of Manufacturing Systems*, *33*(4), pp. 711–718, 2014.

85. Y. Wang, X. Ye, Y. Yang, and W. Zhang, "*Collision-free trajectory planning in human-robot interaction through hand movement prediction from vision*", *2017 IEEE-RAS 17th International Conference on Humanoid Robotics (Humanoids)*, pp. 305–310, 2017.

86. C. Lenz, M. Rickert, G. Panin, and A. Knoll, "*Constraint task-based control in industrial settings*", *2009 IEEE/RSJ International Conference on Intelligent Robots and Systems*, pp. 3058–3063, 2009.

87. E. Matsas, G.C. Vosniakos, and D. Batras, "Prototyping proactive and adaptive techniques for human-robot collaboration in manufacturing using virtual reality", *Robotics and Computer-Integrated Manufacturing*, *50*, pp. 168–180, 2018.

88. K.H. Dinh, O. Oguz, G. Huber, V. Gabler, and D. Wollherr, "*An approach to integrate human motion prediction into local obstacle avoidance in close human-robot collaboration*", *2015 IEEE International Workshop on Advanced Robotics and its Social Impacts (ARSO)*, pp. 1–6, 2015.

3 Machine Learning for Friction Stir Welding

Shivraman Thapliyal
National Institute of Technology Warangal, India

Akshansh Mishra
Politecnico Di Milano, Italy

CONTENTS

3.1 INTRODUCTION

Implementation of artificial intelligence (AI) in various fields such as material design, manufacturing parts, and also in various aerospace applications answers the key question, "Can we automatize reasoning?" The main objective of using AI is to make machines as smart as the human brain, in which the intelligent systems maximize the success of the given goals by using various sets of algorithms. Machine learning can be thought of as a subset of AI, as shown in Figure 3.1, which helps us to solve complex engineering problems.

Machine learning is defined as the domain of computer science which overcomes the disadvantages of the conventional computing method lacking behind to provide the automating solutions to complex engineering problems. Machine learning solves problems in a similar way to a normal human brain. For example, when a student wants to solve the exercise of a Calculus chapter, he goes through some examples that mark the action of training the brain on a given data. After training, it solves for an unknown data, which is the exercise question. Similarly, machine learning algorithms are first trained for the given available dataset and then tested for unknown data.

DOI: 10.1201/9781003202233-4

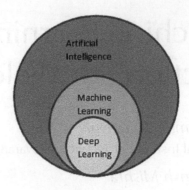

FIGURE 3.1   Representation of AI hierarchy.

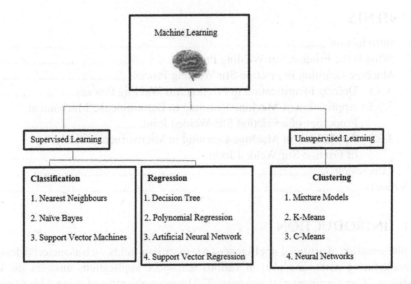

FIGURE 3.2   Division of machine learning algorithms.

For a given large dataset, the machine learning algorithms can detect the pattern and structures for solving tasks such as classification, prediction, and clustering. From a major point of view, machine learning algorithms are subdivided into two models, i.e., supervised machine learning model and unsupervised machine learning model, as shown in Figure 3.2. Supervised learning is a kind of machine learning algorithm where models are trained upon "labeled" training data, in which correct output is mapped to the corresponding input data. This learned data is then used to predict the output for new input variables. In the case of unsupervised learning, the model itself finds the essence of hidden patterns from the given data.

Machine learning algorithms make faster decisions by taking an available chunk of data into account. If any expected results are present, it further uses the computer system to build a program known as a model, as shown in Figure 3.3.

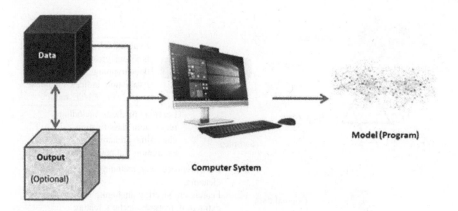

FIGURE 3.3 Working mechanism of machine learning.

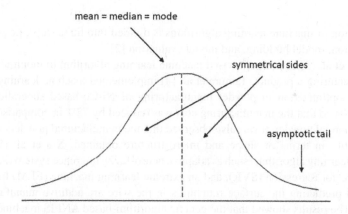

FIGURE 3.4 Normal distribution curve.

Most of the tasks in machine learning generally deal with the finding of maximum likelihood having the highest probability. In order to get the idea of the maximum likelihood, a random variable can be considered which follows the normal distribution curve. Figure 3.4 shows the shape of the probability distribution function of normal distribution.

Mean and standard deviation are the parameters that characterize the given curve. The highest value of the curve is shown by the mean value, while the spread of the curve is represented by the standard deviation value.

In recent days, the manufacturing industries are witnessing the advent of the availability of data [1]. Manufacturing industries are incorporating various mathematical and computer science-based algorithms for meeting research and development needs. Machine learning algorithms usage is highly inculcated in the material science domain too, as shown in Figure 3.5. It is observed from Figure 3.5 that the

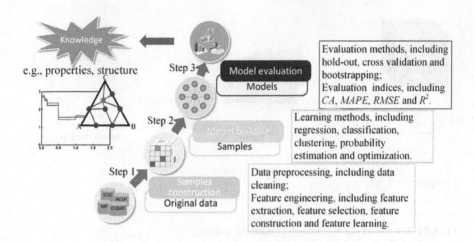

FIGURE 3.5 Construction of machine learning algorithms for material science [2].

construction of machine learning algorithms is divided into three steps, i.e., sample construction, model building, and model evaluation [2].

Wuest et al. [3] used a supervised machine learning algorithm to monitor quality in manufacturing a product. Tamura et al. [4] implemented machine learning-based Bayesian optimization in powder manufacturing of Ni-Co-based superalloy. The results showed that the manufacturing cost was reduced by 72% in comparison with the commercial powder. It was also observed that the manufactured powders were of high quality in terms of shape and microstructure obtained. Xia et al. [5] used machine learning algorithms such as adaptive neuro-fuzzy inference system (ANFIS), support vector regression (SVR), and an extreme learning machine (ELM) for modeling and predicting the surface roughness in the wire arc additive manufacturing process. The results showed that the genetic algorithm-based ANFIS machine learning model yields good accuracy for predicting the surface roughness.

The main objective of this chapter is to discuss the recent application of machine learning algorithms in the friction stir welding process. Section 3.2 will highlight a brief introduction to the friction stir welding process, while the following sections will highlight the application of various machine learning algorithms for defects identification, modeling, and prediction of mechanical properties and the prediction of the microstructure properties of friction stir welded joints.

3.2   WHAT IS THE FRICTION STIR WELDING PROCESS?

Since its invention in 1991, the friction stir welding process has gained wide popularity in the aerospace and automotive manufacturing industries for its high-precision application [6, 7]. The friction stir welding process can be defined as a solid-state welding technology that does not use any consumable electrode to join the workpiece. It is possible to join lightweight materials such as titanium, aluminum, magnesium, etc., by using the friction stir welding process [8–10].

FIGURE 3.6 Friction stir-welding process setup [11].

Figure 3.6 shows the arrangement of the friction stir welding process setup. The setup consists of the (a) two or more similar or dissimilar alloy plates (b) rotating tool, which is harder than the material to be joined (c) backing plate material to support the alloy plates to be joined (d) fixture to hold the plates firmly in order to prevent dislocation during the joining process.

3.3 MACHINE LEARNING IN FRICTION STIR WELDING PROCESS

This section will focus on the application of machine learning algorithms in the friction stir welding process.

3.3.1 DEFECTS IDENTIFICATION IN FRICTION STIR WELDING PROCESS

Improper selection of important input parameters such as tool traverse speed (mm/min) and tool rotational speed (rpm) results in defects being formed due to insufficient heat generation and improper material mixing. It is generally observed that the selection of low tool rotational speed and high tool traverse speed does not result in enough heat generation in order to plasticize the material and hence results in proper intermixing. On the other hand, selection of high tool rotational speed and low tool traverse speed often results in over-plasticization and leads to the formation of intermetallic compounds. Other input parameters such as plunge depth, tool tilt angle, and

offset distance also influence the quality of friction stir welded joints. Flash formation along the border of the weldment occurs if a high tool tilt angle and increased
plunge depth are used.

Du et al. [12] analyzed the condition of void formation by subjecting the machine
learning algorithms such as Bayesian neural network and decision trees on one hun-
dred and eight experimental datasets of three aluminum alloys, AA2024, AA2219,
and AA6061. The schematic representation of the research is shown in Figure 3.7.

Mishra et al. [13] used two texture-based recognition systems, i.e., local binary
pattern (LBP) algorithm and fast Fourier transformation algorithm, to detect surface
irregularities in friction stir welded joints as shown in Figure 3.8 and 3.9. The results
showed that the developed algorithms could detect surface defects such as flash formation and groovy edges. Mishra et al. [14] also used the discrete wavelet transformation method to detect surface defects in the friction stir welded joint, as shown in

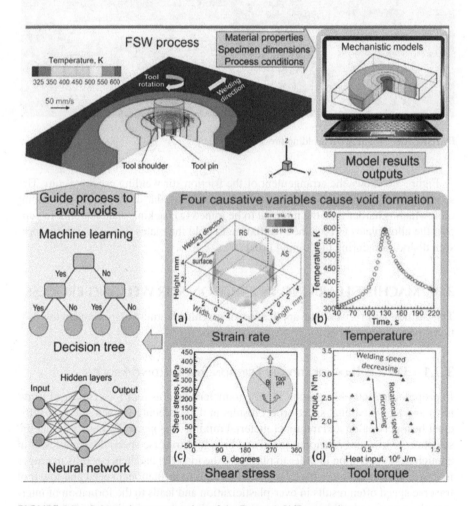

FIGURE 3.7   Schematic representation of the Du et al. [12] research.

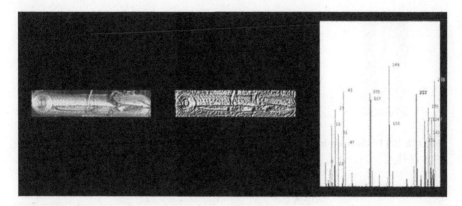

FIGURE 3.8 Surface irregularities analysis by using the LBP algorithm [13].

FIGURE 3.9 Surface irregularities analysis by using fast Fourier transformation method [13].

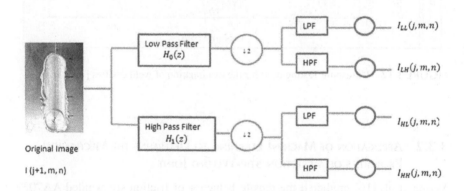

FIGURE 3.10 Discrete wavelet transformation algorithm subjected to friction stir welded joint image [14].

Figure 3.10. It was observed that the surface defects were successfully differentiated by using the developed algorithm, as shown in Figure 3.11.

Hartl et al. [15] developed a convolutional neural network model to identify surface defects present in friction stir-welded joints. The ultrasonic testing was carried out on 120 weld samples, shown in Figure 3.12, to label them as "good" and "defective" samples. The results obtained showed an accuracy of 98.5%.

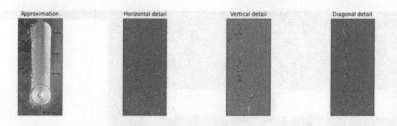

FIGURE 3.11   Surface irregularities analyzed by a discrete wavelet transformation algorithm [14].

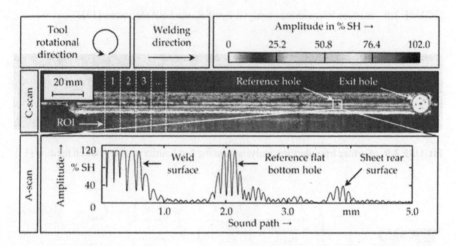

FIGURE 3.12   Ultrasonic Testing used for the examination of weld cavities [15].

3.3.2 APPLICATION OF MACHINE LEARNING TO DETERMINE THE MECHANICAL PROPERTIES OF A FRICTION STIR-WELDED JOINT

Verma et al. [16] predicted the tensile behavior of friction stir-welded AA7039 using machine learning algorithms such as an artificial neural network (ANN), support vector machine (SVM), Gaussian process regression (GPR), and linear regression (LR). Yunus et al. [17] for mathematical modeling of a friction stir-welded joint to predict the joint strength of two dissimilar aluminum alloys by using a new computing method known as genetic programming. Figure 3.13 shows the plot of experimental and predicted values of the tensile strength. Figure 3.14 shows the normal Q-Q plot of error between experimental and genetic programmed data.

The maximum ultimate tensile strength (UTS) of friction stir-welded 6061-T6 aluminum alloy was predicted by Mishra et al. [18] by using an ANN and decision

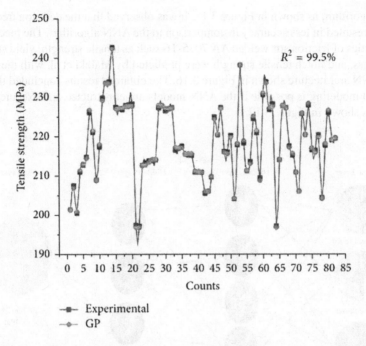

FIGURE 3.13 Plot of experimental and predicted values of the tensile strength [17].

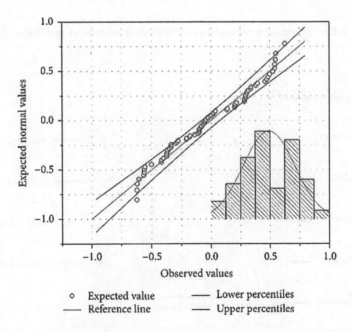

FIGURE 3.14 Normal Q-Q plot of error between experimental and genetic programmed data [17].

tree algorithm, as shown in Figure 3.15. It was observed that the decision trees algorithm resulted in less accuracy in comparison to the ANN algorithm. The mechanical properties of friction stir welded AA 7075-T6 such as tensile strength, yield strength, hardness, and notch tensile strength were predicted by Maleki et al. with the help of the ANN architecture shown in Figure 3.16. The obtained results concluded that successful modeling is possible if the ANN models are constructed and executed properly, as shown in Figure 3.17.

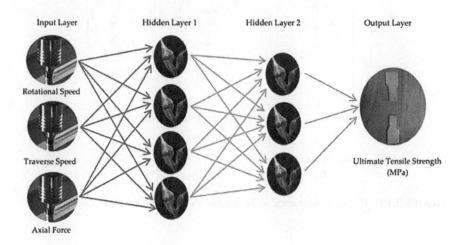

FIGURE 3.15 Schematic representation of artificial neural network architecture [18].

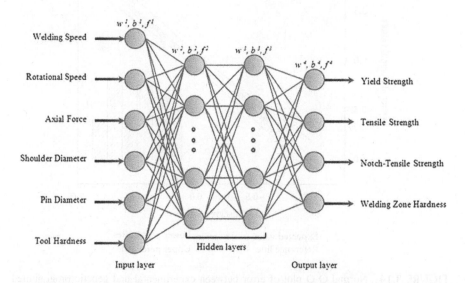

FIGURE 3.16 Neural Network architecture for predicting the mechanical properties [19].

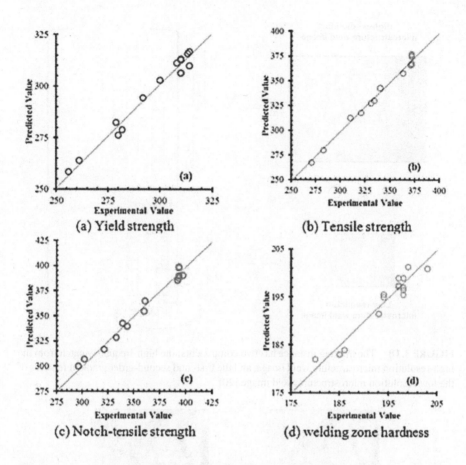

(a) Yield strength

(b) Tensile strength

(c) Notch-tensile strength

(d) welding zone hardness

FIGURE 3.17 Plot of predicted value against experimental values [19].

3.3.3 APPLICATION OF MACHINE LEARNING IN MICROSTRUCTURE STUDY OF FRICTION STIR WELDED JOINT

Unlike the various availability of the research work on the application of machine learning-based algorithms to determine the mechanical property of a friction stir welded joint, there is limited availability of the research work on the application of machine learning in microstructure study. Srinivasan et al. [20] developed a machine learning-based hybrid sparsity model to improve the resolution of the degraded friction stir weld microstructure images, as shown in Figure 3.18. The obtained results are shown in Figure 3.19. Fratini et al. [21] predicted the average grain size in the friction stir welding process by using the ANN approach. Fratini et al. [22] also carried out the study on the modeling of the metallurgical phenomena occurring in friction stir welding processes of AA6082-T6 and AA7075-T6 aluminum alloys by using ANN.

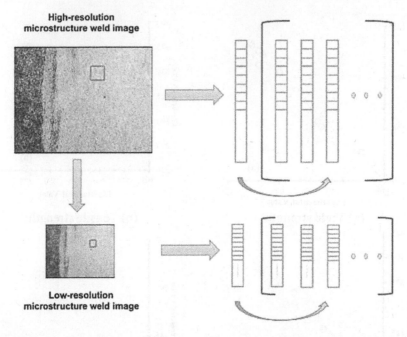

**High-resolution
microstructure weld image**

**Low-resolution
microstructure weld image**

FIGURE 3.18 The training instance has two components: the high-frequency patch from the high-resolution microstructure weld image, and the first- and second-order gradient features of the low-resolution microstructure weld image [20].

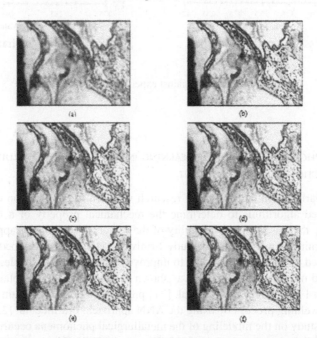

FIGURE 3.19 (a) Low-resolution microstructure weld image; resolution enhancement methods: (a) bi cubic interpolation, (b) K-nearest neighbors, (c) local linear embedding, (d) non-local means, (e) hybrid sparsity model.

3.4 CONCLUSION

In this chapter, an attempt is made to highlight the various application of machine learning algorithms in the friction stir welding process ranging from the determination of the defects to the determination of various mechanical and microstructure properties. It has been observed that mainly supervised learning-based machine learning algorithms such as ANN, regression models, SVM, and decision trees are widely being used in the friction stir welding process. It can be further concluded that the application of machine learning algorithms in the friction stir welding process not only results in good accuracy, but also it leads to the reduction of the time and cost of the experimental work. The future scope can deal with the application of quantum computing-based machine learning algorithms, resulting in higher accuracy and less computing time than conventional machine learning algorithms. Future work can predict the mechanical properties of friction stir welded joints from their microstructure by using convolutional neural network modeling.

REFERENCES

1. Chand, S., and Davis, J. F., 2010, July. *What is smart manufacturing?* Time Magazine.
2. Liu, Y., Zhao, T., Ju, W., and Shi, S., 2017. Materials discovery and design using machine learning. *Journal of Materiomics*, *3*(3), pp. 159–177.
3. Wuest, T., Irgens, C., and Thoben, K. D., 2014. An approach to monitoring quality in manufacturing using supervised machine learning on product state data. *Journal of Intelligent Manufacturing*, *25*(5), pp. 1167–1180.
4. Tamura, R., Osada, T., Minagawa, K., Kohata, T., Hirosawa, M., Tsuda, K., and Kawagishi, K., 2021. Machine learning-driven optimization in powder manufacturing of Ni-Co based superalloy. *Materials & Design*, *198*, p. 109290.
5. Xia, C., Pan, Z., Polden, J., Li, H., Xu, Y., and Chen, S., 2021. Modelling and prediction of surface roughness in wire arc additive manufacturing using machine learning. *Journal of Intelligent Manufacturing*, pp. 1–16. https://link.springer.com/article/10.1007/s10845-020-01725-4.
6. Mishra, R. S., and Ma, Z. Y., 2005. Friction stir welding and processing. *Materials Science & Engineering R: Reports*, 50(1–2), pp. 1–78.
7. Lohwasser, D., and Chen, Z. eds., 2009. *Friction Stir Welding: From Basics to Applications*. Elsevier.
8. Kayode, O., and Akinlabi, E. T., 2019. An overview on joining of aluminium and magnesium alloys using friction stir welding (FSW) for automotive lightweight applications. *Materials Research Express*, 6(11), p. 112005.
9. Kallee, S. W., Thomas, W. M., and Dave Nicholas, E., 2000. Friction stir welding of lightweight materials. *Magnesium Alloys and their Applications*, pp. 173–190. https://doi.org/10.1002/3527607552.ch29.
10. Prater, T., 2014. Friction stir welding of metal matrix composites for use in aerospace structures. *Acta Astronautica*, *93*, pp. 366–373.
11. Jauhari, T. K., 2012. *Development of Multi-Component Device for Load Measurement and Temperature Profile for Friction Stir Welding Process* [M.Sc Thesis]. Penang: Universiti Sains Malaysia; Unpublished.
12. Du, Y., Mukherjee, T., and DebRoy, T., 2019. Conditions for void formation in friction stir welding from machine learning. *npj Computational Materials*, *5*(1), pp. 1–8.
13. Mishra, A., 2020. *Surface Quality Analysis of Friction Stir Welded Joints by Using Fourier Transformation and Local Binary Patterns Algorithms*. Soldagem & Inspeção, p. 25.

14. Mishra, A., 2020. Discrete wavelet transformation approach for surface defects detection in friction stir welded joints. *Fatigue of Aircraft Structures*, *12*, pp. 27–35.

15. Hartl, R., Bachmann, A., Habedank, J. B., Semm, T., and Zaeh, M. F., 2021. Process monitoring in friction stir welding using convolutional neural networks. *Metals*, 11(4), p.535.

16. Verma, S., Gupta, M., and Misra, J. P., 2018. Performance evaluation of friction stir welding using machine learning approaches. *MethodsX*, *5*, pp. 1048–1058.

17. Yunus, M., and Alsoufi, M. S., 2018. Mathematical modelling of a friction stir welding process to predict the joint strength of two dissimilar aluminium alloys using experimental data and genetic programming. *Modelling and Simulation in Engineering*, 2018, pp. 1–18.

18. Mishra, A., 2020. Artificial intelligence algorithms for the analysis of mechanical property of friction stir welded joints by using python programming. *Welding Technology Review*, 92(6), pp. 7–16.

19. Maleki, E., 2015, November. *Artificial neural networks application for modeling of friction stir welding effects on mechanical properties of 7075-T6 aluminum alloy*. In *IOP Conference Series: Materials Science and Engineering* (Vol. 103, No. 1, p. 012034). IOP Publishing.

20. Srinivasan, K., Deepa, N., and Durai Raj Vincent, P. M., 2020. Realizing the resolution enhancement of tube-to-tube plate friction welding microstructure images via hybrid sparsity model for improved weld interface defects diagnosis. *Journal of Internet Technology*, *21*(1), pp. 61–72.

21. Fratini, L., Buffa, G., and Palmeri, D., 2009. Using a neural network for predicting the average grain size in friction stir welding processes. *Computers & Structures*, 87(17–18), pp. 1166–1174.

22. Fratini, L., and Buffa, G., 2008. Metallurgical phenomena modeling in friction stir welding of aluminium alloys: Analytical versus neural network based approaches. *Journal of Engineering Materials and Technology*, *130*(3), pp. 031001.

4 Mathematical and Intelligent Modeling in Tundish Steelmaking

Vipul Kumar Gupta, Pradeep Kumar Jha, and Pramod Kumar Jain
Indian Institute of Technology Roorkee, India

CONTENTS

4.1 INTRODUCTION

In the continuous casting process, a "tundish" refers to an intermediate refractory vessel placed between a ladle and mold to distribute molten steel to different molds. The role of modern continuous casting tundish is not only limited to supply the molten metal. Still, it now acts as a metallurgical reactor to provide clean molten steel with the desired composition. With stringent quality control requirements of the final product and reduced production cost, force foundry engineers give additional responsibilities to tundish, such as refining and processing. Therefore, inclusion flotation and removal, minimize the thermal and chemical losses from molten steel and composition adjustment have become important functions of tundish. Close control of the molten steel flow dynamics is essential to implement these functions. Tundish shape and its flow control device (dams, weirs, baffles, pour pads, etc.) play a pivotal role in modifying the flow. Many researchers studied the effectiveness of these flow control devices (FCDs) [1–5]. The turbulent inhibitor box (TIB) confines the turbulence to near the inlet zone, increasing mean residence time, and thus helps in inclusion removal. Dams, weirs, and baffles guide the molten steel flow, improving flow characteristics and mixing [6–7]. Continuous heat loss to surrounding and external heating/cooling makes the tundish process non-isothermal.

Convection in molten steel is mainly governed by both forced and natural convection [8]. The relative importance of the convection pattern is given by dimensionless number Gr/Re^2. In the middle region of tundish, low velocity, and significant temperature difference ($\approx 2° - 4°$) give rise to dominance of natural convection and therefore, thermal buoyancy significantly modifies the flow [9]. Thermal stratification is also observed near tundish walls. Detailed numerical investigation on the effect of thermal buoyancy on fluid flow is performed [10]. Buoyancy rising helps in inclusion flotation by drifting the flow in the upward direction. Gas bubbling from the curtain at the tundish bottom is another approach to enhance inclusions' flotation and mixing [11–15]. In industry practice, a porous plug is provided at the tundish bottom for gas injection. An optimum flow rate is given to avoid the aperture of slag. Argon gas is injected from the shroud to avoid oxidation due to open eye formation. Powdered flux is used at the tundish top surface to avoid liquid metal exposure to the environment and to prevent oxidation. Slag metal interface stabilization is necessary to avoid excessive slag entrainment.

Flow in tundish is multiphase in nature due to the presence of inclusion particles, gas bubbles, and top slag layer along with molten steel. The schematic of multiphase flow in tundish is shown in Figure 4.1. High molten steel temperature,

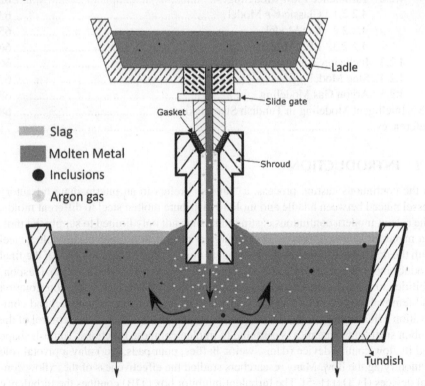

FIGURE 4.1 Tundish schematic.

the large size of the reactor, and visual opacity pose a serious challenge to investigating these multiphase phenomena experimentally. Physical modeling accuracy using a water model is limited to single-phase flow. Therefore, mathematical modeling is a valuable tool to study these complex multiphase flow phenomena. There are multiple mathematics models available to study each phenomenon and reported in previously published work [16]. In this work, the details of the most commonly used models are presented, and their associated governing equations are described. Furthermore, an overview of intelligent modeling in tundish steelmaking is also presented.

4.2 CONSTITUENTS OF MATHEMATICAL MODELING IN TUNDISH

A combination of different mathematical models is used to model multiphase phenomena in the tundish. Fluid flow is modeled using Navier stokes equations for the isothermal case and coupled with energy equation for non-isothermal case. These equations are solved with other equations to study multiphase phenomena. Turbulence in molten steel flow is captured by different turbulent models like k-epsilon, k-omega, large eddy simulation (LES), etc. Inclusion particle modeling and Argon gas bubbling are modeled by Euler–Euler and Euler–Lagrangian approaches. Similarly, magnetohydrodynamics phenomena in tundish are modeled using Maxwell equations. Numerous combinations of models are used to investigate multiphase flow in the tundish. However, it is not possible to cover each model in this chapter. Therefore, the most popular models are described as given in Figure 4.2 and their associated governing equations are presented.

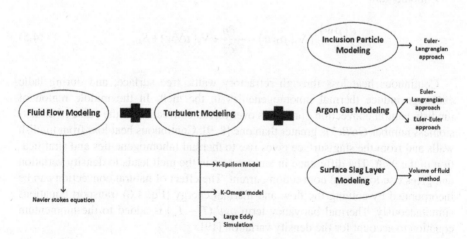

FIGURE 4.2 Constituents of mathematical modeling in tundish.

4.2.1 FLUID FLOW MODELING

Flow dynamics, inclusion motion, entrainment of slag and gas in tundish are mainly governed by fluid flow. The transition from laminar to turbulent flow is given by a dimensionless quantity known as the Reynold number [17] (Eq. 4.1). The characteristically high length of tundish and low kinematic viscosity of molten steel give rise to turbulent flow in the tundish. However, laminar flow is mainly prevalent near the tundish wall. The governing equations of continuity and momentum are fundamental building blocks of fluid flow, given in Eqs. (4.2)–(4.5) [18].

$$Re = \frac{\rho u L}{\mu} \tag{4.1}$$

$$\frac{\partial \rho}{\partial t} + \nabla.(\rho u) = 0 \tag{4.2}$$

Continuity equation

$$\frac{d(\rho u)}{dt} + \nabla.(\rho u u) = -\frac{\partial p}{\partial x} + \nabla.(\mu \nabla u) + S_{Mx} \tag{4.3}$$

X-momentum
Y-momentum

$$\frac{d(\rho v)}{dt} + \nabla.(\rho v u) = -\frac{\partial p}{\partial y} + \nabla.(\mu \nabla v) + S_{My} \tag{4.4}$$

Z-momentum

$$\frac{d(\rho w)}{dt} + \nabla.(\rho w u) = -\frac{\partial p}{\partial z} + \nabla.(\mu \nabla w) + S_{Mz} \tag{4.5}$$

Continuous heat loss through refractory walls, free surface, and during ladle exchange induce thermal inhomogeneities in the melt. In the middle region of tundish, natural convection dominates over forced convection as the value of dimensionless number Gr/Re^2 is greater than one [8, 9]. Continuous heat loss from tundish walls and from the slag surface gives rise to thermal inhomogeneities and stratification in the melt. The difference in temperature in the melt leads to density variation and gives rise to natural convection current. The effect of natural convection can be incorporated by solving the flow and thermal energy (Eq. 4.6) transport equations simultaneously. Thermal buoyancy term $\rho g \beta_T (T - T_\infty)$ is added to the momentum equation to account for the density variation [19].

$$\frac{d(\rho i)}{dt} + \nabla.(\rho i u) = -p\nabla.u + \nabla.(k\nabla T) + \Phi + S_i \tag{4.6}$$

4.2.1.1 Flow Characteristics

Tundish operations should ensure that desired steel composition, cleanliness and inclusion-free steel is delivered to mold. For that, close control of flow dynamics in the tundish is necessary. A schematic diagram of the fluid flow in the tundish is shown in Figure 4.3. Various combination of flow control devices (FCDs) is used to control flow in the tundish. It is observed that the inclusion removal tendency increases with the use of dams. For a particular position of the dam, inclusion removal increases with dam height [1]. The pouring chamber or impact pad helps in turbulent suppression and inclusion flotation [4]. The optimal size of baffle and impact pad increase uniformity of molten steel [3]. An experimental technique known as the stimulus response technique is used for finding the distribution of <u>residence</u> time of fluid in the vessel. In this technique, radioactive material or tracer is added into the fluid stream through a tundish nozzle. Step input and pulse input are the two most commonly used methods for tracer injection. Tracer concentration at exit is monitored and results are plotted in the form of dimensionless concentration v/s dimensionless time curve, also known as the residence time distribution (RTD) curve, and normally used to characterize the flow in mixing (Plug and Well mixed) volume and dead volume in the tundish.

If V is the volume of tundish and Q is the volumetric flow rate, then the average time spent by molten steel in the tundish or theoretical average residence time will be:

$$\bar{t} = \frac{V}{Q}$$

Dimensionless time can be obtained by dividing any time with theoretical average residence time:

$$\theta = \frac{t}{\bar{t}}$$

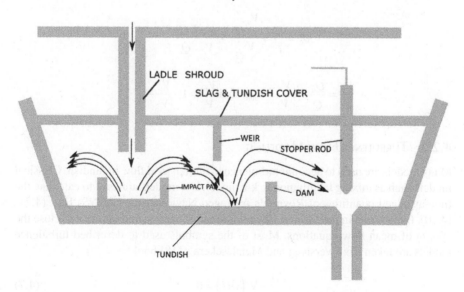

FIGURE 4.3 Flow in tundish.

Similarly, the dimensionless concentration for pulse input of tracer C can be calculated by dividing any concentration of tracer at the exit with the average concentration:

$$C = \frac{c}{q/V}$$

where q is the quantity of tracer.

Flow systems in the tundish may be classified by the type of flow they exhibited and can be divided into three categories [20, 21].

Plug Flow: In this type of flow system, all fluid elements have the same speed and identical residence time. Fluid elements preserve their identity during the passage. In the plug flow region, longitudinal mixing is nonexistent, however, there may be transverse mixing to any extent.

Well Mixed Flow: In this case, the tracer is dispersed immediately after introduced at the inlet. Maximum mixing possible in the vessel. Therefore tracer concentration at exit is equal to that in the vessel [21].

Dead Flow: It is also called inactive volume. Fluid elements spend longer time in the vessel. This type of behavior indicates inefficient use of the reactor volume and should be avoided or minimized in practice.

Quantitative analysis of these flow regions is used to investigate the tundish performance, and is as follows [20].

$$\overline{\theta}_c = \frac{Mean\ residence\ time\ upto\ \theta = 2}{Mean\ residence\ time} = \frac{\overline{t_c}}{\overline{t}}$$

$$\overline{\theta}_c = \frac{\overline{t_c}}{\overline{t}} = \frac{V_a/Q_a}{V/Q} = \frac{V_a}{V} * \frac{Q}{Q_a}$$

$$\frac{V_a}{V} = \frac{Q_a}{Q}\overline{\theta}_c, \frac{V_p}{V} = \theta_{min}\ and\ \frac{V_m}{V} = 1 - \frac{V_p}{V} - \frac{V_a}{V}$$

4.2.2 TURBULENCE FLOW MODELING

Many models are used to account turbulence in molten steel flow in tundish. Classical models such as mixing length model, k-ε model, k-ω models are used to calculate the time-averaged quantities of Reynolds Averaged Navier stokes (RANS) Eqs. (4.7)–(4.10). Extra terms in the form of Reynolds stresses need to find out how to close the system of mean flow equations. Most of the symbols used to described turbulence models are taken from Versteeg and Malalasekera's [18] book.

$$\frac{\partial \overline{\rho}}{\partial t} + \nabla.\left(\overline{\rho}\tilde{U}\right) = 0 \tag{4.7}$$

Continuity
Reynolds Equations

$$\frac{\partial\left(\bar{\rho}\tilde{U}\right)}{\partial t}+\nabla.\left(\bar{\rho}\tilde{U}\tilde{\mathbf{U}}\right)=-\frac{\partial\bar{P}}{\partial x}+\nabla.\left(\mu\nabla\tilde{U}\right)+\left[-\frac{\partial(\overline{\bar{\rho}u'^{2}})}{\partial x}-\frac{\partial(\overline{\bar{\rho}u'v'})}{\partial y}-\frac{\partial(\overline{\bar{\rho}u'w'})}{\partial z}\right]+S_{Mx}$$

(4.8)

$$\frac{\partial\left(\bar{\rho}\tilde{V}\right)}{\partial t}+\nabla.\left(\bar{\rho}\tilde{V}\tilde{\mathbf{U}}\right)=-\frac{\partial\bar{P}}{\partial y}+\nabla.\left(\mu\nabla\tilde{V}\right)+\left[-\frac{\partial(\overline{\bar{\rho}u'v'})}{\partial x}-\frac{\partial(\overline{\bar{\rho}v'^{2}})}{\partial y}-\frac{\partial(\overline{\bar{\rho}v'w'})}{\partial z}\right]+S_{My}$$

(4.9)

$$\frac{\partial\left(\bar{\rho}\tilde{W}\right)}{\partial t}+\nabla.\left(\bar{\rho}\tilde{W}\tilde{\mathbf{U}}\right)=-\frac{\partial\bar{P}}{\partial z}+\nabla.\left(\mu\nabla\tilde{W}\right)+\left[-\frac{\partial(\overline{\bar{\rho}u'w'})}{\partial x}-\frac{\partial(\overline{\bar{\rho}v'w'})}{\partial y}-\frac{\partial(\overline{\bar{\rho}w'^{2}})}{\partial z}\right]+S_{Mz}$$

(4.10)

In the above equations, six additional stress terms are introduced and known as Reynold stresses. Turbulent flow can only be computed if these stress terms are known. For that, Boussinesq proposed that Reynold stress (τ_{ij}), as given in Eq. 4.11, might be proportional to mean rates of deformation:

$$\tau_{ij}=-\rho\overline{u_i'u_j'}=\mu_t\left(\frac{\partial U_i}{\partial x_j}+\frac{\partial U_j}{\partial x_i}\right)-\frac{2}{3}\rho k\delta_{ij}$$

(4.11)

k is the turbulent kinetic energy per unit mass. Depending on the values of subscript i and j values of Reynold stresses can be computed. One unknown μ_t is necessary to evaluate to close the system of mean flow Eqs. (4.7)–(4.10). In order to evaluate μ_t, two equation-based models are proposed, as follows.

4.2.2.1 Classic k-ε Model

This is a two-equation model. Two transport equations are used to represent turbulent properties of flow, one for turbulent kinetic energy and one for dissipation rate [18–22]. Eddy viscosity is determined from single turbulent length scale.

Governing equations of turbulent kinetic energy and dissipation rate are given below (Eq. 4.12 and 4.13):

$$\mu_t=\rho C_\mu\frac{k^2}{\varepsilon}$$

$$\frac{\partial\left(\rho k\right)}{\partial t}+\nabla.\left(\rho k\mathbf{U}\right)=\nabla.\left[\frac{\mu_t}{\sigma_k}\nabla k\right]+2\mu_t S_{ij}.S_{ij}-\rho\varepsilon$$

(4.12)

$$\frac{\partial(\rho\varepsilon)}{\partial t}+\nabla.(\rho\varepsilon\mathbf{U})=\nabla.\left[\frac{\mu_t}{\sigma_\varepsilon}\nabla\varepsilon\right]+C_{1\varepsilon}\frac{\varepsilon}{k}2\mu_t S_{ij}.S_{ij}-C_{2\varepsilon}\rho\frac{\varepsilon^2}{k} \qquad (4.13)$$

Where

$$C_\mu=0.09, \sigma_k=1, \sigma_\varepsilon=1.3, C_{1\varepsilon}=1.44, C_{2\varepsilon}=1.92$$

k-ε model is found to be more accurate for free shear flow and for small pressure gradient. However, at near wall, due to the dominance of viscous stress, wall damping needs to be applied. The constants C_μ, $C_{1\varepsilon}$, and $C_{2\varepsilon}$ are multiplied with the following functions [18, 23]:

$$f_\mu=\left[1-\exp(-0.0165 Re_y)\right]^2\left(1+\frac{20.5}{Re_t}\right)$$

$$f_1=\left(1+\frac{0.05}{f_\mu}\right)^3, f_2=1-\exp(-Re_t^2)$$

Variation in standard k-e model

Realizable k- ε model [24, 25] is a variant of the standard k-ε model. It contains a new formulation of turbulent viscosity and a new transport equation for the dissipation rate, ε, has been derived from an exact equation for the transport of the mean-square vorticity fluctuation. The dissipation rate can be expressed as in Eq. 4.14:

$$\frac{\partial\rho\varepsilon}{\partial t}+\frac{\partial}{\partial x_j}(\rho\varepsilon u_j)=\frac{\partial}{\partial x_j}\left[\left(\mu+\frac{\mu_t}{\sigma_\varepsilon}\right)\frac{\partial\varepsilon}{\partial x_j}\right]+\rho C_1 S\varepsilon-\rho C_2\frac{\varepsilon^2}{k+\sqrt{v\varepsilon}} \qquad (4.14)$$

Where

$$G_k=\mu_t\frac{\partial u_j}{\partial x_i}(\frac{\partial u_i}{\partial x_j}+\frac{\partial u_j}{\partial x_i})$$

$$C_1=\max.\left[0.43,\frac{\eta}{\eta+5}\right],$$

$$\eta=S\frac{k}{\varepsilon}, S=\sqrt{2S_{ij}S_{ij}}$$

$$\mu_t=\rho C_\mu\frac{k^2}{\varepsilon}$$

$$C_\mu=\frac{1}{A_O+A_S\dfrac{kU^*}{\varepsilon}}$$

$$U^*=\sqrt{S_{ij}S_{ij}+\hat{\Omega}_{ij}\hat{\Omega}_{ij}}$$

$$\hat{\Omega}_{ij} = \Omega_{ij} - 2\varepsilon_{ijk}\omega_k; \Omega_{ij} - \bar{\Omega}_{ij} - \varepsilon_{ijk}\omega_k$$

$\bar{\Omega}_{ij}$ is the mean rate of rotation tensor viewed in a rotating reference frame with the angular velocity ω_k

The model constants A_O and A_s are given by

$$A_O = 4.04, A_s = \sqrt{6}\cos\varnothing$$

$$\varnothing = \frac{1}{3}\cos^{-1}\sqrt{6}W, W = \frac{S_{ij}S_{jk}S_{ki}}{\hat{S}^3}, \hat{S} = \sqrt{S_{ij}S_{ij}}, S_{ij} = \frac{1}{2}\left(\frac{\partial u_j}{\partial x_i} + \frac{\partial u_i}{\partial x_j}\right)$$

Another variation in the standard k-ε model is the renormalization group (RNG) k-ε model [26]. In this model, small scale motion in governing equations is expressed into large scale motion and modified viscosity. The mathematical formulation of RNG k-ε model is as follows (Eq. 4.15 and 4.16)

RNG k-ε model

$$\frac{\partial(\rho k)}{\partial t} + \nabla.(\rho k U) = \nabla.(\alpha_k \mu_{eff}\nabla k) + \tau_{ij}.S_{ij} - \rho\varepsilon \tag{4.15}$$

$$\frac{\partial(\rho\varepsilon)}{\partial t} + \nabla.(\rho\varepsilon U) = \nabla.(\alpha_\varepsilon \mu_{eff}\nabla\varepsilon) + C_{1\varepsilon}^*\frac{\varepsilon}{k}\tau_{ij}.S_{ij} - C_{2\varepsilon}\rho\frac{\varepsilon^2}{k} \tag{4.16}$$

$$\tau_{ij} = -\rho\overline{u_i'u_j'} = 2\mu_t S_{ij} - \frac{2}{3}\rho k\delta_{ij},$$

$$\mu_{eff} = \mu + \mu_t, \mu_t = \rho C_\mu\frac{k^2}{\varepsilon}$$

$$C_\mu = 0.0845, \alpha_k = \alpha_\varepsilon = 1.39, C_{1\varepsilon} = 1.42, C_{2\varepsilon} = 1.68$$

and

$$C_{1\varepsilon}^* = C_{1\varepsilon} - \frac{\eta\left(1 - \dfrac{\eta}{\eta_0}\right)}{(1 + \beta\eta^3)}, \eta = \frac{k}{\varepsilon}\sqrt{2S_{ij}.S_{ij}}, \eta_0 = 4.377, \beta = 0.012$$

4.2.2.2 k-ω Model

It is also called low Re model. It gives better results in the case of low Reynold number, adverse pressure gradient, thick boundary layer compared to k-ε model. This model can be directly used for near wall treatment without applying damping functions.

$$\mu_t = \frac{\rho k}{\omega}$$

Transport equations for turbulent kinetic energy and specific turbulent dissipation rate(ω) are as follows (Eq. 4.17 and 4.18):

$$\frac{\partial(\rho k)}{\partial t} = \nabla.(\rho k U) = \nabla.\left[\left(\mu + \frac{\mu_t}{\sigma_k}\right)\nabla k\right] + P_k - \beta * \rho k \omega \tag{4.17}$$

Where

$$P_k = \left(2\mu_t S_{ij}.S_{ij} - \frac{2}{3}\rho k \frac{\partial U_i}{\partial x_j}\delta_{ij}\right)$$

$$\frac{\partial(\rho \omega)}{\partial t} = \nabla.(\rho \omega U) = \nabla.\left[\left(\mu + \frac{\mu_t}{\sigma_\omega}\right)\nabla \omega\right] + \gamma_1\left(2\rho S_{ij}.S_{ij} - \frac{2}{3}\rho \omega \frac{\partial U_i}{\partial x_j}\delta_{ij}\right) - \beta_1 \rho \omega^2 \tag{4.18}$$

Where $\sigma_k = 2$, $\sigma_\omega = 2$, $\gamma_1 = 0.553$, $\beta_1 = 0.075$, $\beta* = 0.09$

4.2.2.3   LES Model

Turbulent flow involves a wide range of eddies, in terms of their length and time scales. The scale of the largest eddies is comparable to the characteristic length of mean flow. Dissipation of turbulent kinetic energy normally occurs due to smaller eddies. In theory, it is possible to resolve a whole range of turbulent scale using direct numerical simulation (DNS). But high computational cost put limitation on this [27]. In LES, large eddies are resolved directly, while small eddies are modeled. Thus, LES falls between RANS and DNS in terms of the fraction of resolve eddies. Instead of time averaging, LES uses spatial filtering operations to separate larger and small eddies. The detailed mathematical formulation of LES can be found in many textbooks and research articles [28–34].

4.2.3   Inclusion Transport Modeling

A well-designed tundish should promote inclusion flotation and removal. Terminal rise velocity of particles is given by stokes relation (Eq. 4.19).

$$V_s = \frac{g(\rho - \rho_p)d^2}{18\mu} \tag{4.9}$$

Inclusion particles in tundish can be removed by several mechanisms. Thermal buoyancy helps inclusion to move upwards. Due to high turbulence, inclusion coalescence occurs near the inlet zone, bigger size particles form and rise upwards due to inertial buoyancy. FCDs aid in inclusion removal by changing flow dynamics. Argon gas bubbling from the bottom wall of tundish force inclusion toward slag layer [35]. In this method, the continuous phase is treated as continuum by solving Navier stokes equations, while the dispersed phase is solved by tracking many particles, bubbles, or droplets through the calculated flow field. The dispersed phase can exchange momentum, mass, and energy with the fluid phase. Inclusion transport in molten steel flow can be modeled using the Euler–Langrangian approach and governed by following the force balance in Eq.4.20. First term on the right side of the equation account drag force, the second incorporate buoyancy force and the third term $\vec{F}$ is for extra force under special circumstance. These can be Virtual force, Lift force, Brownian force, and Magnus lift force [36–40].

$$m_p \frac{d\overrightarrow{u_p}}{dt} = m_p \frac{\vec{u} - \overrightarrow{u_p}}{\tau_r} + m_p \frac{\vec{g}\left(\rho_p - \rho\right)}{\rho_p} + \vec{F} \tag{4.20}$$

Particles relaxation time calculated by

$$\tau_r = \frac{\rho_p d_p^2}{18\mu} \frac{24}{C_d Re}$$

Relative Reynold number is defined as

$$Re \equiv \frac{\rho d_p \left|\overrightarrow{u_p} - \vec{u}\right|}{\mu}$$

To model the chaotic effect of turbulent eddies, on particle motion a discrete random walk model is applied during inclusion trajectory calculations in which a random velocity vector u_p' is added to the calculated time-averaged vector, $\bar{u}_p$ to obtain the inclusion velocity u_p at each time step as it travels through the established flow field. Each random component of the inclusion velocity is proportional to the local turbulent kinetic energy level, k according to the following equation

$$u_p' = \zeta_i \sqrt{u_i'^2} = \zeta_i \sqrt{\frac{2k}{3}}$$

Where ζ_i is a random number, normally distributed between -1 and 1 that changes at each time step. Inclusions are injected computationally at many different locations distributed homogeneously over the inlet plane. Each trajectory is calculated through the constant steel flow field until the inclusion either is trapped or exits the tundish outlet.

4.2.4 SLAG MODELING

Slag cover helps by protecting the molten metal from surrounding exposure and therefore minimize the chances of oxidation. It also aids inclusion removal by absorbing the foreign particles and hence improves steel cleanliness. The volume of fluid (VOF) model is used to track the metal slag interface. In this model, single set of Navier stokes equations are used to tracking the volume fraction of each fluid throughout the domain. This model is based on the condition that the two or more fluids are not interpenetrating. For each additional phase, a variable called volume fraction is introduced: the volume fraction of the phase in the computational cell. In each control volume, the sum of volume fractions of all the phases equal to unity. The variable and properties of flow in each cell will be dependent on phases which are present in that cell and also affected by their volume fraction values. In summary

$$\alpha_q = 0 : The\ cell\ is\ empty\left(of\ the\ q^{th}\ fluid\right)$$

$$\alpha_q = 1 : The\ cell\ is\ full\left(of\ the\ q^{th}\ fluid\right)$$

$$0 < \alpha_q < 1 : The\ cell\ contains\ interface\ between\ the\ q^{th}\ fluid\ or\ one\ or\ more\ fluid$$

4.2.5 ARGON GAS MODELING

Gas bubbling as a flow modifier is used to control the flow pattern in tundish. It also enhances the uniformity of molten metal in terms gas shrouding is also used to protect the incoming metal from air contact, thus avoiding oxidation. Inclusion flotation increase in presence of gas bubbling, as bubble particle attachment aid in their upwards rising. Many studies are reported on modeling gas curtain [11, 12, 14, 15, 41, 42]. Both Euler–Lagrangian and Euler–Euler approaches are used to model gas bubbling. In Euler–Euler modeling separate Navier stokes equations and energy equations are solved for both liquid and gas phase. A single pressure is shared by all phases. The volume of phase q, v_q is defined as

$$V_q = \int \alpha_q dV$$

Where $\sum_{q=1}^{n} \alpha_q = 1$

The effective density of phase q is

$$\hat{\rho}_q = \alpha_q \rho_q$$

4.3 INTELLIGENT MODELING IN TUNDISH STEELMAKING

All the relevant phenomena occur in tundish steelmaking is difficult to be model by mathematical modeling due to two reasons. One, apart from process models presented above, little knowledge is available for modeling of other metallurgical phenomena such as oxidation deoxidation of molten steel, phase transformation, chemical reactions within the liquid metal and at slag metal interface. Second, the high computational cost associated with multiphase modeling put another limitation in coupling the different models. Therefore, other powerful modeling techniques

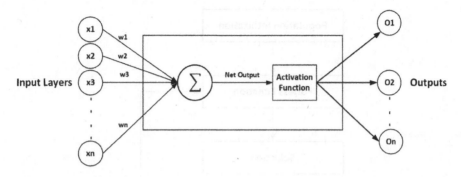

FIGURE 4.4 Basic structure of ANN.

such as artificial neural networks (ANN), genetic algorithm (collectively referred to as artificial intelligence (AI) or artificial intelligence methods) are getting popular [19]. ANN is increasingly used in the process control of steelmaking. Application area includes flow control, flow optimization, parametric optimization etc. A typical structure of ANN is shown in Figure 4.4. Input data is received in different categories of input variables ($x1, x2....xn$). These data are processed by their respective weightage or strength ($w1, w2, ...wn$) and summed up. After that, processed input transformed into output using the activation function. Depending on the type of problem, various functions are used, such as sigmoid function, threshold function, hyperbolic tangent function, etc. Different parameters such as input flow rate, temperature, tundish depth, number of outlets etc. can be input variables. Output can be mixing volume fraction inclusion removal rate etc. fraction, plug volume fraction.

Optimization method, Genetic algorithm (GA) is based on natural evolution process. GA simply manipulates various values of input variables in search space till best solution is obtained. The basic structure of GA is shown in Figure 4.5. It starts with initialization, wherein the initial population of candidates is generated. In this step, values of initial variables, which satisfying given constraints are taken as initial populations. Fitness function took candidate solution as input and produce output. The function should possess two characteristics, one it should fast, second, it should quantitatively measure the accuracy of the solution. In the selection step, a pool of parents chosen based on a good fitness score. Crossover creates two individuals from the existing parent population. There are multiple ways in which, this is performed. Examples are: one point crossover, multipoint crossover, uniform crossover. In mutation, idea is to keep random genes to maintain diversity in population. Unlike crossover, mutation produces offspring from single parent string. Different types of termination conditions are used to determine when GA will be end, like, when improvements in population stopped in iterations when the objective function value reaches a predetermined value. Fluid flow optimization study in tundish is performed by Amit et al. [43] using GA. They have taken the velocity of injection and injection depth as input variables, and the inverse of stress as a fitness function. Significant computation burden is reduced as convergence obtained after 50 iterations compared to 2000–5000 required CFD.

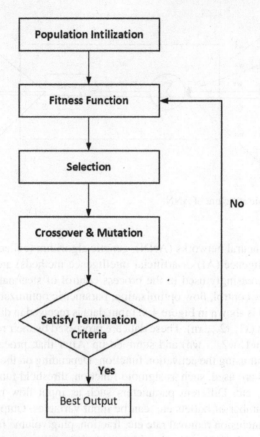

FIGURE 4.5 Basic structure of GA.

NOMENCLATURES

u	
u, v, w	Flow velocity in x, y, and z-direction
p	Pressure
t	Time
g	Gravitational acceleration (9.81 m/s^2)
C_1, C_2, C_μ	Turbulent constants
F_D	Drag force
k	Turbulent kinetic energy
ε	Dissipation rate
T	Temperature
μ	Dynamic viscosity of the fluid
S_M	Momentum source term
v_a, v_p, v_d	Active volume, plug volume, and dead volume
ν	Kinematic viscosity
ϕ	Dissipation function

k	Thermal conductivity
U	Mean component of velocity (U, V, W)
u'	Fluctuating components of velocity (u', v', w')
P	Mean flow pressure
p'	Fluctuating flow pressure
$\bar{U}$	Time-averaged variable
$\tilde{U}$	Density averaged variable
μ_t	Turbulent viscosity
δ_{ij}	Kronecker delta
S_{ij}	Rate of deformation of the fluid element in a turbulent flow
u_p	Inclusion particle velocity
ρ_p	Inclusion particle density
d_p	Inclusion particle diameter
R_e	Relative Reynold number
Re_t	Turbulent Reynold number
t_a	Theoretical mean residence time
$\bar{t}$	Actual residence time
$\bar{t_c}$	Two times the theoretical mean residence time
t_{min}	Minimum residence time
t_{peak}	Time taken to attain peak concentration of tracer at outlet
t_p	Plug flow residence time

REFERENCES

1. Jha PK, Rao PS, Dewan A Effect of height and position of dams on inclusion removal in a six strand tundish. *ISIJ Int.* 2008;48(2):1–4.

2. Morales RD, Torres-Alonso E, Ramirez-Lopez P, Garcia-Demedices L, Najera-Bastida A Dissipation of turbulent kinetic energy in a tundish by inhibitors with different designs. *steel Res int.* 2007;164(c):318–326.

3. Zhong L Cai, Wang M An, Chen B Yu, Wang C Rong, Zhu Y Xiong Flow control in six-strand billet continuous casting tundish with different configurations. *J Iron Steel Res Int.* 2010;17(7):7–12. doi:10.1016/S1006-706X(10)60148-3

4. Tripathi A, Ajmani SK Effect of shape and flow control devices on the fluid flow characteristics in three different industrial six strand billet caster tundish. *ISIJ Int.* 2011;51(10):1647–1656. doi:10.2355/isijinternational.51.1647

5. Tripathi A, Ajmani SK Numerical investigation of fluid flow phenomenon in a curved shape tundish of billet caster. *ISIJ Int.* 2005;45(11):1616–1625. doi:10.2355/isijinternational.45.1616

6. Jha PK, Dash SK Effect of outlet positions and various turbulence models on mixing in a single and multi strand tundish. *Int J Numer Methods Heat Fluid Flow.* 2002;12(5):560–584. doi:10.1108/09615530210434296

7. Palafox-Ramos J, De J, López-Ramírez S, Morales RD Melt flow optimisation using turbulence inhibitors in large volume tundishes. *Ironmak Steelmak.* 2001;28(2):101–109. doi:10.1179/030192301678000

8. Sheng DY, Jonsson L Two-fluid simulation on the mixed convection flow pattern in a nonisothermal water model of continuous casting. *Metall Mater Trans B.* 2000;31 (August):867–875.

9. Sheng DY, Jonsson L Investigation of transient fluid flow and heat transfer in a continuous casting tundish by numerical analysis verified with nonisothermal water model
   experiments. *Metall Mater Trans B Process Metall Mater Process Sci.* 1999;30(5):979–
   985. doi:10.1007/s11663-999-0103-9

10. Lopez Ramirez S, Morales RD, Serrano JAR Numerical simulation of the effects of
    buoyancy forces and flow control devices on fluid flow and heat transfer phenomena of
    liquid steel in a tundish. *Numer Heat Transf Part A Appl.* 2000;37(1):69–85.
    doi:10.1080/104077800274424

11. Liu Y, Ersson M, Liu H, Jönsson P, Gan Y Comparison of euler-euler approach and
    euler–lagrange approach to model gas injection in a ladle. *Steel Res Int.* 2019;90(5):1–
    13. doi:10.1002/srin.201800494

12. Holzinger G, Thumfart M Flow interaction in continuous casting tundish due to bubble
    curtain operation. *Steel Res Int.* 2019;90(6):1–11. doi:10.1002/srin.201800642

13. Neves L, Tavares RP Analysis of the mathematical model of the gas bubbling curtain
    injection on the bottom and the walls of a continuous casting tundish. *Ironmak Steelmak.*
    2017;44(8):559–567. doi:10.1080/03019233.2016.1222122

14. Morales RD, Ramos-Banderas A, Sánchez-Perez R Mathematical simulation and mod-
    eling of steel flow with gas bubbling in trough type tundishes. *AISTech - Iron Steel
    Technol Conf Proc.* 2004;2(5):867–877.

15. Zhong LC, Li LY, Wang B, Zhang L, Zhu LX, Zhang QF Fluid flow behaviour in slab
    continuous casting tundish with different configurations of gas bubbling curtain.
    Ironmak Steelmak. 2008;35(6):436–440. doi:10.1179/174328108X318365

16. Yang H, Vanka SP, Thomas BG Mathematical modeling of multiphase flow in steel
    continuous casting. *ISIJ Int.* 2019;59(6):956–972. doi:10.2355/isijinternational.
    ISIJINT-2018-743

17. Reynolds O XXIX. An experimental investigation of the circumstances which determine
    whether the motion of water shall be direct or sinuous, and of the law of resistance in
    parallel channels. *Philos Trans R Soc London.* 1883;174:935–982. doi:10.1098/
    rstl.1883.0029

18. Versteeg HK, Malalasekera M *An Introduction to Computational Fluid Dynamics.* 2
    Edition. PHI; 2007.

19. Mazumdar D, Evans JW *Modeling of Steelmaking Processes.* CRC Press; 2010.
    doi:10.1007/978-981-15-2437-0_14

20. Yogeshwar S, Emi T Melt flow characterization in continuous casting tundishes. *ISIJ
    Int.* 1996;72(6):293–300.

21. Szekely J, Ilegbusi J, Olusegun J *The Physical and Mathematical Modeling of Tundish
    Operations.* Springer-Verlag doi:10.1007/978-1-4613-9626-0

22. Launder BE, Spalding DB The numerical computation of turbulent flows. *Comput
    Methods Appl Mech Eng.* 1974;269. doi:10.1007/JHEP10(2012)057

23. Patel VC, Rodi W, Scheuerer G Turbulence models for near-wall and low reynolds num-
    ber flows – a review. *AIAA J.* 1985;23(9):1308–1319. doi:10.2514/3.9086

24. Shih T-H, Liou WW, Shabbir A, Yang Z, Zhu I A new k-epsilon eddy-viscosity model
    for high reynolds number turbulent flows. *Comput Fluids.* 1995;24(3):227–238.
    doi:10.1016/0045-7930(94)00032-T

25. Gupta VK, Jha PK, Jain PK Transient numerical simulation of solidification in continu-
    ous casting slab caster. In: Phanden RK, Mathiyazhagan K, Kumar R, Paulo Davim J,
    eds. *Advances in Industrial and Production Engineering.* Singapore: Springer Singapore;
    2021:637–649.

26. Yakhot V, Orszag SA, Thangam S, Gatski TB, Speziale CG Development of turbulence models for shear flows by a double expansion technique. *Phys Fluids A*. 1992;4(7):1510–1520. doi:10.1063/1.858424

27. Inc. ANSYS. ANSYS Manual 12.0. Ansys Documents.

28. Jin K, Vanka SP, Thomas BG. Large eddy simulations of the effects of embr and sen submergence depth on turbulent flow in the mold region of a steel caster. *Metall Mater Trans B Process Metall Mater Process Sci*. 2017;48(1):162–178. doi:10.1007/s11663-016-0801-z

29. Cho SM, Thomas BG, Kim SH Transient two-phase flow in slide-gate nozzle and mold of continuous steel slab casting with and without double-ruler electro-magnetic braking. *Metall Mater Trans B Process Metall Mater Process Sci*. 2016;47(5):3080–3098. doi:10.1007/s11663-016-0752-4

30. Singh R, Thomas BG, Vanka SP Large eddy simulations of double-ruler electromagnetic field effect on transient flow during continuous casting. *Metall Mater Trans B Process Metall Mater Process Sci*. 2014;45(3):1098–1115. doi:10.1007/s11663-014-0022-2

31. Liu Z, Li B, Jiang M, Zhang L, Xu G Large eddy simulation of unsteady argon/steel two phase turbulent flow in a continuous casting mold. *Jinshu Xuebao/Acta Metall Sin*. 2013;49(5):513–522. doi:10.3724/SP.J.1037.2012.00760

32. Li B, Liu Z, Qi F, Wang F, Xu G Large eddy simulation for unsteady turbulent flow in thin slab continuous casting mold. *Jinshu Xuebao/Acta Metall Sin*. 2012;48(1):23–32. doi:10.3724/SP.J.1037.2011.00464

33. Yuan Q, Thomas BG, Vanka SP Study of transient flow and particle transport in continuous steel caster molds: Part I. Fluid flow. *Metall Mater Trans B Process Metall Mater Process Sci*. 2004;35(4):685–702. doi:10.1007/s11663-004-0009-5

34. Ji CB, Li JS, Yang SF, Sun LY Large eddy simulation of turbulent fluid flow in liquid metal of continuous casting. *J Iron Steel Res Int*. 2013;20(1):34–39. doi:10.1016/S1006-706X(13)60041-2

35. Zhang L Fluid flow, heat transfer and inclusion motion in a four-strand billet continuous casting tundish. *Steel Res Int*. 2005;76(11):784–796. doi:10.1002/srin.200506097

36. Oesterlé B, Bui Dinh T Experiments on the lift of a spinning sphere in a range of intermediate Reynolds numbers. *Exp Fluids*. 1998;25(1):16–22. doi:10.1007/s003480050203

37. Tsuji Y, Oshima T, Morikawa Y Numerical simulation of pneumatic conveying in a horizontal pipe. *KONA Powder Part J*. 1985;3(3):38–51. doi:10.14356/kona.1985009

38. Rubinow SI, Keller JB The transverse force on a spinning sphere moving in a viscous fluid. *J Fluid Mech*. 1961;11(3):447–459. doi:10.1017/S0022112061000640

39. Gosman AD, Ioannides E Aspects of computer simulation of liquid-fuelled combustors. *AIAA Pap*. 1981;7(6):482–490. doi:10.2514/6.1981-323

40. Saffman PG The lift on a small sphere in a slow shear flow. *J Fluid Mech*. 1965;22(2):385–400. doi:10.1017/S0022112065000824

41. Chang S, Zhong L, Zou Z Simulation of flow and heat fields in a seven-strand tundish with gas curtain for molten steel continuous-casting. *ISIJ Int*. 2015;55(4):837–844. doi:10.2355/isijinternational.55.837

42. Cwudziński A Numerical and physical modeling of liquid steel flow structure for one strand tundish with modern system of argon injection. *Steel Res Int*. 2017;88(9):1–14. doi:10.1002/srin.201600484

43. Kumar A, Chakraborty S, Chakraborti N Fluid flow in a tundish optimized through genetic algorithms. *Steel Res Int*. 2007;78(7):517–521. doi:10.1002/srin.200706242

5 Analysis of Inclusion Behavior In-Mold During Continuous Casting

Rajneesh Kumar
Indian Institute of Technology Roorkee, India

Ambrish Maurya
National Institute of Technology Patna, India

Pradeep Kumar Jha
Indian Institute of Technology Roorkee, India

CONTENTS

5.1 INTRODUCTION

The continuous casting process, also referred to as strand casting, is a manufacturing industry process to cast a continuous metal length. The shape of the mold cavity determines the shape of the finished product. The casting is traveling downward, its size increasing with time. New molten metal is supplied continuously to the mold to solidify casting at precisely the correct rate.

There are several reasons for the popularity of continuous casting over the ingot casting process. These include increased productivity, higher yield, and reduced costs, directly affecting any steel-making operation's bottom line. Thus, continuous casting is a dominant process for the production of finished and semi-finished steel. The increase in popularity of the continuous casting process has occurred concurrently with the tremendous advancement in the development of new materials with comparable and, in many instances, superior mechanical properties to steel. In reality, new materials such as metal matrix composites, structural ceramics, and ceramic composites have long been expected to replace steel quickly. But steel is still the most cost-effective material for structural applications. However, a more significant reason for the continued popularity of steel has been the tremendous improvement in the quality of steel products, possibly driven by composite materials' development over the last several years.

For steel to remain a competitive material, it is necessary to improve further the quality of steel produced in the steel industries. Reducing cost via quality improvement can be possible with reduced wastage of material, energy, and human resources. Further improvements in the quality of steel produced would demand an even closer control over the composition and reduction of defects.

The molten steel flow behavior inside the mold significantly affects the quality of steel. It depends on the many complex phenomena such as surface-level fluctuations, inclusion transport, and superheat [1]. Figure 5.1 shows the schematic

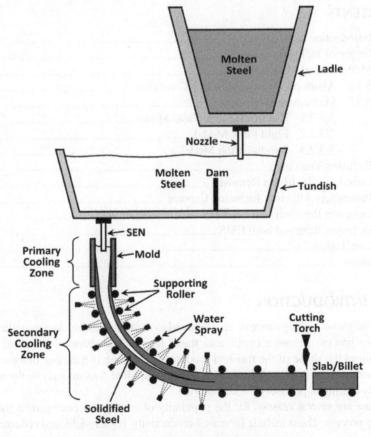

FIGURE 5.1   Schematic of the continuous casting process [2].

of the continuous casting process. Molten steel is poured from the ladle into the tundish in the continuous casting process, continuously filling the mold through a submerged entry nozzle (SEN). The mold is water-cooled to remove enough heat to solidify a thick enough shell to prevent the solidified shell from breaking due to Ferro-static pressure. Below the mold, heat is further removed by the strand surface by spraying the water, and the strand finally becomes fully solid when it reaches the "metallurgical length".

The inclusions trapped in solid steel can decrease the surface and subsurface quality of the final products. Inclusion removal in the metallurgical reactor, on the other hand, is a complex process that involves fluid flow, inclusion collision and coalescence, flotation, wall adhesion, and gas bubbles [3]. Inclusions in the mold arise due to (a) chemical reactions in the mold during freezing, (b) coming with incoming molten steel, and (c) entrapment of mold powder below the surface, which causes several kinds of internal and surface defects in steel products. Non-metallic inclusions in liquid steel can significantly influence steel properties, such as decreased ductility, decreased fracture toughness, induced cracks, poor polishability, reduced resistance to corrosion, and lower resistance hydrogen-induced cracks [4].

5.2   ORIGIN OF INCLUSIONS

The continuously casting unit essentially consists of three metallurgical reactors – the ladle, the tundish, and the mold, as shown in Figure 5.1. The ladle is a batch reactor, whereas the tundish and the mold are continuous reactors. The defects that arise in a cast slab (or billet) are intimately related to fluid flow, heat transfer, and mass transfer phenomena occurring in each of these reactors. Cracking that occurs during the solidification of the strand is influenced by thermal stresses. The mass transfer influences macro-segregation during the solidification of billets and slabs. The mass transfer also influences the distribution of alloying additions made in the ladle. Its final inclusion content determines the cleanliness of steel products.

Non- metallic inclusions are generated during the various processing steps of the steel-making processes, primarily during the deoxidation of steel by adding deoxidants such as aluminum, silicon, and manganese. Inclusions may also be generated by the erosion of the vessels' refractory lining, by reoxidation of the steel during its transfer from the ladle to tundish, and by the slag's entrapment in the ladle, tundish, or the mold. Fluid flow in continuous casting strongly influences the final inclusion content [5]. Appropriate fluid flow patterns help in the inclusion removal by aiding flotation of inclusions; fluid flow can also be the source of inclusion generation in the cast product by the erosion of the refractory linings and its influence slag entrapment.

In the mold, various physicochemical processes involve viscous flow, interfacial phenomena, mass transfer, etc., at high temperatures. Figure 5.2 shows mold fluid flow and dynamic multi-phase physicochemical reactions [6]. Chemical reactions cause inclusions to form during solidification. Oxides, sulfides, oxysulfides, nitrides, and carbides are the typical forms of inclusions that form during the solidification of molten steel. Homogeneous distributions of small size (<40 μm) inclusions are being used to control the products' microstructure. Most of the large size oxides and nitrites

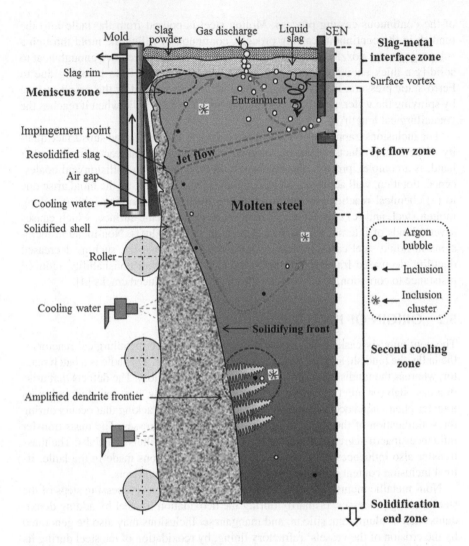

FIGURE 5.2 Mold fluid flow and dynamic multi-phase physicochemical reactions (From Liu, Z., Li, B., *Powder Technology*, 323, 403–415, 2018. With permission.)

form when steel is in the molten state. It can cause defects in casting materials, such as processing difficulties and failures, reducing productivity, reducing product properties, and reducing yield if it is not controlled or removed from the steel [7].

Fluid flow in continuous casting plays a vital role in deciding the quality of the cast slab. Realizing these, several researchers have attempted to study fluid flow in various stages of the continuous casting process [8–11]. The study of fluid flow in continuous casting is made difficult by the process's temperature and the molten metal's opacity, preventing direct visual observation of molten metal flow in the mold. To overcome these difficulties, researchers have used water modeling techniques [6, 12–14] and mathematical modeling [15–18].

5.3 MATHEMATICAL MODELING

The mathematical model involves describing certain phenomena in the real world through a single or system of mathematical equations – algebraic, and differential, and integral. The aim of using such a mathematical description is usually to obtain a quantitative estimate of various parameters of the process in question without resorting to expensive and cumbersome experimentation. In some instances, mathematical modeling may provide the only means of estimating specific process parameters.

In a continuous casting system, fluid flow, heat transfer, solidification, slag-metal reactions, inert gas purging of mold, nozzle clogging, inclusion, inclusion generation, and removal are just some of the many processes that occur in the system. While developing a system model, it is not necessary to include every phenomenon occurring within the system. A model that incorporates every one of these phenomena, though useful, would only complicate the system's description without providing any further insight into the nature of the various phenomena. Therefore, it is necessary to employ reasonable simplification to better understand the phenomenon under consideration without introducing unnecessary complications. Simultaneously, it should ensure that the model is sufficiently realistic so that the model results can provide an excellent qualitative insight into the phenomenon occurring in the existing system.

5.3.1 ASSUMPTIONS AND BOUNDARY CONDITIONS

The fluid flow in the continuous casting mold with an electromagnetic stirrer (EMS) interacts with the magnetic field and solidification, a combined solution of Maxwell's equation, the momentum conservation equation, solidification equations, and the solute transport equation needed to represent transport phenomena in the mold quantitatively.

The mathematical model for fluid flow, solidification, and inclusion removal with electromagnetic forces is based on the following assumptions and boundary conditions:

- As compared to the longitudinal direction, the temperature gradient along the transverse direction is slightly higher. As a consequence, heat conduction along the longitudinal direction is ignored to save time in the calculation.
- The molten melt fluid is Newtonian, incompressible, and viscous, with an unsteady fluid flow.
- Since the presence of flux powder at the meniscus reduces heat loss to a marginal degree, heat loss from the meniscus is presumed to be adiabatic.
- Despite the fact that the thermo-physical properties of steel's solid and liquid phases are influenced by temperature and composition, they were considered constant.
- Any shrinkage caused by solidification or wall shear stresses is not taken into account.
- Since the mold was considered stationary, the effect of mold oscillation on the variation of mold wall temperature near the meniscus was negligible.

- A linear mixture rule describes the dependence of both liquidus and solidus temperatures on the alloying elements in the heat transfer and solidification model.
- The free surface is assumed to be smooth, and the effect of slag on fluid flow is ignored.
- During the solidification process, induced current's Joule heating produced is minimal compared to the latent heat released, which is ignored in the modeling.
- The effects of inclusion movement on fluid flow are not taken into account.
- The influence of fluid flow on the electromagnetic field is ignored due to the low Reynolds number of M-EMS. The electromagnetic field is considered quasi-static.
- The solidified steel shell is set to move along the casting direction at a pulling velocity equal to the casting speed.
- The inclusions are spherical and move independently of one another.

Since various researchers incorporate parameters and boundary conditions differently for evaluating the effectiveness and improving the overall degree of accuracy, the above assumptions and boundary conditions are not applicable for all cases. As a result, depending on the modeling methodology used, the boundary conditions are chosen, the form of analysis and the number of parameters considered during modeling, a few of the above assumptions and boundary conditions may change.

5.3.2 Mathematical Model

5.3.2.1 Electromagnetic Field Model

An electromagnetic stirrer aims to create a rotational flow in molten steel by generating the Lorentz force that increases the tangential velocity around the mold's perimeter. Maxwell's equations can be used to calculate the electromagnetic field are defined as follows [19]:

Gauss's law:

$$\nabla . \vec{D} = q \tag{5.1}$$

Gauss's law for magnetism:

$$\nabla . \vec{B} = 0 \tag{5.2}$$

Faraday's law of induction:

$$\nabla \times \vec{E} = -\frac{\partial \vec{B}}{\partial t} \tag{5.3}$$

Ampere's circuital law (with Maxwell's addition):

$$\nabla \times \vec{H} = \vec{J} + \frac{\partial \vec{D}}{\partial t} \tag{5.4}$$

Where $\vec{H}$ and $\vec{E}$ are the magnetic and electric field intensities. $\vec{B}$ and $\vec{D}$ are magnetic and electric field flux densities. Electric charge density is represented by q, and electric current density is represented by $\vec{J}$. The field's displacement in space is defined by Equations (5.1) and (5.2), while Equations (5.3) and (5.4) show how it moves in the domain. Ohm's law describes the electric current density in a moving fluid as [19]:

$$\vec{J} = \sigma\left(\vec{E} + \vec{U} \times \vec{B}\right) \tag{5.5}$$

The induction equation is derived from Ohm's law and Maxwell's equations as follows:

$$\frac{\partial \vec{B}}{\partial t} + \left(\vec{U}.\nabla\right)\vec{B} = \frac{1}{\mu\sigma}\nabla^2\vec{B} + \left(\vec{B}.\nabla\right)\vec{U} \tag{5.6}$$

The applied magnetic field $\vec{B}_0$ is added to the induced magnetic field $\vec{b}$ to form the magnetic field.

$$\vec{B} = \left(\vec{B}_0 + \vec{b}\right) \tag{5.7}$$

The electromagnetic force, $\vec{F}$, that varies over time as a result of the interaction between an electric current and a magnetic field also added in the momentum conservation Equation (5.10), can be calculated using

$$\vec{F} = \vec{J} \times \vec{B} = \vec{J} \times \left(\vec{B}_0 + \vec{b}\right) \tag{5.8}$$

5.3.2.2 Fluid Flow Model

Continuity (mass balance) and momentum conservation (time-dependent Navier-Strokes) equations for solving the turbulent fluid flow are shown as follows:

$$\Delta.\vec{U} = 0 \tag{5.9}$$

$$\frac{\partial}{\partial t}\rho\vec{U} + \rho\nabla.\left(\vec{U}\vec{U}\right) = -\nabla P + \nabla.\left[\mu_{eff}\left(\nabla.\vec{U}\right)\right] + \rho\vec{g} + S + \vec{F} \tag{5.10}$$

The liquid's effective dynamic viscosity μ_{eff} in the momentum conservation Equation (5.10) is the addition of turbulent viscosity and dynamic viscosity and enthalpy-porosity technique is being used for solidification modeling to define each cell's porosity in the domain in the form of a liquid fraction (f_l). The porous zone is treated in liquid-solid mushy with porosity equal to the fraction of liquid. When the molten metal is fully solidified, the liquid fraction becomes zero in that region and fluid velocities also become zero. The momentum sink (S) in the momentum conservation Equation (5.10) is well-defined as:

$$S = \frac{\left(1 - f_l\right)^2}{\left(f_l^3 + \xi\right)} A_{mush}\left(\vec{U} - \vec{U}_{pull}\right) \tag{5.11}$$

A small positive number ξ in the above equation is used to avoid zero. Fluid flow in the mushy region is affected by the mushy constant, A_{mush}. Many researchers reported a value of A_{mush} between 10^4 to 10^8 for various solidification processes. The momentum sink term's value is significant for solidified material, which dominates in the Navier–Strokes Equation (5.10) that extinguishes these regions' velocities. The newly solidified material moves out from the mold with $\vec{U}_{pull}$ velocity set to be equal to casting speed [19].

The realizable k-ε turbulence model is coupled with a solidification model to determine the fluid flow's turbulence. It avoids the singularity in the mushy zone region, which arises due to the low Reynolds number. Superior performance is provided for rotation flows, boundary layers under adverse solid pressure gradients, separation, and recirculation. Equations for turbulent kinetic energy (k) and dissipation rate (ε) are represented as:

$$\rho\frac{\partial k}{\partial t}+\nabla.\left(\rho k\vec{U}\right)=\nabla.\left[\left(\mu_l+\alpha_k\mu_t\right)\nabla k\right]+G-\rho\varepsilon+S_k \tag{5.12}$$

$$\rho\frac{\partial\varepsilon}{\partial t}+\nabla.\left(\rho\varepsilon\vec{U}\right)=\nabla.\left[\left(\mu_l+\alpha_\varepsilon\mu_t\right)\nabla\varepsilon\right]+C_{1\varepsilon}\frac{\varepsilon}{k}G-C_{2\varepsilon}\rho\frac{\varepsilon^2}{k+\sqrt{v\varepsilon}}+S_\varepsilon \tag{5.13}$$

Where $C_1\varepsilon=1.44$ and $C_2\varepsilon=1.92$ are the model parameters. α_k and α_ε are the inverse turbulent Prandtl numbers for k and ε. S_k and S_ε are user-defined source terms. G represents the generation of turbulence kinetic energy due to buoyancy.

5.3.2.3 Solidification Model
Enthalpy formulation by means of energy conservation equation is:

$$\rho\frac{\partial H}{\partial t}+\rho\nabla.\left(\vec{U}H\right)=\nabla.\left(k_{eff}\nabla T\right)+Q_L \tag{5.14}$$

H is the enthalpy of the material is the sum of the sensible enthalpy h, and the latent heat, ΔH.

$$h=h_{ref}+\int_{T_{ref}}^{T}c_p dT \tag{5.15}$$

ρ is the melt density, h_{ref} is the reference enthalpy, and c_p is the specific heat, $\vec{U}$ the melt's velocity, and the liquid's effective conductivity k_{eff} is the sum of the material's thermal conductivity (k) and turbulent thermal conductivity (k_t).

The source term Q_L in Equation (5.14) is expressed as:

$$Q_L=\rho L\frac{\partial(1-f_l)}{\partial t}+\rho L\vec{U}_{pull}.\nabla(1-f_l) \tag{5.16}$$

The liquid fraction, f_l, can be represented as [19]

$$f_l = \begin{cases} 0 & if \ T < T_{solidus} \\ \dfrac{\left(T - T_{solidus}\right)}{\left(T_{liquidus} - T_{solidus}\right)} & if \ T_{solidus} < T < T_{liquidus} \\ 1 & if \ T > T_{liqidus} \end{cases} \qquad (5.17)$$

The temperature solution is essentially an iteration of the energy Equation (5.14) and the liquid fraction Equation (5.17). Using Equation (5.17) to change the liquid fraction directly causes the energy equation to converge slowly. The liquid fraction is modified using the method proposed by Voller and Swaminathan [20]. For pure metals with equivalent $T_{solidus}$ and $T_{liquidus}$, a specific heat-based method is used instead of the earlier method by Voller and Prakash [21].

5.4   INCLUSION TRACKING

Stochastic tracking model used for simulation the motions of inclusions (spherical) during continuous steel casting by solving the following equation [22]:

$$m_p \frac{d\vec{u}_p}{dt} = \frac{1}{8} \pi d_p^3 \rho_l C_D \left| \vec{u}_l - \vec{u}_p \right| \left(\vec{u}_l + \vec{u}_l' - \vec{u}_p \right) + \frac{1}{6} \pi d_p^3 \rho_p \vec{g} + \vec{F}_L$$
$$+ \frac{\rho_l \pi d_p^3}{12} \left(\frac{D\vec{u}_l}{Dt} - \frac{d\vec{u}_p}{dt} \right) + \frac{d_p^3 \pi}{6} \rho_p \frac{D\vec{u}_l}{Dt} \qquad (5.18)$$

The steady-state drag force, gravitational force, lift force, added mass force, pressure gradient force, and stress gradient force are the terms on the right-hand side (RHS) of Equation (5.18). The melt and inclusion are denoted by the superscripts l and p, respectively. The diameter is d, the mass is m, the gravity is g, and the fluctuating velocity is $\vec{u}_l'$ due to turbulence [22]. Thomas and Coworkers [11] give details about $\vec{F}_L$ and C_D.

5.5   CRITERIA FOR INCLUSION REMOVAL

The reflection angle of inclusion is supposed to be equal to the incidence angle on the liquidus iso-surface to model inclusion pushing at the solidification front. Inclusion capture criteria begin when the melt temperature is lower than the liquidus temperature, as described by Pfeiler [23]. The inclusion capture criteria are shown in Figure 5.3. The shell is modeled to entrap smaller inclusion particles than the primary dendrite arm spacing (PDAS). Many computational simulation studies [6, 11, 22] have considered the various forces operating on an inclusion near a dendritic front, which are combined to form the resultant force. Depending on the orientation of the resultant force, the inclusion will be engulfed, pushed away, or roll along the front [11, 22]. In another work [6], forces acting on inclusion along the solidification front are shown in Figure 5.4(a). In this model, if the inclusions

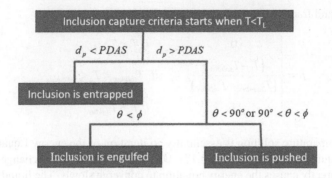

FIGURE 5.3　Flow chart of the inclusion capture criteria [23].

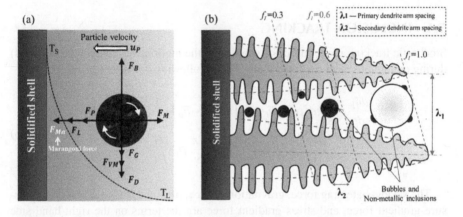

FIGURE 5.4　(a) Force balance on the inclusion along a dendritic solidification front and (b) diagram of the region where solids and liquids coexist (From Liu, Z., Li, B., *Powder Technology*, 323, 403–415, 2018. With permission.)

are greater than PDAS, the dendrites can push it instead of trapping it. When they move to the zone, $q_2(0.3 < f_l \leq 0.6)$ as shown in Figure 5.4(b), the inclusions are modeled to be capture. Different researchers used different inclusion removal criteria [5, 14, 15, 24–30]. Steel sheets with non-metallic inclusions develop surface defects. The injection of gas assists in the elimination of inclusions [31]. In order to speed up the removal of non-metallic inclusions, magnetic forces were used in the mold [17, 32].

5.6　PARAMETERS AFFECTING INCLUSION CAPTURE

The force balance capture criterion considers many data about inclusion entrapment behavior at a metal solidification front and the probability of a fraction of inclusions trapped under a given set of process conditions.

The forces on inclusion, force balance on inclusion, and the conditions for inclusion entrapment are all affected by the following parameters [33]:

- *Inclusion diameter*: If inclusion diameter is smaller than the PDAS enters the solidification front, it is caught regardless of fluid velocity. If the cross-flow velocity around the dendrites is zero, larger inclusions are also caught.
- *Inclusion density*: The effect of inclusion density on the inclusion buoyancy force has a major impact on inclusion capture. Increasing the inclusion density reduces its buoyancy force, decreasing the probability of the inclusion drifting upwards.
- *PDAS*: The shell entraps inclusions that are smaller than the PDAS. Depending on the resulting force's direction, an inclusion near a dendritic solidification front that is larger than the PDAS can be engulfed, pulled backward, or rolled along the front.
- *Sulfur concentration*: By causing a gradient in surface tension and a corresponding force toward the solidification front, sulfur concentration in the solidifying steel affects inclusion capture. Inclusion capture is increased when the melt contains more sulfur.
- *Dendrite tip radius*: The radius of the dendrite tip varies depending on the solidification conditions, such as front velocity, composition, liquidus, and solidus temperature.
- *Solidification front velocity*: The solidification velocity influences lubrication force, which acts where the dendrite tips nearly contact the inclusion. The lubrication force increases as the solidification velocity increases.
- *Solidification front angle*: Inclusion capture is greatly influenced by lowering the solidification front. The buoyancy component encourages inclusions to move upwards, toward the inner radius, at a lower solidification front angle.
- *Cross-flow velocity*: The difference between the actual fluid velocity at the solidification front and the casting speed is the cross-flow velocity. Increasing the cross-flow velocity helps the inclusion rotate more easily around the dendrite tip, allowing it to prevent capture.

5.7   INCLUSION REMOVAL WITHOUT EMS

The molten steel flow behavior dramatically influences the quality of steel within the mold. Surface-level fluctuations, inclusion transport, and superheat are among the many complex phenomena influenced by the flow pattern. Gutierrez and coworkers reported that the casting speed and inclusion size were critical factors for removing inclusions [34]. As the casting speed increases, more drag force is produced, resulting in a higher percentage of inclusions being removed. To predict entrapment of inclusion in the full length of a billet caster, Zhang and Wang [35] developed two approaches (full solidification approach and sink term approach). Ho and Hwang

[15] created the model to investigate the effects of nozzle designs and conditions for operation on the removal of inclusions that had already been held to the mold. A Lagrangian trajectory tracking approach has been used by Yuan and Coworkers [7] to model inclusion capture and motion in continuous slab casters based on time-dependent flow fields obtained by "large-eddy simulations". So long as the inclusions aren't entrapped in the solidifying shell, smaller bubbles are more effective at eliminating inclusions by bubble flotation. Bubble flotation is more effective at removing inclusions when the gas flow rate is higher [31]. Zhang and coworkers [36] found that more inclusion entrainment occurs from the top surface as turbulence energy and surface velocity are increase. Smaller inclusions are simpler to entrap into the steel than larger sizes.

5.8 INCLUSION REMOVAL WITH EMS

With the rise in demand for high-quality steel, the significance of high-cleanliness steel has risen. Stirring molten steel to facilitate coagulation and coarsening of inclusions has long been thought to be critical for the improved removal of non-metallic inclusions. The model developed by Li and Tsukihashi [30] on the motion of inclusion particles in slab caster was used to study the paths and velocities of inclusion particles by considering the effects of argon gas injection and magnetic field application. Electromagnetically induced flow in the mold prevents the trapping of large alumina clusters in the solidified shell [17]. According to the numerical simulation [32], magnetic flux intensity, inclusion particle size, and melt inlet velocity all affected inclusion trajectory and removal efficiency. Ambrish and Jha [19] investigated how different stirrer positions in a billet mold affected fluid flow and solidification, but its effect on inclusion removal is not considered. Li and coworkers [16] investigated the effect of EMS location on the flow field and inclusion removal fraction in a slab casting mold. When EMS is applied, the tangential velocity of the molten steel around the mold's perimeter rises [19, 37], washing the inclusion particles far away from the dendritic solidification front and reducing inclusion entrapment in the solidified shell. And compared to when there is no EMS, the upper flow pattern varies dramatically. Below the meniscus, a horizontal recirculating flow zone is formed. Figure 5.5 depicts the removal fraction of inclusions for various EMS locations [16]. The removal fraction increases as the EMS location decreases. Since the formation of the top recirculating flow, which is a crucial factor in deciding the number of inclusions that are being removed by the free surface, is prevented as EMS is added above the SEN port in case 1, the removal fraction decreases on increasing EMS currents. In case 3, however, when EMS is applied at a lower stage of the SEN port, the removal of inclusions is more due to the EMS enhanced effect on the upper stage recirculating flow. The proper application of EMS also refined the equiaxed type of grain structure because of the homogenization of the melt and formation of more possible nucleation sites [37], along with the removal of inclusions. This increases the quality of the finalized product.

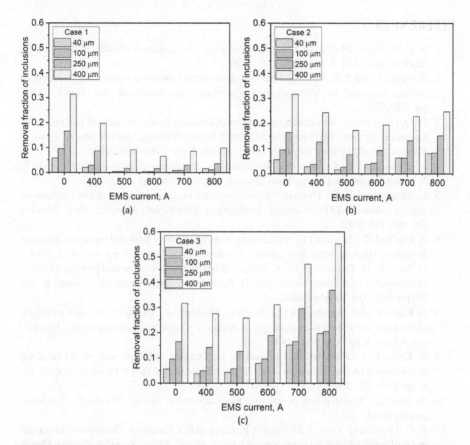

FIGURE 5.5 Removal fraction of inclusions with different inclusion diameters with EMS if stirrer installed (a) above the SEN port, (b) near the SEN port, (c) below the SEN port [16].

5.9 CONCLUSION

Nowadays, steelmakers seek to manufacture steel with minimal contaminants from alumina and sulfide inclusions. Although larger inclusions are simpler to float and remove, smaller inclusions (50 μm or less) are more difficult to remove. Calcium is poured into steel to help remove alumina and sulfide inclusions. Slag at the meniscus prevents oxygen from surrounding entering into the liquid steel and also absorbs non-metallic inclusions that rise to the surface due to buoyancy. The cast product's inclusion distribution is influenced by the flow of fluid pattern of liquid steel in the mold. High-level fluid flow fluctuation at the slag and liquid steel interface leads to mold flux entrapment into the liquid steel pool [38]. Therefore, SEN design is essential; the proper design of SEN outlets has a regulating influence. In order to facilitate inclusion flotation, the flow should have an upward component. Electromagnetic forces are helpful when appropriately applied to the mold to remove the inclusion and improve the microstructure of the cast products.

REFERENCES

1. B. G. Thomas, "Modeling of continuous casting defects related to mold fluid flow," *Iron Steel Technol.*, vol. 3, no. 5, pp. 1–6, 2006.
2. A. Maurya, and P. K. Jha, "Analysis of solidification kinetics in mold during continuous casting process," in *Manufacturing Techniques for Materials*, no. March, 2018, pp. 539–557.
3. D. Q. Geng et al., "Simulation on decarburization and inclusion removal process in the Ruhrstahl–Heraeus (RH) process with ladle bottom blowing," *Metall. Mater. Trans. B Process Metall. Mater. Process. Sci.*, vol. 46, no. 3, pp. 1484–1493, 2015.
4. L. Zhang, and B. G. Thomas, "*Inclusions in continuous casting of steel*," in *XXIV National Steelmaking Symposium, Morelia, Mich, Mexico*, pp. 1–42, 2003.
5. L. Zhang, and B. G. Thomas, "*Fluid flow and inclusion motion in the continuous casting strand*," *XXIV National Steelmaking Symposium, Morelia, Mich, Mexico*, pp. 184–196, 2003.
6. Z. Liu, and B. Li, "Effect of vertical length on asymmetric flow and inclusion transport in vertical-bending continuous caster," *Powder Technol.*, vol. 323, pp. 403–415, 2018.
7. Q. Yuan, B. G. Thomas, and S. P. Vanka, "Study of transient flow and particle transport in continuous steel caster molds: Part II. Particle transport," *Metall Mater Trans B*, vol. 35, pp. 703–714, August, 2004.
8. N. Kubo, T. Ishii, J. Kubota, and T. Ikagawa, "Numerical simulation of molten steel flow under a magnetic field with argon gas bubbling in a continuous casting mold," *ISIJ Int.*, vol. 44, no. 3, pp. 556–564, 2004.
9. S. Koric, L. C. Hibbeler, R. Liu, and B. G. Thomas, "Multiphysics model of metal solidification on the continuum level," *Numer. Heat Transf. Part B Fundam.*, vol. 58, no. 6, pp. 371–392, 2010.
10. A. Maurya, "Investigation of in-mold electromagnetic stirring process in continuous casting mold," 2017.
11. B. G. Thomas, Q. Yuan, S. Mahmood, R. Liu, and R. Chaudhary, "Transport and entrapment of particles in steel continuous casting," *Metall. Mater. Trans. B Process Metall. Mater. Process. Sci.*, vol. 45, no. 1, pp. 22–35, 2014.
12. Y. Kwon, J. Zhang, and H. G. Lee, "Water model and CFD studies of bubble dispersion and inclusions removal in continuous casting mold of steel," *ISIJ Int.*, vol. 46, no. 2, pp. 257–266, 2006.
13. K. T. Zhang, J. H. Liu, and H. Cui, "Effect of flow field on surface slag entrainment and inclusion adsorption in a continuous casting mold," *Steel Res. Int.*, vol. 1900437, no. 30, pp. 1–7, 2019.
14. J. Pötschke, and V. Rogge, "On the behaviour of foreign particles at an advancing solid-liquid interface," *J. Cryst. Growth*, vol. 94, no. 3, pp. 726–738, 1989.
15. Y. H. Ho, and W. S. Hwang, "Numerical simulation of inclusion removal in a billet continuous casting mold based on the partial-cell technique," *ISIJ Int.*, vol. 43, no. 11, pp. 1715–1723, 2003.
16. B. Li, H. Lu, Y. Zhong, Z. Ren, and Z. Lei, "Numerical simulation for the influence of EMS position on fluid flow and inclusion removal in a slab continuous casting mold," *ISIJ Int.*, vol. 60, no. 6, pp. 1204–1212, 2020.
17. W. Yamada, A. Kiyose, J. Fukuda, H. Tanaka, and J.-I. Nakashima, "Simulation of coagulation of non-metallic inclusions in tundish and their trapping into solidified shell in continuous casting mould," *Ironmak. Steelmak.*, vol. 30, no. 2, pp. 151–157, 2003.

18. D. Geng, H. Lei, and J. He, "Numerical simulation for collision and growth of inclusions in ladles stirred with different porous plug configurations," *ISIJ Int.*, vol. 50, no. 11, pp. 1597–1605, 2010.

19. A. Maurya, and P. K. Jha, "Influence of electromagnetic stirrer position on fluid flow and solidification in continuous casting mold," *Appl. Math. Model.*, vol. 48, pp. 736–748, 2017.

20. V. R. Voller, and C. R. Swaminathan, "General source-based method for solidification phase change," *Numer. Heat Transf. Part B*, vol. 19, pp. 175–189, 1991.

21. V. R. Voller, and C. Prakash, "A fixed grid numerical modelling methodology for convection-diffusion mushy region phase-change problems," *Int. J. Heat Mass Transf.*, vol. 30, no. 8, pp. 1709–1719, 1987.

22. C. Pfeiler, B. G. Thomas, M. Wu, A. Ludwig, and A. Kharicha, "Solidification and particle entrapment during continuous casting of steel," *Steel Res. Int.*, vol. 79, no. 8, pp. 599–607, 2008.

23. C. Pfeiler, "*Modeling of turbulent particle / gas dispersion in the mold region and particle entrapment into the solid shell of a steel continuous caster,*" 2008.

24. Y. Miki, H. Kitaoka, T. Sakuraya, and T. Fujii, "Mechanism for separating inclusions from molten steel stirred with a rotating electro-magnetic field," *ISIJ Int.*, vol. 32, no. 1, pp. 142–149, 1992.

25. D. Shangguan, S. Ahuja, and D. M. Stefanescu, "Insoluble particle and an advancing solid / liquid interface," *Metall. Trans. A*, vol. 23A, pp. 669–680, 1992.

26. D. M. Stefanescu, F. R. Juretzko, B. K. Dhindaw, A. Catalina, S. Sen, and P. A. Curreri, "Particle engulfment and pushing by solidifying interfaces: Part II. Microgravity experiments and theoretical analysis," *Metall. Mater. Trans. A Phys. Metall. Mater. Sci.*, vol. 29, no. 6, pp. 1697–1706, 1998.

27. J. K. Kim, and P. K. Rohatgi, "An analytical solution of the critical interface velocity for the encapturing of insoluble particles by a moving solid/liquid interface," *Metall. Mater. Trans. A Phys. Metall. Mater. Sci.*, vol. 29, no. 1, pp. 351–358, 1998.

28. D. M. Stefanescu, and A. V. Catalina, "Calculation of the critical velocity for the pushing/engulfment transition of nonmetallic inclusions in steel," *ISIJ Int.*, vol. 38, no. 5, pp. 503–505, 1998.

29. Z. Wang, K. Mukai, and I. J. Lee, "Behavior of fine bubbles in front of the solidifying interface," *ISIJ Int.*, vol. 39, no. 6, pp. 553–562, 1999.

30. B. Li, and F. Tsukihashi, "Numerical estimation of the effect of the magnetic field application on the motion of inclusion in continuous casting of steel," *ISIJ Int.*, vol. 43, no. 6, pp. 923–931, 2003.

31. L. Zhang, J. U. N. Aoki, and B. G. Thomas, "Inclusion removal by bubble flotation in a continuous casting mold," *Metall. Mater. Trans. B*, vol. 37, pp. 361–379, 2006.

32. M. Reza Afshar, M. Reza Aboutalebi, R. I. L. Guthrie, and M. Isac, "Modeling of electromagnetic separation of inclusions from molten metals," *Int. J. Mech. Sci.*, vol. 52, no. 9, pp. 1107–1114, 2010.

33. S. Mahmood, "Modeling of flow asymmetries and particle entrapment in nozzle and mold during continuous casting of steel slabs," 2006.

34. H. Arcos-Gutierrez, C. A. Espinosa, G. Barrera-Cardiel, H. Guillermo, and C. Garcidueñas, "Behavior and removal of inclusions by means of the use of mathematical and physical simulations as well as the measured vibrations with an accelerometer in a funnel mold in thin slab continuous casting," *ISIJ International*, vol. 55, no. 5. pp. 1017–1024, 2015.

35. L. Zhang, and Y. Wang, "Modeling the entrapment of nonmetallic inclusions in steel continuous-casting billets," *JOM J. Miner. Met. Mater. Soc.*, vol. 64, no. 9, pp. 1063–1074, 2012.

36. L. Zhang, Y. Wang, and X. Zuo, "Flow transport and inclusion motion in steel continuous-casting mold under submerged entry nozzle clogging condition," *Metall. Mater. Trans. B*, vol. 39B, pp. 534–550, 2008.

37. A. Maurya, and P. K. Jha, "Numerical investigation of in-mold electromagnetic stirring process for fluid flow and solidification," *COMPEL – Int. J. Comput. Math. Electr. Electron. Eng.*, vol. 36, no. 4, pp. 1106–1119, 2017.

38. A. Maurya, and P. K. Jha, "Two-phase analysis of interface level fluctuation in continuous casting mold with electromagnetic stirring," *Int. J. Numer. Methods Heat Fluid Flow*, vol. 28, no. 9, pp. 2036–2051, 2018.

6 Modeling of Inclusion Motion Under Interfacial Tension in a Flash Welding Process

Md Irfanul Haque Siddiqui
King Saud University, Saudi Arabia

Ambrish Maurya
National Institute of Technology Patna, India

Masood Ashraf
Prince Sattam Bin Abdulaziz University, Saudi Arabia

Fisal Asiri
Department of Mathematics, Taibah University, Saudi Arabia

CONTENTS

6.1 INTRODUCTION

Due to its high efficiency and good welding accuracy, flash butt welding (FBW) is usually used in the automotive industry, pipeline construction, sheet welding, hot rolled coils, rail welding, ship structure welding, and ship mooring equipment chains, etc. [1, 2]. In this welding process, the end of the part to be welded is connected to the transformer's secondary circuit, and a flash is generated by supplying current. For

DOI: 10.1201/9781003202233-7

efficient operation, one end is clamped by a movable plate, while the other is fixed by a clamping mechanism. When two parts are brought together under a high current supply, resistive heat is produced. After that, enough heat is produced to melt the metal and form a weld pool between the two parts. The surface of the movable piece is butted against another fixed surface with a higher force in the final phase of the operation. The extrusion of molten metal oxides and other impurities onto the outer surfaces is aided by this disturbing operation, which improves weld efficiency [3].

In recent years, various higher-strength steel materials have been used in various industry forays, making it more difficult to preserve the durability and mechanical properties of flash welded joints. The micro inclusion cannot be visible with the normal eye, but these must be seen under the microscopes. Inclusions are normally in size range of one to ten microns. The inclusion sizes, their distribution, chemistry, and type of inclusions influence the grain size and phases. In certain cases, it acts as the nuclei for other phases to form and influence the mechanical properties, such as decreasing the toughness ductility and strength. Previous studies indicated that welding parameters affected the mechanical properties and microstructures of micro-alloy (HSLA) steel [4, 5]. It was also suggested that oxide inclusions produced during the welding process have a negative effect on the mechanical properties of welded joints, such as resilience, hardiness, fatigability, and surface appearance, among other things. For example, Lu et al. [6] studied the surface morphology of HSLA 590CL flash butt welded joints in wheel rims and discovered non-metallic inclusions on the fracture surface, concluding that micro-cracks formed between inclusions and steel matrix. The presence of inclusions lowered the hardness quality of a flash weld joint, according to Shajan et al. [7, 8]. Inclusions were shown to be responsible for crack formation in welded joints by Yu et al. [9] and Joo et al. [9, 10]. As a result, removing oxide inclusions from welded joints is a crucial step in the flash welding process.

The oxide inclusion distribution in FBW joints under various welding parameters should be examined to ensure good efficiency. As a result, it is important to look into the effects of upsetting force (upsetting rate), inclusion scale, and flickering temperature on the inclusion distribution in weld joints. Oxides containing Si, Mn, and Al were generated in the sheet edge during the flashing process and persisted in the weld surface without being expelled during the upset force application process, according to Ichiyama and Kodama [11]. They discovered that steels with higher inclusion particles had lower hardness. They discovered that steels with higher inclusion particles had lower hardness. Ichiyama and Kodama [11] investigated the expulsion of oxide inclusions during disturbing activity by adjusting current density in another study. They concluded that a higher unsettling current promotes molten layer extrusion and inclusion-exclusion from the weld joint. The different parameters involved in the FBW process were analyzed by Kim et al. [3]. They found that even at a low upset rate, oxides were not discharged from the welded joint. Furthermore, Lu et al. [12] demonstrated that increasing the upset pressure will improve the extrusion of oxide inclusions from FBW joints. The oxide inclusions are formed during flashing in the presence of oxygen in the atmosphere [9, 11, 13, 14]. The amount of oxides generated is thought to increase as the flashing time lengthens [15]. Xi et al. [16] found that an unnecessary flash allowance caused oxide inclusions to get trapped in the weld joint. The removal and entrapment of oxide inclusions in the weld zone were influenced greatly by disrupting pressure and length parameters [17, 18]. Despite these attempts, the distribution of inclusions in FBW joints is thought to be incomplete.

Non-metallic particles embedded in the weld metal or weld interface are called inclusions. Insufficient welding technology, poor joint contact, or both may produce inclusions. The sharp notches between the weld boundaries or between the weld beads help slag inclusion. When using a coated electrode, a slag layer will form on the top of the weld, which must be cleaned up after welding. The hollows and sharp corners seem to capture slag fragments. Slag removal involves chisel hammers or wire brushes, and the difficulty varies depending on the type of electrode coating. If the welding spatter is not removed properly, welding spatter inclusions may appear in the weld. Slag is the deoxidation product of the reaction between the flux, air, and surface oxides because it is the residue of the flux coating. If two adjacent welds are buried without sufficient overlap and a gap is formed, the slag will get stuck in the weld. The entrained slag will not be removed until the next layer is deposited. Excessive undercut on the weld toe or irregular surface profile of the previous weld may even trap the slag in the cavity in the multi-pass weld. Single particles or longer inclusion lines are possible. Failure to properly remove slag in one weld pass and then produce another weld pass is the most common source of slag inclusions. Effective work strategies will help reduce risks. In closed joints, it is also important to use the correct welding process to avoid the use of thick electrodes. It is also important to prevent undercut during welding.

On the other side, the motion of inclusion particles is dependent upon various parameters. One of the important parameters is surfactant concentration. Any change and any modification in surfactant concentration affect the interfacial tension in the liquid metal. Subsequently, interfacial tension affects the Marangoni forces applied on the non-metallic particle. It has been seen that surfactant concentration has affected the interfacial tension. There may be various types of surfactant in the molten pool, for example, sulfur nitrogen-oxygen. During the solidification, the solid and liquid boundary removes surfactant remove surfactant and thus, there is the concentration of surfactant at the solidification boundary. Since the increase in surfactant concentration reduces the interfacial tension at the solid-liquid boundary. Since in a welding pool, there are several types of non-metallic inclusions, and during the solidification of the weld melt pool, non-metallic inclusion is affected by the interfacial tension. It would be great if we can see how inclusion particles are affected by the surfactant concentration. Several studies have looked into the engulfment or pushing of gas bubbles or inclusion particles by an advancing solid at the solid-liquid interface. Shibata et al. looked at the action of inclusions on the metal surface immediately in front of the solid-liquid interface [19]. Several researchers have pointed out that the Marangoni effect, which is induced by a temperature gradient or a surface-active element concentration gradient, will affect bulk flow in microgravity experiments [20, 21].

Also, with very low oxygen and sulfur concentrations, Marangoni flow was very high during the solidification of the steel melt, according to Yin and Emi [21]. Interfacial stress gradients at the solid-liquid interface boundary layer cause pressing and engulfment of fine bubbles at the solid-liquid interface, according to Mukai et al.

[22–24]. Mukai et al. discovered alumina inclusions in stainless steel slabs with a related propensity. The inclusion growth and dissolution in the weld pool are numerically predicted by Hong et al. [25]. They looked at the structure, size distribution, and number density of oxide inclusions in the welded region. It is also worth looking at whether Marangoni flow has a major impact on the FBW operation.

The purpose of this work is to understand how FBW parameters (such as upsetting rate, initial temperature, and inclusion size) affect inclusion movements during weld pool solidification. The kinematic characteristics of the solidified thin strip weld of the SPFH590 steel plate are considered to be based on the two-dimensional numerical representation of computational fluid dynamics (CFD). The multiphase VOF numerical model was combined with plate dynamic motion, discrete phase, and melt solidification. At the solid-liquid interface, the inclusion pressing and engulfment phenomena were also investigated.

6.2 EXPERIMENTS

The measures involved in the flash welding process are shown in Figure 6.1. AC FBW was used on SPFH590 micro-alloyed steel plates in this research. Table 6.1 shows the chemical composition of SPFH590 steel. The sample had a thickness of

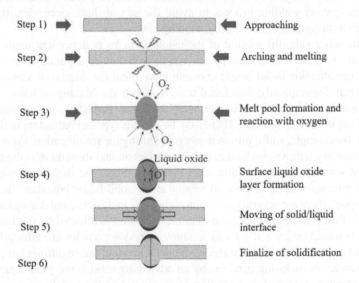

Step 1) ➡️ ⬅️ Approaching

Step 2) ➡️ ⬅️ Arching and melting

Step 3) ➡️ O_2 ⬅️ Melt pool formation and reaction with oxygen

O_2

Liquid oxide

Step 4) [O] Surface liquid oxide layer formation

Step 5) Moving of solid/liquid interface

Step 6) Finalize of solidification

FIGURE 6.1 Representation of AC flash butt welding process.

TABLE 6.1
Chemical composition of SPFH590 steel (wt%)

Element	C	Mn	Si	P	Al	Nb
Concentration	≤0.1	≤3.0	≤0.5	≤0.1	≤0.1	≤0.1

2.4 mm and a width of 3.0 mm. The samples were prepared in the welding direction after flash welding. Backscatter electron (BSE) images taken with a field emission scanning electron microscope were used to analyze the samples (FE-SEM, Hitachi, S-4300 model). The samples were assembled and polished up to 1 m polishing paper using 1 m diamond suspension for SEM analysis. The wheel rim is joined by AC flash welding in Figure 6.2. The distribution of inclusions along the weld joint core was investigated, and various types of oxides were discovered in the weld joint. Table 6.2 lists the sample's thermophysical properties. The simulation feedback has been taken from the data in Tables 6.1 and 6.2. Jeong et al. [26] experimentally developed the surface tension of SPFH590 steel and the interfacial tension between SPFH590 steel and an alumina inclusion. The surface tension of SPFH590 steel is calculated as follows:

$$\sigma_L = \left(1510.59274 + 0.08277\ T\right) - \left(1040.95926 - 0.51563\ T\right) \\ \left\{\ln\left[1 + \exp\left(-3.58336 + 19845.87952/T\right)\left(weight\ \%\ S\right)\right]\right\} \quad (6.1)$$

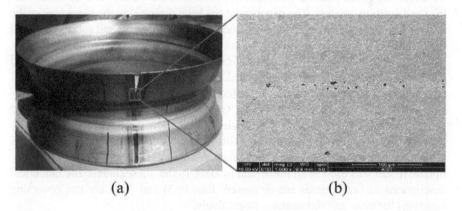

(a)   (b)

FIGURE 6.2 (a) Wheel rim joined by flash butt welding (b) inclusions in the welded joint.

TABLE 6.2
Physical properties of the sample

Parameters	Data
Solidus temperature	1781 K
Liquidus temperature	1798 K
Density of molten steel	ρ (kg m^{-3}) = 8621.17–0.88T
Viscosity of molten steel	μ (mPa s) = A exp(B/(RT))
Surface tension (σ_L) & Interfacial tension (σ_{PL})	Eqs. 6.1 and 6.2
Specific heat	750 J kg^{-1} K^{-1}
Thermal conductivity	41 m^{-1} K^{-1}

The following equation describes the interfacial stress between SPFH590 steel and an alumina inclusion.

$$\sigma_{PL} = \{3050.50999 - 131437.96734 \times (weight\,\%\,S)$$
$$- 1.54448 \times 10^{7} (weight\,\%\,S)^{2} - 3.37816 \times 10^{9} (weight\,\%\,S)^{3}\}$$
$$+ \{(-0.84975 - 79.73919 \times (weight\,\%\,S) + 7655.0561$$
$$\times (weight\,\%\,S)^{2} + 1.96183 \times 10^{6} \times (weight\,\%\,S)^{3}\} T \qquad (6.2)$$

6.3   NUMERICAL MODELING

6.3.1   THE GOVERNING EQUATIONS

There are three stages to AC FBW. The two conjugate faces are first pushed forward by a process. After that, an AC arc creates a proper flash between the joints. Finally, the disturbing of the plates is performed. A portion of the weld pool is extruded during the disturbing process. The experiment has been designed in the same way as the flash welding process. Moving plates (squeezing the weld pool) and successive weld pool solidification are used in the CFD model to simulate the disturbing operation. Besides, the distribution of inclusions has been tracked during the unsettling and solidification processes. Two-dimensional, multiphase numerical models were created in this research to investigate the effects of disturbing parameters on the inclusion distribution in the FBW process. On both macro- and micro-scales, CFD simulations were performed. The inclusion distribution during unsettling activity was examined for the macro-scale analysis. The micro-scale research looked at the pressing and engulfment of alumina inclusions at the solid-liquid interface. The multiphase model was used in all numerical models. In the 3.3 segments, the numerical descriptions of both phases are discussed. Eqs. (6.3) and (6.4) are the governing equations for mass and momentum, respectively.

$$\frac{\partial \rho}{\partial t} + \nabla \cdot (\rho \vec{v}) = S_m \qquad (6.3)$$

$$\frac{\partial}{\partial t}(\rho \vec{v}) + \nabla \cdot (\rho v \vec{v}) = -\nabla p + \nabla \cdot (\bar{\bar{\tau}}) + \rho \vec{g} + \vec{F} \qquad (6.4)$$

where p is the static pressure, $\bar{\bar{\tau}}$ is the stress tensor and $\rho \vec{g}$ and $\vec{F}$ are gravitational body force external body forces, respectively. The stress tensor, $\bar{\bar{\tau}}$ is given by Eq. (6.5).

$$\bar{\bar{\tau}} = \mu \left[\{\nabla \vec{v} + \nabla \vec{v}^{T}\} - \frac{2}{3} \nabla \cdot \vec{v} I \right] \qquad (6.5)$$

where μ is the molecular viscosity, I is the unit tensor, and the second term on the right-hand side is the effect of volume dilation. The energy equation is expressed in Eq. (6.6):

$$\frac{\partial}{\partial t}(\rho E) + \nabla \cdot \{\vec{v}(\rho E + p)\} = \nabla \cdot \left\{ k_{eff} \nabla T - \sum_j h_j \vec{J}_j + \left(\bar{\tau}_{eff} \cdot \vec{v} \right) \right\} + S_h \tag{6.6}$$

where k_{eff} is the effective conductivity ($k+k_t$, where k_t is the turbulent thermal conductivity, defined according to the turbulence model being used), and $\vec{J}_j$ is the diffusion flux of species j. S_h is a volumetric heat source term.

The enthalpy-porosity procedure was used to measure the solidification of the weld pool. The mushy area (liquid fraction between 0 and 1) is treated as a porous medium in this process. The liquid fraction of each cell is referred to as the cell's porosity. The solidified cells are called non-porous cells with a porosity of one. As a result, velocities of totally solidified cells are assumed to be 0. The mushy zone is often modeled as a "pseudo" porous medium, with a porosity of 1 to 0 as the substance solidifies [27]. The material's enthalpy is calculated by adding the sensible enthalpy, h, and the latent gas, ΔH:

$$H = h + \Delta H \tag{6.7}$$

where,

$$h = h_{ref} + \int_{T_{ref}}^{T} C_p dT$$

and h_{ref} is reference enthalpy, T_{ref} is reference temperature, C_p is the specific heat at constant pressure.

Further, liquid fraction, β, can be defined as:

$$\beta = 0 \quad if \; T < T_{solidus}$$

$$\beta = 1 \quad if \; T > T_{solidus}$$

$$\beta = \frac{T - T_{solidus}}{T_{liquidus} - T_{solidus}} \quad if \; T_{solidus} < T < T_{liquidus}$$

The weld pool's latent heat content is given as L, $\Delta H = \beta$. For both solids and liquids, the latent heat content may range from 0 to 1. Furthermore, the energy equation for solidification/melting problems is written as:

$$\frac{\partial}{\partial t}(\rho H) + \nabla \cdot (\rho \vec{v} H) = \nabla \cdot (k \nabla T) + S \tag{6.8}$$

where, H is enthalpy and S is the source term.

The solution of a convection-diffusion equation for the ith species predicts the local mass fraction of molten steel, sulfur material, Y_i. The following is the conservation equation for all liquid phases:

$$\frac{\partial}{\partial t}(\rho Y_i) + \nabla \cdot (\rho \vec{v} Y_i) = -\nabla \cdot \vec{J}_j \qquad (6.9)$$

In the case of sulfur diffusion in iron, the diffusion coefficient can be calculated using the following equation [28]:

$$D = \frac{kT}{2\pi\mu d}\left[\frac{m_1 + m_2}{2m_2}\right]^{\frac{1}{2}} \qquad (6.10)$$

where d is a metallic diameter, m_1 and m_2 are an atomic mass of solute and solvent, respectively. T is the temperature of the melt, μ is the viscosity of the molten metal, and k is Boltzmann's constant (1.38×10^{-3} J/K).

The volume of fluid (VOF) model allows for multiphase computations with precisely specified immiscible incompressible fluid interfaces. Molten steel and alumina incorporation was used as two immiscible fluids in this study. Pressure and velocity are factors that are shared by all processes and correspond to volume-averaged values. The flow equations are directly volume-averaged to produce a single set of equations, and the fluid interface is monitored using the color function α, which is defined as:

- $\alpha = 1 \Rightarrow$ control volume is filled only with phase 1
- $\alpha = 0 \Rightarrow$ volume is filled only with phase 2
- $0 < \alpha < 1 \Rightarrow$ interface present

The flow front is advanced by solving the following transport equation of the fluid:

$$\frac{\partial F}{\partial t} + u \cdot \nabla F = 0 \qquad (6.11)$$

Here F, is the volume fraction of the fluid in a cell and u is the flow velocity vector.

Geo-reconstruct scheme advection schemes were used in this research. The discretization process of governing equations is well known to have a major impact on device representation. As a result, the explicit scheme with the geo-reconstruct interface interpolation scheme yielded the solution to the two-dimensional problem. Normal finite-difference interpolation schemes are extended to the volume fraction values computed at the previous time point in the explicit method.

$$\frac{\alpha_q^{n+1}\rho_q^{n+1} - \alpha_q^n\rho_q^{n+1}}{\Delta t}V + \sum_f\left(\rho_q U_f^n \alpha_{q,f}^n\right) = \left[\sum_{p=1}^n\left(\dot{m}_{pq} - \dot{m}_{qp}\right) + S_{\alpha_q}\right]V \qquad (6.12)$$

6.3.2 Numerical Details

CFD simulations were used to predict the inclusion distribution during the AC flash welding process. During the first step of the simulation, the distribution of inclusions was tracked during the unsettling process. The effect of three parameters has been investigated: upsetting rate, alumina particle size, and the weld pool's initial temperature. In this simulation, a two-dimensional, multiphase numerical model was created to investigate the distribution of inclusions during the disturbing activity in the FBW process. The dynamic mesh model was combined with the discrete step model from Ansys Fluent to initialize the inclusion and plate motion (Academic version: 18.0) (upsetting operation). Multiphase simulation has also been done using the volume of the fluid system. A species model was used to predict the local sulfur content in molten steel. Molten iron, ambient air, and isolated particles have been chosen as three phases in this simulation. Because of the size of geometry and viscosity of molten iron, the disturbing process is believed to be laminar [29]. The domain of the first simulation scenario is depicted in Figure 6.3.

The revolving side of a dynamic wall is called a static wall. A convective atmospheric domain was found for the outer domain. The thermal conductivity of the steel plate is stable. It was thought that the internal domain was a deforming body. The outer walls were seen as a stationary body that served as a pressure inlet and outlet. Seventeen thousand (17,000) tetrahedra meshes were used to discretize the model domain. In this simulation, the domain's maximum cell size was 30 microns. The thermophysical properties of interest are described in Table 6.1. Equation 6.1 shows the experimental association for surface tension between molten iron and air. Equation 6.2 expresses the interfacial association of molten steel and alumina inclusion used in the current numerical model. Pushing and engulfment of inclusions particles is investigated in the simulation's second process. The second simulations aim to accurately predict alumina inclusions under the Marangoni power, which is caused by a sulfur concentration and temperature gradient. For the simulation, a complete realm of welded joint was initially assumed. The geometric domain considered for

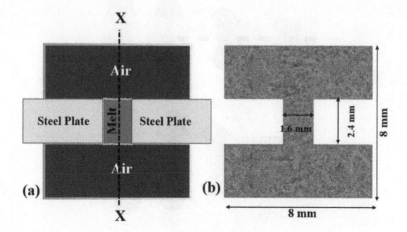

FIGURE 6.3   (a) Modeling of flash welding (b) domain meshing.

moving and engulfment of alumina inclusions during solidification is shown in Figure 6.8. The numerical model consists of two phases: molten steel and alumina inclusions. The Ansys Fluent species model was used to achieve the sulfur concentration in the domain. Besides, the solidification model was used to investigate the solidification mechanism. The association between interfacial stress and other thermophysical principles persisted unchanged. In both simulation scenarios, the transient state solution was found. At each time level, the species continuity equation was solved.

6.4 RESULTS AND DISCUSSION

6.4.1 EFFECT OF FLASH BUTT WELDING PARAMETERS AND INCLUSION SIZE

In figure 6.4, it can be seen here that inclusion particles are moving because of the motion of the plates. In this work, three different cases of velocity have been studied. In the first case, the velocity is 19.3 meters per second, which is the highest one, and we can see here tach inclusion particles are moving rapidly in the direction of the outer word. For the second case, we can see here that the movement of particles is not as rapid as in the previous case. In the third case, where the velocity was lesser, it can be seen noted that particles are settled in a welded joint. Hence, we can say that the high upsetting rate affects the motion. In other words, inclusion motions were

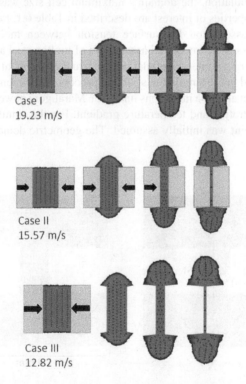

Case I
19.23 m/s

Case II
15.57 m/s

Case III
12.82 m/s

FIGURE 6.4   Alumina inclusion particles distribution after upsetting operation.

affected by the upsetting rate, and it is expected that inclusions are forced in an outward direction. In FBW, the upsetting rate is critical. Three cases of unsettling pace, 19.3 m/s, 15.57 m/s, and 12.82 m/s, have been investigated in this study. The higher disturbing rate necessitates a greater force to pass the plates, which affects the weld pool's flow dynamics. In comparison to the lower upsetting average, the high upsetting rate washed out most of the alumina inclusion particles from the weld joint. Figure 6.4 depicts the expected passage of inclusion particles at various disturbing rates. When a strong force is applied to the weld region, alumina particles are displaced from the weld zone.

It should be remembered that as the unsettling rate increases, the average velocity of inclusion particles increases. Additionally, a higher upsetting intensity improves the separation of alumina inclusions from the weld region. In the case of a low disturbing volume, more particle density can be found near the middle of the welding field. When a higher upsetting rate is used, however, a higher concentration of inclusions can be seen on the outside side of the weld joint.

In Figure 6.5 the velocity distribution of increased particles has been shown. The velocity distribution illustrates the movement and velocity of increased particles add different time steps. The plot was created from a set of inclusion particles whose motion has been tracking concerning time and position. Here we can see that different article has different motion at a specific time. It also depicts that various particles have a different velocity than the corresponding time step.

Effect of inclusion diameters: The function of inclusion diameter was investigated in another study. The inclusion motions were shown to be greatly influenced by the diameter of the inclusion particles. The effect of inclusion particle size on platen movement was investigated using four different sizes of inclusion particles in this analysis. Inclusion particles of smaller sizes are displaced rather than inclusion

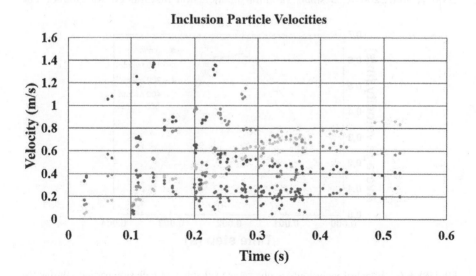

FIGURE 6.5 Predicted velocities of inclusion particles during upsetting operation.

particles of larger sizes. Figure 6.5 depicts the expected effects, with smaller inclusion particles being displaced from the welded zone.

Figure 6.7 depicts the predicted velocity of alumina inclusion particles as a function of particle size (diameters). Except for the inclusion particles with a diameter of 400 microns, the total particle velocity remains approximately constant for all sizes. The huge scale of alumina inclusions is impossible to displace from the weld zone during disrupting activity. The average velocity of large-size inclusion particles (especially those with a diameter of 400 microns) is also observed to be lower in magnitude, even at 0.0025s. From the Figure 6.6, it can be noted the different diameter of inclusion particle has shown a unique average velocity. Especially, we can see here that a larger diameter of particles has a higher average velocity. However, inclusion particles having small diameter has no different average velocities. But here we can keep in mind that these are the average velocity. If we see the velocity of individual particles as shown in Figure 6.5, we can realize that each particle has a specific velocity at each time step. Thus, we can say that inclusion velocity is very important and significant to its behavior and distribution at each time step.

The effect of the initial weld pool temperature on alumina inclusion distribution has been studied. The original temperature of the welding tank will be changed by an arc flash. Figure 6.8 shows the influence of the initial flash/weld pool temperature on the distribution of alumina inclusions. Inclusion distributions were shown to be limited to the weld joint at low weld pool temperatures (1808 K).

Effect of temperature: The effect of the initial weld pool temperature on alumina inclusion distribution has been investigated. An arc flash will change the initial temperature of the welding tank. The effect of the initial flash/weld pool temperature on the distribution of alumina inclusions can be seen in Figure 6.7. At low weld pool temperatures, inclusion distributions were found to be confined to the weld joint (1808 K). When the original flash temperature was raised in the other two scenarios, 2000 K and 2200 K, a small variation in inclusion distribution was found. The

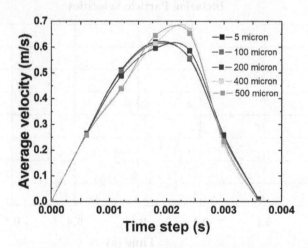

FIGURE 6.6 Effect of inclusion size on average velocities of inclusion particles during the flash welding process.

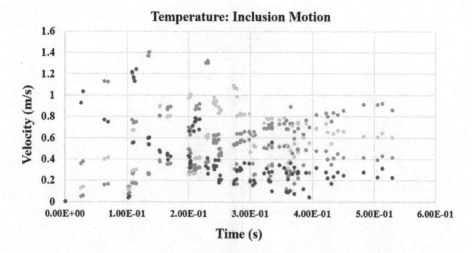

FIGURE 6.7   Predicted velocities of particles under different initial flashing temperatures.

predicted velocity of alumina inclusions during the unsettling phase is seen in Figure 6.7. On three separate flickering temperatures, the average velocity was expected. Up to 0.0018 seconds, the average velocity remained constant. Furthermore, as the temperature dropped, the average velocity dropped. However, a small change in average velocity can be observed at the end of the unsettling process.

6.4.2 PUSHING AND ENGULFMENT OF INCLUSIONS

Oxide inclusion particles are affected by interfacial tension force during weld pool solidification. As a result, a shift in the motion of inclusions is predicted. Inclusion particle engulfment and pushing near the solid-liquid interface are influenced by interfacial stress. Because of the difference in interfacial tension caused by surfactant concentration and temperature gradient, inclusions are subject to the Marangoni power. The action of alumina inclusion particles at the solidification interface has been studied due to sulfur content in the molten weld pool in this phase of the research. While considering the interfacial tension mediated force acting on the inclusion particles at the solid/liquid interface, the results for alumina inclusion engulfment and pushing are also discussed. The molten steel weld pool's surface energy is reduced by a limited volume of sulfur surfactant concentration.

A 2d, two-phase numerical simulation was used to investigate the action of alumina inclusions during the solidification process in this analysis. The full scale of geometry has been assumed in this case study, as seen in Figure 6.8. The model's length and width are 2.4 mm × 0.1 mm, respectively. The alumina inclusion particles (20 m in diameter) were strewn around the domain at random to be studied. Radiative and convective boundaries are considered on the domain's upper and lower sides. A conductive wall runs down the left and right sides of the domain.

At an initial flickering temperature of 1808 K, the simulation calculation began. Figures 6.9(a) and (b) depict the instantaneous location of inclusions during weld

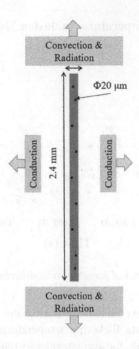

FIGURE 6.8   Geometric domain considered for simulation.

pool solidification at two separate sulfur content levels of 10 ppm and 64 ppm, respectively. It also depicts the solidification process's temperature contour. The movement of alumina inclusions was affected by the Marangoni force caused by sulfur content and temperature gradient. At this size of the domain, however, inclusion motion during solidification is undetectable.

The generated interfacial strain does affect alumina inclusions, but pressing and engulfment are restricted to a few microns. Another cause for minute inclusion particle displacement may be the weld pool's rapid solidification rate. According to the simulation, the weld pool reached solidus temperature in 0.013 seconds. The quick solidification of the weld pool, on the other hand, gives the alumina inclusions little time to pass. As a result, the movement of alumina inclusions under the Marangoni force was restricted to 30 microns. As a result, a more complex and accurate numerical simulation was needed. As a result, a new CFD model was created on a small scale (200 m long and 100 m wide) to investigate the engulfment and pressing of alumina inclusions in the solid/liquid front during weld pool solidification. Figure 6.11 depicts the smaller model's geometric domain. This domain's minimum cell size is on the order of 1 micron. The domain's boundary status was changed from its original state to capture the pressing and engulfment phenomena.

The domain's upper wall is believed to be convective, while the rest of the domain's walls are assumed to be adiabatic. As unidirectional heat transfer was predicted further pushing and engulfment phenomena can be caught precisely by adding these boundary conditions. As the cell temperature exceeds solidus temperature, the alumina inclusions particles velocities are thought to stop (1781 K). Figure 6.12

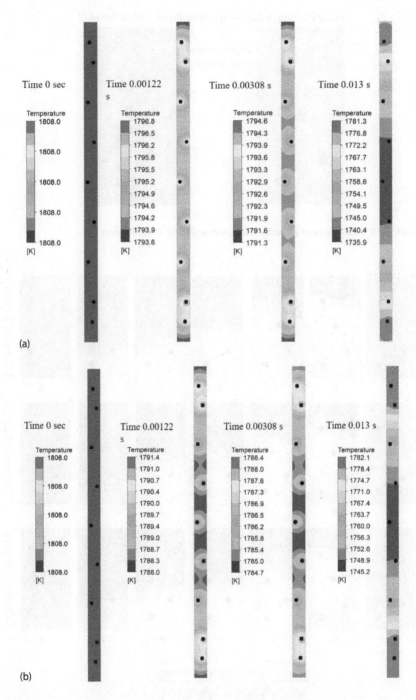

FIGURE 6.9 (a) Prediction of inclusion particle movement along with temperature contour (S 10 ppm). (b) Prediction of inclusion particle movement along with temperature contour (S 64 ppm).

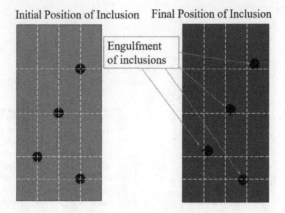

FIGURE 6.10 Engulfment of inclusion particles during solidification.

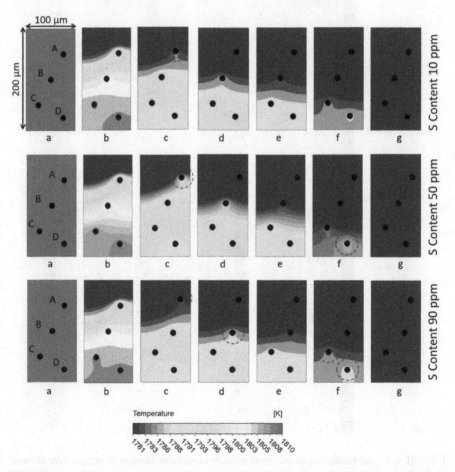

FIGURE 6.11 Pushing and engulfment of alumina inclusion particles during solidification of the weld pool.

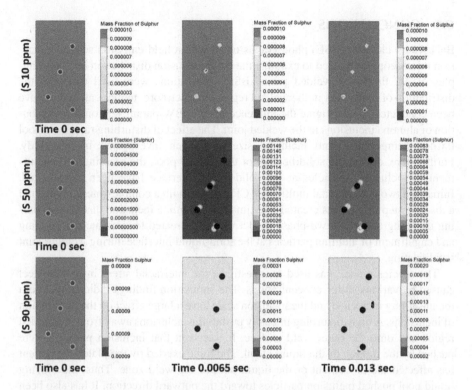

FIGURE 6.12 Contours of sulfur concentration during solidification of molten welding pool (up to solidus temperature 1781 K).

depicts the pushing and engulfment of inclusions during the solidification period. The curve depicts an instantaneous snapshot of inclusion displacement as a function of sulfur concentration. The temperature of the cells is represented by the color band. The solidus temperature contour line is depicted by the blue color band. The interfacial tension formed between the molten steel and alumina process affects inclusion particles, as can be seen here. Nonetheless, the phenomenon of pulling and engulfment is insignificant. Where the sulfur content is between 10 and 90 parts per million, particles are forced into the molten region.

The solid-liquid propagating front, on the other hand, engulfs alumina inclusions. The simulation findings show that interfacial tension force caused by sulfur concentration and temperature gradient affects alumina inclusions in weld pools. Due to the weld pool's rapid solidification, the alumina inclusions' motion activity was only registered for a few seconds. The sulfur concentration contour is seen in Figure 6.12 during the weld's solidification. Concentration contours up to solidus temperature, as seen in this snapshot. In the case of 10 ppm, it can be shown that the sulfur concentration is uniform. As the sulfur level is raised to 50 ppm and 90 ppm, sulfur accumulation is higher around the alumina inclusion particles during solidification. As a result, but on a smaller scale, the engulfment of alumina inclusion particles in flash welding operations is caused by surfactant concentration (in this case, sulfur) and temperature gradient.

6.5 CONCLUSIONS

Backscatter electron (BSE) photographs taken with a field emission scanning electron microscope were used to examine the oxide inclusion distribution in a few samples of AC flash butt welded joints. BSE observations were used to assess the distribution of inclusions in the welded region. Furthermore, numerical studies have been conducted to investigate the influence of AC FBW parameters on the distribution of alumina inclusions in the welded joint. The effect of disturbing rate, weld pool flashing temperature, and inclusion size have been investigated in this study. Furthermore, during the solidification of the welded pool, the pressing and engulfment of inclusions particles at the solid-liquid interface have been investigated. Initially, a two-dimensional multiphase CFD model with a complex mesh model and a discrete step model was created to estimate the alumina inclusion distribution during upsetting action. A two-phase model was also created to simulate the pressing and engulfment of alumina particles at the solid-liquid interface during welding joint solidification.

The species model was used to investigate the interfacial stress in molten steel caused by various sulfur concentrations. The simulation findings indicate that both the unsettling incidence and the inclusion scale have a large effect on the distribution of inclusions. A high disturbing intensity pushes the inclusions away from the welded region and onto the outer weld surface. It was seen that inclusion particles were pushed by the motion of the liquid metal. The force exerted by the platen movement has affected the movement of the liquid pool of the weld zone. Thus, the moving liquid pool pushed inclusion particles toward the outward direction. It has also been seen that some inclusion particles were found in the middle zone of the weld. Hence, we can say that the initial movement of inclusion particles during the upsetting rate is significantly affected by the liquid metal pool motion. When the diameter of inclusion particles is increased, a related result is observed. Also, the initial flash temperature affects inclusion motion. When the initial flashing temperature was increased from 1808 K to 2000 K, a change in alumina particle distribution was observed. However, raising the flickering temperature to 2200 K has no discernible effect on the distribution of inclusions. In practice, however, the initial flash temperature affects weld consistency and inclusion pollution. It has been seen from the experiment and the numerical results that inclusion movement is dependent upon various factors. Firstly, the upsetting rate temperature and inclusions diameter play an important role. Further, we have seen that surfactant concentration is very important. The surfactant concentration affects the inclusion motion behavior. Moreover, surfactant concentration affects the interfacial tension, and subsequently, inclusions particles are affected by Marangoni forces. The present work shows that the inclusion particle's motion behavior was significantly affected by surface consultation. This work noted that sulfur concentration in micro-alloyed steel had affected the inclusion motion at the solid-liquid interface. It was also noted that inclusions particles were pushed and engulfed by the moving solid little interface. The findings are solely concerned with the distribution of alumina inclusions already present in the weld pool under the conditions described. A computational investigation was carried out to specifically evaluate the motion activity of alumina particles at the solid-liquid interface of the weld pool under the effect of interfacial stress. The sulfur concentration and temperature differential caused the Marangoni power in the welding zone. Under the effect of a higher surface tension gradient between inclusions and melt, the expected findings show that inclusions were vulnerable to engulfment by the solidification front. The more sulfur content was increased, the more alumina inclusion particles became engulfed. A rise in sulfur content (to 100 ppm) had no discernible effect. Despite this, due to the rapid solidification rate, the displacement of inclusions under the influence of surface tension gradient was minimal.

REFERENCES

1. W. Wang, Y. Shi, Y. Lei, and Z. Tian, "FEM simulation on microstructure of DC flash butt welding for an ultra-fine grain steel," *J. Mater. Process. Technol.*, vol. 161, no. 3, pp. 497–503, 2005.
2. Y. Ichiyama, and S. Kodama, "Flash-butt welding of high strength steels," *Nippon Steel Tech.*, vol. 95, pp. 81–87, 2007.
3. D. C. Kim, W. J. So, and M. J. Kang, "Effect of flash butt welding parameters on weld quality of mooring chain," *Arch. Mater. Sci. Eng.*, vol. 38, no. 2, pp. 112–117, 2009.
4. C. Çetinkaya, and U. Arabaci, "Flash butt welding application on 16MnCr5 chain steel and investigations of mechanical properties," *Mater. Des.*, vol. 27, no. 10, pp. 1187–1195, 2006.
5. S. Bhattacharyya, M. Adhikary, M. B. Das, and S. Sarkar, "Failure analysis of cracking in wheel rims - material and manufacturing aspects," *Eng. Fail. Anal.*, vol. 15, no. 5, pp. 547–554, 2008.
6. P. Lu, Z. Xu, K. Jiang, F. Ma, and Y. Shu, "Influence of flash butt welding parameters on microstructure and mechanical properties of HSLA 590CL welded joints in wheel rims," *J. Mater. Res.*, vol. 32, no. 4, pp. 831–842, 2017.
7. N. Shajan, K. S. Arora, V. Sharma, and M. Shome, "Effect of upset pressure on texture evolution and its correlation to toughness in flash butt joints," *Sci. Technol. Weld. Join.*, vol. 23, no. 5, pp. 434–440, 2018.
8. N. Shajan, K. S. Arora, B. Asati, V. Sharma, and M. Shome, "Effects of post-weld heat treatment on the microstructure and toughness of flash butt welded high-strength low-alloy steel," *Metall. Mater. Trans. A*, vol. 49, no. 4, pp. 1276–1286, 2018.
9. X. Yu, L. Feng, S. Qin, Y. Zhang, and Y. He, "Fracture analysis of U71Mn rail flash-butt welding joint," *Case Stud. Eng. Fail. Anal.*, vol. 4, pp. 20–25, 2015.
10. M. S. Joo, K.-M. Noh, W.-K. Kim, J.-H. Bae, and C.-S. Lee, "A study of metallurgical factors for defect formation in electric resistance welded API steel pipes," *Metall. Mater. Trans. E*, vol. 2, no. 2, pp. 119–130, 2015.
11. Y. Ichiyama and T. Saito, "Factors affecting flash weldability in high strength steel – A study on toughness improvement of flash welded joints in high strength steel," *Weld. Int.*, vol. 18, no. 6, pp. 436–443, 2004.
12. P. Lu, Z. Xu, Y. Shu, and F. Ma, "Microstructure and failure analysis of flash butt welded HSLA 590CL steel joints in wheel rims," *JOM*, vol. 69, no. 2, pp. 135–143, 2017.
13. T. Taka, K. Kunishige, N. Yamauchi, and N. Nagao, "Hot-rolled steel sheet with excellent flash weldability for automotive wheel rim use," *ISIJ Int.*, vol. 29, no. 6, pp. 503–510, 1989.
14. L. B. Godefroid, G. L. Faria, L. C. Cândido, and T. G. Viana, "Failure analysis of recurrent cases of fatigue fracture in flash butt welded rails," *Eng. Fail. Anal.*, vol. 58, pp. 407–416, 2015.

15. T. Saito, and Y. Ichiyama, "Correlation between welding phenomena and weld defects before and after the start of upsetting: Welding phenomena and process control in flash welding of steel sheet (2nd Report)," *Weld. Int.*, vol. 10, no. 3, pp. 173–180, 1996.

16. C. Xi, D. Sun, Z. Xuan, J. Wang, and G. Song, "Microstructures and mechanical properties of flash butt welded high strength steel joints," *Mater. Des.*, vol. 96, pp. 506–514, 2015.

17. C. W. Ziemian, M. M. Sharma, and D. E. Whaley, "Effects of flashing and upset sequences on microstructure, hardness, and tensile properties of welded structural steel joints," *Mater. Des.*, vol. 33, no. 1, pp. 175–184, 2012.

18. M. I. H. Siddiqui, D. D. Geleta, G. Bae, and J. Lee, "Numerical modeling of the inclusion behavior during AC flash butt welding," *ISIJ Int.*, vol. 60, no. 11, pp. 1–9, 2020.

19. H. Shibata, H. Yin, S. Yoshinaga, T. Emi, and M. Suzuki, "In situ observation of engulfment and pushing of nonmetallic inclusions in steel melt by advancing melt/solid interface," *ISIJ Int.*, vol. 38, no. 0, pp. 49–56, 1998.

20. P. R. Scheller, J. Lee, T. Yoshikwa, and T. Tanaka, "Treatise on process metallurgy volume 2: Process phenomena," in *Treatise on Process Metallurgy*, vol. 2, S. Seetharaman, Ed. Oxford, UK: Elsevier, 2013, pp. 119–139.

21. H. Yin, and T. Emi, "Marangoni flow at the gas/melt interface of steel," *Metall. Mater. Trans. B*, vol. 34, no. 5, pp. 483–493, 2003.

22. Z. Wang, K. Mukai, and J. Lee, "Behavior of fine bubbles in front of the solidifying interface," *ISIJ Int.*, vol. 39, no. 6, pp. 553–562, 1999.

23. K. Mukai, and M. Zeze, "Motion of fine particles under interfacial tension gradient in relation to continuous casting process," *Steel Res.*, vol. 74, no. 3, pp. 131–138, 2003.

24. T. Matsushita, K. Mukai, and M. Zeze, "Correspondence between surface tension estimated by a surface thermodynamic model and number of bubbles in the vicinity of the surface of steel products in continuous casting process," *ISIJ Int.*, vol. 53, no. 1, pp. 18–26, 2013.

25. T. Hong, T. Debroy, S. S. B. Abu, and S. A. David, "Modeling of inclusion growth and dissolution in the weld pool," *Metall. Mater. Trans. B*, vol. 31, no. 1, pp. 161–169, 2000.

26. J. Jeong et al., "Interfacial tension between SPFH590 microalloyed steel and alumina," *Metall. Mater. Trans. B*, vol. 51, no. 2, pp. 690–696, 2020.

27. *ANSYS FLUENT Theory Guide, 18.2.* no. August. Canonsburg, PA: ANSYS Inc. USA, 2017.

28. T. A. Engh, C. J. Simensen, and O. Wijk, *Principles of Metal Refining*. Oxford University Press, 1992.

29. M. I. H. Siddiqui, H. Alshehri, J. Orfi, M. A. Ali, and D. Dobrotă, "Computational fluid dynamics (CFD) simulation of inclusion motion under interfacial tension in a flash welding process," *Metals*, vol. 11, p. 1073, 2021.

Section B

Mechanical Design Engineering

7 A Robust Approach for Roundness Evaluation

D. S. Srinivasu
Indian Institute of Technology Madras, India

N. Venkaiah
Indian Institute of Technology Tirupati, India

CONTENTS

7.1 INTRODUCTION

Parts generally deviate from their nominal size and shape due to systematic and random errors during manufacturing. The deviations present on the parts influence their performance. Deviations from the nominal shape include long-wavelength components called form error, apart from short (roughness) and medium (waviness) wavelength components. Roughness is the characteristic process marks. Instabilities produce waviness in the machining process. Form error is the deviation from the desired surface shape, caused by factors such as the part being held too firmly or not firmly enough, inaccuracies in guideways and spindles of machine tools, and uneven wear in machining equipment.

Measurement of form deviation is necessary to control assembly criteria and/or general performance, as the parts produced are bound to deviate from the nominal shapes. In form measurement, instrument/setup used, sampling interval, etc., determine the wavelengths captured. Generally, the shorter wavelength components are ignored. The majority of engineering components include some features with rotational symmetry. A circular feature is one such example, and it may exist in the form of a hole or boss. Circularity is a condition wherein all the measured data points are

DOI: 10.1201/9781003202233-9

equidistant from a center and is influenced by the overall form of the manufactured part. Various types of filters with micro and nano-sized hole arrays in bio-filters (Miller, 2004; Zhu et al., 2009; Biswas, Kuar, Biswas, & Mitra, 2010, Biswas, Kuar, Sarkar, & Mitra, 2010; Kai et al., 2010; Liu, 2010; Jahana, Rahman, & Wong, 2011; Maity & Singh, 2012) are some of the applications, wherein the accuracy level required is very high. Hence, the measurement and evaluation of circularity with a high degree of accuracy is of utmost importance. The circularity errors are estimated for those components whose height is small when compared to diameter, as in bearing races. ISO specifies that an ideal or reference feature must be established from the measured profile such that deviation between it and the measured profile is the least possible value (ISO 1101-1983, 1983; ISO/TS 12181-1, 2003; ISO/TS 12181-2, 2003). Generally, the ideal feature is taken to be a straight line for straightness, a plane for flatness, a circle for circularity, a cylinder for cylindricity, etc.

7.1.1 Definition of Circularity

Form measurement takes into account the overall shape of the surface to be measured. The terms roundness and circularity refer to the same form and are, therefore, synonymously used. A coordinate measuring machine (CMM) is used to generate the roundness data, which includes size, position, and orientation details apart from the form details. Measured data points of a circular profile (Figure 7.1) in a plane perpendicular to the axis are P_1, P_2,...., P_{12}. The coordinates (x_i, y_i), where $i = 1, 2,...., N$ (N = size of the dataset), are the Cartesian coordinates. A reference circle is established to the measurement profile using an appropriate fitting algorithm. Deviations of the measured points are computed from the reference circle.

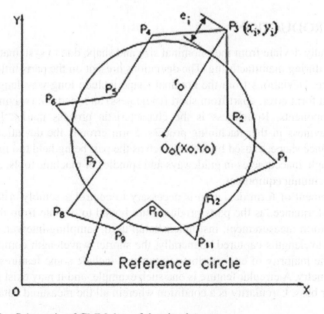

FIGURE 7.1 Schematic of CMM data of the circular part.

Distance between the data point and the center (x_0, y_0) of the reference circle is,

$$r_i = \sqrt{(x_i - x_0)^2 + (y_i - y_0)^2} \tag{7.1}$$

The deviation between the measured point and the reference circle is,

$$e_i = r_i - r_0 \tag{7.2}$$

where r_0 is the radius of the reference circle and is calculated as the mean of the radial distances from the measured points to the center coordinates of the reference circle. Let $e_{max} = $ maximum (e_i), $i = 1, 2....,N$ $e_{min} = $ minimum (e_i), $i = 1, 2....,N$. Now, the objective function for the circularity problem can be posed as: Circularity error,

$$\Delta = |e_{max}| + |e_{min}| \tag{7.3}$$

7.1.2 Various Approaches in Roundness Evaluation

There are four approaches to assess circularity, such as (i) least-squares (LS), (ii) minimum circumscribing (MC), (iii) maximum inscribing (MI), and (iv) minimum zone (MZ). In the LS method, a feature is fitted to the data points such that the sum of the squares of the deviations of the actual profile from the fitted feature is the minimum (Figure 7.2(a)). The largest peak to valley deviation from this feature is then computed to give the roundness error. This is the most straightforward in the sense that all the profile data is used to establish the center. This method is fast, easy to implement, and gives a unique solution. However, the problem with this method is that it does not conform to standards and generally overestimates the error. As a result, good parts may be rejected during the inspection. In the MC method, a feature is drawn around all the data and is shrunk until it is constrained by three peaks, as shown in Figure 7.2(b). The roundness error is simply the largest valley measured from this feature. The MI method is the opposite of the previous method. A feature is drawn within all the data and is expanded until it is constrained by three valleys, as shown in Figure 7.2(c). The roundness error is simply the largest peak measured

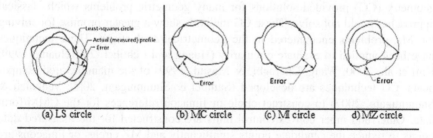

| (a) LS circle | (b) MC circle | (c) MI circle | d) MZ circle |

FIGURE 7.2 Various reference features for circularity assessment (a) LS circle (b) MC circle (c) MI circle (d) MZ circle.

from this feature. The MC and the MI methods are useful for function-oriented evaluations. In the MZ method, two concentric features; one outside the data and the other inside, are drawn as shown in Figure 7.2(d). These features are then brought together such that zone (difference in radius of two features) is the minimum. This difference in the radii of the two features is the roundness error. As far as the actual roundness value is concerned, the MZ method by definition results in the smallest and conforms to the ISO guidelines.

7.1.3 Various Computational Methods for Evaluation of Circularity

Several computational methods to evaluate the circularity error are classified as: (i) numerical methods; (ii) computational geometry-based methods; (iii) search-based methods; and (iv) hybrid methods.

(i) *Numerical methods*

Shunmugam (1986) suggested the median technique, which gives a lesser error value than that of the LSM. Xiong (1990) developed a general mathematical model to solve different profiles, including circularity where the LP and exchange algorithm are used. An approach closely related to Chebyshev approximation for the efficient approximation of finite point sets in R^2 and R^3 by circles and cylinders has been proposed by Drieschner (1993). It is shown that the approximation criterion used is best modeled mathematically through the concept of a parallel body. An optimization theory is formulated analytically to evaluate roundness error (Kaiser & Morin, 1994). A strategy for the MZ evaluation of circles is proposed by Lai and Chen (1996) that employs a nonlinear transformation to convert a circle into a line. 'Control line rotation scheme' is then applied, and then a straightness evaluation scheme was employed to find the MZ value for the feature concerned. A series of inverse transformation procedures are then carried out to attain desired feature parameters. It is found that the method can yield a smaller error value than that of the LS method. The convergence of the heuristic algorithm reported by Rajagopal and Anand (1999) is not guaranteed.

(ii) *Computational geometry-based methods*

Through the development of efficient data structures and algorithms, computational geometry (CG) provided solutions for many geometric problems which classical approaches could not solve. These CG methods show a greater promise for solving the MZ problems encountered in the geometrical evaluations. These techniques have been applied to evaluate circularity (Huang and Lehtihet, 2001Huang, 1999; Kim et al., 2000; Wang, Cheraghi, & Masud, 1999) of the manufactured components. CG techniques are developed (Samuel & Shunmugam, 2000; Venkaiah & Shunmugam, 2007) to construct circle or limacon references for the CMM/form data. Outer and inner convex/control hulls are constructed for the measured data points to reduce the candidate points significantly and MZ circles or limacons are constructed using the concept of equidistant lines. Circularity error is reported as the radial distance between these MZ circles or limacons.

(iii) *Search-based methods*

While dealing with a multi-variable optimization problem, the conventional optimization algorithms based on unidirectional search, direct search, and gradient methods evaluate the derivatives of the objective function. Therefore, these algorithms suffer from the issue of getting stuck in local minima. However, the non-conventional approaches such as genetic algorithm (GA), simulated annealing (SA), and particle swarm optimization (PSO) search for the optimal solution simultaneously at multiple locations, thereby addressing the issue of local minima. However, the PSO updates the velocity assigned to each potential solution instead of crossover and mutation operators (Kovvur, Ramaswami, Anand, & Anand, 2008; Sun, 2009). Attempts are also made to apply GA for circularity evaluations (Chen, 2000; Liao & Yu, 2001; Lu & Lu, 1999; Sharma, Rajagopal, & Anand, 2000; Wen, Xia, & Zhao, 2006). However, the GA alone gives a near-optimal solution only. Limitations of existing algorithms and the need for the parts with tighter tolerances for advanced technologies motivated the researchers in the field of form metrology to develop more efficient algorithms (Li & Liu, 2019; Li, Zhu, Guo, & Liu, 2020; Rhinithaa et al., 2018; Zheng, 2019).

7.2 SCOPE OF THE CHAPTER

Although researchers in the past proposed various computational techniques, the majority of the works compare the obtained results against that of the LS method. ISO recommends general guidelines on circularity evaluation. However, a specific method to establish the reference feature is not suggested as there is no consensus upon a particular method. We have developed a hybrid approach to address these issues, and accuracy and robustness, by exploiting the best features of both conventional and non-conventional algorithms. A detailed discussion on *modeling, experimentation, evaluation,* and *validation* against various computational methods is presented.

7.3 PROPOSED HYBRID METHOD

Following the ISO standards, a novel hybrid strategy (Figure 7.3) is presented in this chapter to establish the MZ circles to the measured dataset. This strategy aims to

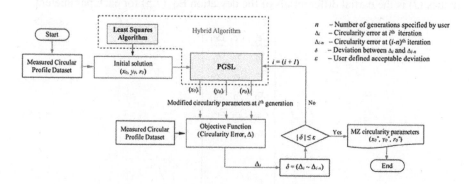

FIGURE 7.3 Hybrid global search algorithm for MZ solution (Srinivasu & Venkaiah, 2017).

address the issues such as accuracy, local minima, computational complexity, non-linearity, and robustness. The strategy comprises (i) a hybrid approach, which is a combination of the LS method and probabilistic global search Lausanne (PGSL) for identification of the global minima, i.e., establishing the MZ circles, and (ii) an effective stopping criterion to terminate the search. While finding the global optimal solution, the LS method and PGSL complement each other. For obtaining the global optimal solution, most of the search approaches end up in near-optimal solutions. This is because these approaches need to deal with larger search space throughout the search process and do not have a fine search feature in them. However, the present hybrid strategy employs the LS method to identify the small potential zone. PGSL does the fine search within this zone by exploiting its unique features like *global and focused searches*. Because of the complementing nature of the LS method and PGSL, the proposed strategy can find a better solution without getting stuck at the near-optimal solution.

PGSL, being a search technique, needs an initial solution to start with. The initial solution for the circularity problem comprises the center coordinates of the reference circle (x_0, y_0). An initial solution, which is closer to the global optimal solution, reduces the computational effort significantly. Hence, a method that can provide a suboptimal solution is highly desirable to reduce the search space by reducing the search time and computational complexity. The literature found that the LS method with its sound mathematical basis can identify such a suboptimal solution. Although the solution from the LS method overestimates the error and does not conform to the ISO standards, it serves well to act as the initial solution. The deviation of a measured point from the reference circle is given by Eq. (7.2). LS method is based on minimizing the sum of the squares of the deviations. The objective function is

$$\text{Minimize} \sum_{i-1}^{N} e_i^2 \tag{7.4}$$

Since Eq. (7.4) is nonlinear, a nonlinear LS method is employed to solve this problem. This method needs a Jacobian matrix to be formed. The required Jacobian matrix (J) is the partial differentials of the deviation Eq. (7.5) for each parameter.

$$J = \begin{bmatrix} \dfrac{\partial e_1}{x_0} & \dfrac{\partial e_1}{y_0} & \dfrac{\partial e_1}{r_0} \\ \dfrac{\partial e_2}{x_0} & \dfrac{\partial e_2}{y_0} & \dfrac{\partial e_2}{r_0} \\ \dfrac{\partial e_3}{x_0} & \dfrac{\partial e_3}{y_0} & \dfrac{\partial e_3}{r_0} \\ . \\ . \\ . \end{bmatrix} \tag{7.5}$$

where

$$\frac{\partial e_1}{x_0} = \frac{x_o - x_i}{\sqrt{x_i^2 - 2x_0 x_i + y_i^2 - 2y_0 y_i + x_0^2 + y_0^2}}$$

$$\frac{\partial e_1}{y_0} = \frac{y_o - y_i}{\sqrt{x_i^2 - 2x_0 x_i + y_i^2 - 2y_0 y_i + x_0^2 + y_0^2}}$$

$$\frac{\partial e_i}{r_0} = -1$$

This is an iterative process; therefore, an initial solution for the unknowns (x_o, y_o, r_o) is needed and the same may be obtained as follows:

$$x_0 = \frac{1}{N} \sum x_i$$

$$y_0 = \frac{1}{N} \sum y_i \qquad (7.6)$$

$$r_0 = \sqrt{(x_1 - x_0)^2 + (y_1 - y_0)^2}$$

Where N represents the total number of measured points.

The vector of residuals, K is given by

$$k = \begin{bmatrix} 0 - \left(\sqrt{(x_1 - x_0)^2 + (y_1 - y_0)^2} - r_0 \right) \\ 0 - \left(\sqrt{(x_2 - x_0)^2 + (y_2 - y_0)^2} - r_0 \right) \\ 0 - \left(\sqrt{(x_3 - x_0)^2 + (y_3 - y_0)^2} - r_0 \right) \\ \cdot \\ \cdot \\ \cdot \end{bmatrix} \qquad (7.7)$$

It is essential to solve for adjustment values, ΔV follows:

$$\Delta V = \begin{bmatrix} \Delta x_0 \\ \Delta y_0 \\ \Delta r_0 \end{bmatrix}$$

where,

$$\Delta V = (J'J)^{-1} J'K \qquad (7.8)$$

On solving the above equation, the circle parameters are then adjusted as

$$x_0' = x_0 + \Delta x_0; \quad y_0' = y_0 + \Delta y_0 \qquad ; r_0' = r_0 + \Delta r_0 \qquad (7.9)$$

This new solution is now used to compute a new J matrix and a new residual vector, which are then used to get a new adjustment vector ΔV. This procedure is continued until the adjusted values are closer to zero. Even today, most of the CMMs use the LSM to report the final solution. A further fine search was carried out using the PGSL to improve this solution.

Probabilistic global search method

The second stage, i.e., PGSL (Figure 7.4) was employed to carry out the fine search for the MZ circles in the reduced search space. The assumption in the PGSL is that better solutions are found in the neighborhood of good ones (Raphael & Smith, 2000; Raphel & Smith, 2003). PGSL finds the global optimal solution based on four nested cycles (Figure 7.4): *sampling, probability updating, focusing*, and *subdomain cycles*. The *sampling cycle* helps in exploring the whole search space; the *probability updating cycle* and the *focusing cycles* refine the search space, and the *subdomain cycle* directs the search toward convergence. These cycles are implemented by a global and selective sampling of the search space according to a probability distribution function (PDF) that is varied dynamically during the search process.

In the global search by the PGSL, the potential search space for the center coordinates (x_0, y_0) was defined to lie within the limits $(x_{min}, x_{max}) = (x_0 - 0.5x_0, x_0 + 0.5x_0)$,

```
Sub-domain cycle
for: 1 to NSDC
{
        Focusing cycle
        for: 1 to NFC
        {
                Probability updation cycle
                for: 1 to NPUC
                {
                        Sampling cycle
                        {
                                Sampling resolution = (xmax - xmin)/N
                                Evaluate NS samples
                                Select BS
                        }
                        Refine1: Probability_Update = Probability_Old*1.2|
                                         (for interval containing BS)
                }
                Find CBEST
                Refine2: (i) Divide interval containing CBEST into NDIV
                         (ii) Assign 50% probability for interval containing CBEST + 50% for remaining intervals
        }
        Refine3: Realize the narrowed down search space by centering around CBEST
}
```

FIGURE 7.4 Pseudo code for probabilistic global search method (Srinivasu & Venkaiah, 2017).

and $(y_{min}, y_{max}) = (y_0-0.5y_0, y_0+0.5y_0)$. These ranges are safely set based on the LS solution obtained from stage 1 and its closeness to the MZ solution. The PDF of each parameter (X- and Y- coordinates of the center: x_0, y_0) was created by assuming a uniform distribution (for x_0: x_{min} to x_{max}; for y_0: y_{min} to y_{max}). The two probability distribution functions ($PDFx_0$ and $PDFy_0$) have intervals of constant width ($x_{max} - x_{min}$ and $y_{max} - y_{min}$), at the beginning, which was divided into a fixed number of intervals, N. In the *sampling cycle*, a population of two points was randomly generated according to the PDFs of the variables. As the search progresses, probabilities and intervals are dynamically updated. The search space is gradually narrowed down to achieve convergence.

The strategy followed for this purpose is presented as follows: The suitability of each solution (i.e., x_0 and y_0) for fitting the MZ circles to the dataset was evaluated by determining circularity error (Δ_i). This process was repeated for all the points and the best point (BS) was selected. This completes the *sampling cycle* (Figure 7.4), which was invoked for several probability updating cycle times (NPUC), that is 1. In each iteration, the PDFs of x_0 and y_0 are modified using the probability updating algorithm. In this process, the interval containing the value of the interim best solution (BS) was identified first. The probability of the interval was multiplied by a factor of 1.2, and the PDF is normalized. The probability updating cycle (Figure 7.4) was repeated for several focusing cycle times (NFC), which is 400. The search was increasingly focused on the interval containing the current best point. This was done by subdividing the interval containing the value of the best point of each variable (x_0 and y_0). A 50% probability was assigned to the best interval, and this probability was divided uniformly. This completes a *focusing cycle*, which was repeated for several subdomain cycles (NSDC) times, that is, 400. In the subsequent iteration, the current search space was modified by changing the limits of the parameters to be optimized. This way, the size of the subdomain decreases gradually, and the solution converges. This completes a *subdomain cycle*. The iterative process is terminated based on variation (10^{-8}) in the solution in the last ten iterations. Following this strategy, the MZ circles were fitted to the circularity data, yielding the minimum circularity error, and the corresponding circularity parameters (x_0^*, y_0^*, r_0^*) are the optimal global parameters.

7.4   RESULTS AND DISCUSSION

Table 7.1 presents the circularity parameters and MZ values obtained for different datasets. It may be seen that the error value based on the present approach is significantly smaller than that of the LS method. Results obtained using the present approach are found to be either matching with or lesser than the reported ones. In the case of dataset 2, no improvement was noticed from the reported solution; this might be since the reported solution happens to be the MZ solution. The improvements obtained for datasets 1 and 3 using the present hybrid approach seem to be marginal. However, the order of improvements proves crucial in the case of miniaturized parts, such as micro-manufactured components. Hence the present method is robust in yielding accurate solutions.

TABLE 7.1
Circularity results (Srinivasu & Venkaiah, 2017)

Data	MZ Circle Parameters			Circularity Error, Δ (mm)	
	x_0 (mm)	y_0 (mm)	r_0 (mm)	LS Method	MZ Method
Dataset 1[XZ]	0.03561497	−0.05292948	1.0002	0.0092	0.00854
(39 points)					0.0086[XZ]
Dataset 2[XZ]	0.00534671	0.00790906	1.4888	0.9854	0.95742
(100 points)					0.95742[XZ]
Dataset 3[Y]	82.9909414	97.00838754	30.02972621	39.1017	38.2304
(24 points)					38.2326[Y]

X. Z. Xiuming and Zhaoyao (2010); Y. Ye Ding et al. (2007).

7.5 SUMMARY

A simple yet efficient hybrid approach, discussed in this chapter, combines the LS and PGSL methods to accurately evaluate the circularity error. This approach possesses both global and local search features. The proposed approach has been tested on different sizes of benchmark datasets, and it was found to give accurate results, demonstrating its robustness. The results conform to the MZ criterion specified by ISO. As this strategy is simple to understand, practitioners can easily implement and test it. This development is significant and has an edge in today's scenario, demanding tighter geometrical tolerances, especially in advanced and high-value-added applications. Although the present strategy deals with the 2D geometric feature, it can also be extended to 3D features, such as sphericity, cylindricity, etc.

ACKNOWLEDGMENT

The authors would like to thank the *International Journal of Advanced Manufacturing Technology* and Springer for permitting us to use some of the content published by the same authors.

REFERENCES

Biswas R, Kuar AS, Biswas SK, and Mitra S (2010) Characterization of hole circularity in pulsed Nd:YAG laser micro-drilling of TiN–Al2O3 composites, *International Journal of Advanced Manufacturing Technology*, 51, 983–994.

Biswas R, Kuar AS, Sarkar S, and Mitra S (2010) A parametric study of pulsed Nd:YAG laser micro-drilling of gamma-titanium aluminide, *Optics & Laser Technology*, 42, 23–31.

Chen M (2000) Roundness inspection strategies for machine visions using nonlinear programs and genetic algorithms, *Journal of Production Research*, 38(13), 2967–2988.

Drieschner R (1993) Chebyshev approximation to data by geometric elements, *Numerical Algorithms*, 5, 509–522.

Huang J (1999) An exact solution for the roundness evaluation problems, *Precision Engineering*, 23, 2–8.

Huang J, and Lehtihet EA (2001) Contribution to the minimax evaluation of circularity error, *International Journal of Production Research*, 39(16), 3813–3826.

ISO 1101-1983 (1983) *Technical drawings: tolerancing of form, orientation, location, and runout—generalities, definitions, symbols, indications on drawing*, ISO, Geneva.

ISO/TS 12181-1 (2003) *Geometrical product specifications (GPS)—roundness—part 1: vocabulary and parameters of roundness*.

ISO/TS 12181-2 (2003) *Geometrical product specifications (GPS)—roundness—part 2: specification operators* [1].

Kaiser MJ, and Morin TL (1994) Centers, out-of-roundness measures and mathematical programming, *Computers and Industrial Engineering*, 26 (1), 35–54.

Kim K, Lee S, and Jung H-B (2000) Assessing roundness errors using discrete voronoi diagrams, *International Journal of Advanced Manufacturing Technology*, 16, 559–563.

Kovvur Y, Ramaswami H, Anand RB, and Anand S (2008) Minimum zone form tolerance evaluation using particle swarm optimization, *International Journal of Systems Technologies and Applications*, 4(1/2), 79–96.

Lai JY, and Chen IH (1996) Minimum zone evaluation of circles and cylinders, *International Journal of Machine Tools and Manufacture*, 36(4), 435–451.

Li X, and Liu Y (2019) A quick algorithm for evaluation of minimum zone circles from polar coordinate data, *The Review of Scientific Instruments*, 90, 125114.

Li X, Zhu H, Guo Z, and Liu Y (2020) Simple and efficient algorithm for the roundness error from polar coordinate measurement data, *The Review of Scientific Instruments*, 91, 025105.

Liao P, and Yu SY (2001) A calculating method of circle radius using genetic algorithms, *Metrology Transaction China*, 22, 87–89.

Lu P, and Lu JS (1999) A calculating method of circle error using genetic algorithms, *Trans Nanjing University Aeronaut Astronaut*, 31(4), 393–397.

Rajagopal K, and Anand S (1999) Assessment of circularity error using a selective data partition approach, *International Journal of Production Research*, 37, 3959–3979.

Raphael B, and Smith IFC (2000) *A probabilistic search algorithm for finding optimally directed solutions, Proceedings Construction Information Technology, Icelandic Building Research Institute*, Reykjavik, Iceland, 708–721.

Raphel B, and Smith IFC (2003) A direct stochastic algorithm for global search. *Applied Mathematics and Computation*, 146, 729–758.

Rhinithaa PT, Selvakumar P, Sudhakaran N, Anirudh V, Lawrence KD, and Mathew J, Comparative study of roundness evaluation algorithms for coordinate measurement and form data, *Precision Engineering*, 51 (2018), 458–467.

Samuel GL, and Shunmugam MS (2000) Evaluation of circularity from coordinate and form data using computational geometric techniques. *Precision Engineering*, 24, 251–263.

Sharma R, Rajagopal K, and Anand S (2000) Genetic algorithm based approach for robust evaluation of form tolerances, *Journal of Manufacturing Systems*, 19(1), 46–57.

Shunmugam MS (1986) On assessment of geometric errors. *International Journal of Production Research* 24(2), 413–425.

Srinivasu DS, and Venkaiah N (2017) Minimum zone evaluation of roundness using hybrid global search approach, *The International Journal of Advanced Manufacturing Technology*, 92, 2743–2754.

Sun T-H (2009) Applying particle swarm optimization algorithm to roundness measurement, *Expert Systems with Applications*, 36(2), Part 2, 3428–3438.

Venkaiah N, and Shunmugam MS (2007) Evaluation of form data using computational geometric techniques, part I: circularity error. *International Journal of Machine Tools and Manufacture* 47(7–8), 1229–1236.

Wang M, Cheraghi SH, and Masud ASM (1999) Circularity error evaluation: Theory and algorithm, *Precision Engineering*, 23, 164–176.

Wen X, Xia Q, and Zhao Y (2006) An effective genetic algorithm for circularity error unified evaluation, *International Journal of Machine Tools and Manufacture*, 46, 1770–1777.

Xiong YL (1990) Computer-aided measurement of profile error of complex surfaces and curves: Theory and algorithm, *International Journal of Machine Tools and Manufacture*, 30, 339–357.

Zheng Y (2019) A simple unified branch-and-bound algorithm for minimum zone circularity and sphericity errors, *Measurement Science and Technology*, 31, 045005.

Egashira K, Morita Y, and Hattori Y (2010) Electrical discharge machining of submicron holes using ultrasmall-diameter electrodes, *Precision Engineering*, 34, 139–144.

Jahana MP, Rahman M, and Wong YS (2011) A review on the conventional and micro-electrodischarge machining of tungsten carbide, *International Journal of Machine Tools & Manufacture*, 51, 837–858.

Liu H-T (2010) Waterjet technology for machining fine features pertaining to micromachining, *Journal of Manufacturing Processes*, 12.

Maity KP, and Singh RK (2012) An optimisation of micro-EDM operation for fabrication of micro-hole, *International Journal of Advanced Manufacturing Technology*, 61, 1221–1229.

Miller DS (2004) Micromachining with abrasive waterjets, *Journal of Materials Processing Technology*, 2004 (149), 37–42.

Zhu D, Qu NS, Li HS, Zeng YB, Li DL, and Qian SQ (2009) Electrochemical micromachining of microstructures of micro hole and dimple array, *CIRP Annals – Manufacturing Technology*, 58, 177–180.

8 Computational Techniques for Predicting Process Parameters in the Magnetorheological Fluid-Assisted Finishing Process

Atul Singh Rajput, Sajan Kapil, and Manas Das
Indian Institute of Technology Guwahati, India

CONTENTS

DOI: 10.1201/9781003202233-10

8.1   INTRODUCTION

Manufacturing has a considerable influence on the development of human lives. The main concern during the manufacture of a product is that how well the product is made. Many developments are made to enhance product build quality to meet the consumer's anticipations. For better manufacturing systems, processes and materials, different techniques have been developed. The combined effort of all these processes turns raw material into the desired product. The quality of a product mainly defines how the finished product is good compared to its specifications. The most common index used to measure the product's surface quality is surface roughness (R_a). However, various surface finishing processes were developed to enhance the product's surface quality [1, 2]. The magnetorheological fluid-assisted finishing (MFAF) process is one of the advanced surface finishing techniques [3–7]. This process utilizes Magnetorheological (MR) fluid, consisting of ferromagnetic particles, abrasives, carrier fluid, and stabilizers, producing a multipoint, unbonded polishing tool. This improved polishing tool helps to provide a uniform surface quality on the free-form surfaces. The MR fluid's rheological properties can be controlled through the external magnetic field as per the surface finish requirement [8]. Different modes through which MR fluid is utilized for finishing purposes in the MFAF process are shown in Figure 8.1.

Flow mode: Utilization of constant velocity flow of MR fluid to enhance the surface quality of the workpiece: e.g., Magnetorheological Abrasive Flow Finishing (MRAFF) and R-MRAFF [9, 10].

Squeeze mode: Concentrated and focused magnetic field is used to squeeze the MR fluid between the magnet and workpiece and is utilized as a finishing tool, e.g., ball end magnetorheological finishing (BEMRF) [11].

Along with the magnetic field, several other process parameters, like finishing time, working distance, MR fluid constituent composition, and concentration, are

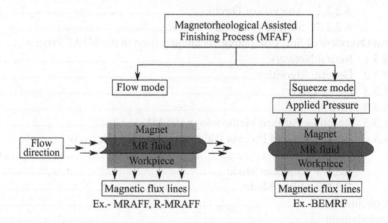

FIGURE 8.1   Different types of Magnetorheological Fluid Assisted Finishing (MFAF) processes.

vital for determining the workpiece's surface quality [12]. Optimizing such critical process parameters requires an explicit computational model to describe the finished component's surface characteristics. These computational methods work based on artificial intelligence (AI), where different algorithms are developed to mimic the human decision-making system [13]. These well-designed algorithms are capable of solving nonlinear, complex, multidimensional problems. The different soft computing techniques that are used to optimize the process parameters of MFAF processes are JAYA, response surface methodology (RSM), fuzzy logic (FL), neural networks (NN), genetic algorithms (GA), etc. [14, 15]. This chapter aims to analyze various computational techniques and theoretical models applied to the MFAF processes to obtain a surface roughness as a function of multiple process parameters that influence the finishing performance.

8.2 ANALYTICAL ANALYSIS

The development of the MFAF process provides uniform surface quality on the free-form surfaces without affecting the material's topography. Perhaps, the quantitative analysis of the different forces acting on the workpiece helps get a deep insight into the physics involved during the finishing operation. The following section describes the analytical model to determine the workpiece's surface roughness during the MFAF process (flow and squeeze mode). The following assumptions were made to analyze the impact of forces on the workpiece's surface during the finishing operation [16].

- The workpiece surface is isotropic with a uniform roughness profile.
- The roughness peaks are triangular and spread all over the workpiece surface, as shown in Figure 8.2.
- Abrasive and *Carbonyl Iron (CI)* particles are spherical with uniform size.
- Debris produced during finishing operation does not affect the process capability.
- The magnetic field does not vary with the time throughout the finishing process.

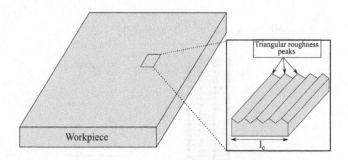

FIGURE 8.2 Triangular roughness peaks on the workpiece surface.

8.2.1 FLOW MODE

The flow mode of the MFAF process utilizes the constant velocity flow of MR fluid to enhance the surface quality of the workpiece, e.g., MRAFF and R-MRAFF. Herein, the constant flow of the MR fluid is generated through the reciprocation motion of the piston. MR fluid is stored between the two inline pistons, moves with the same velocity. The forces generated during the process are explained in the subsequent section.

8.2.1.1 Analysis of Forces

During the surface finishing process, the magnetic field generates the normal force on abrasive particles embedded in between the ferromagnetic particles, i.e., carbonyl iron particles (CIP). The magnitude of the magnetic force on the CI particles can be determined by Eq. (8.1) [17].

$$F_m = m\mu_0\chi_m H\nabla H \tag{8.1}$$

Where mass and mass-susceptibility of CI particles is represented by m and χ_m, respectively. The intensity and gradient of the magnetic field are named H and ∇H, respectively. The permeability of the free space is represented as μ_0. Apart from the magnetic force (F_m), radial force (F_r) (a function of medium viscosity and applied extrusion pressure on the piston) and centrifugal force (F_{cen}) (produced due to magnets' rotational motion) are generated on the abrasive particles. These forces act in the normal direction on the abrasives and indent into the workpiece's surface, as shown in Figure 8.3. Hence, the total normal force (F_n) acting on the abrasive can be calculated from Eq. (8.2).

$$\overrightarrow{F_n} = \overrightarrow{F_m} + \overrightarrow{F_r} + \overrightarrow{F_{cen}} \tag{8.2}$$

The reciprocating motion of the MR fluid generates the axial force on the abrasive (F_a), along with the action of the tangential force (F_t) produced due to the rotation of the magnets, as shown in Eq. (8.3).

$$F_t = 2m(\omega \times v) \tag{8.3}$$

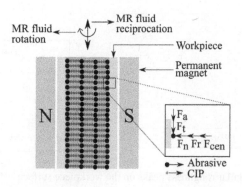

FIGURE 8.3   Different forces acting on the abrasive particle during MFAF (flow mode).

Where ω is angular velocity and ν linear velocity of the abrasive particles, these axial and tangential forces help to provide the shear action (F_s) of the abrasives on the workpiece surface as shown in Eq. (8.4).

$$\overrightarrow{F_s} = \overrightarrow{F_t} + \overrightarrow{F_a} \tag{8.4}$$

Hence, the normal force acting on the abrasives indents it into the workpiece, whereas the shear force helps to remove the indented abrasives from the workpiece, as shown in Figure 8.3. The combined action of these forces erodes materials from the workpiece and enhances the surface quality. The abrasive's indentation thickness (t) into the workpiece can be determined from Eq. (8.5) through the geometrical analysis shown in Figure 8.4.

$$t = \frac{D_1}{2} - \frac{1}{2}\sqrt{D_1^2 - D_2^2} \tag{8.5}$$

The depth of indentation is denoted by t, the diameter of abrasive and indentation is represented by D_1 and D_2, respectively. However, the indentation diameter of the abrasives varies with workpiece hardness and can be calculated through the *BHN* as shown in Eq. 8.6.

$$BHN = \frac{F_n}{\frac{\pi D_1}{2}\left(D_1 - \sqrt{D_1^2 - D_2^2}\right)} \tag{8.6}$$

Now, the area of impact (A_p) due to the indention of the abrasive on the workpiece can be calculated from Eq. (8.7).

$$A_p = \frac{(D_1)^2}{4}\left[\cos^{-1}\left(1 - \frac{2t}{D_1}\right)\right] - \left(\left(\frac{D_1}{2} - t\right)\sqrt{t(D_1 - t)}\right) \tag{8.7}$$

The analysis of the different forces generated during the MFAF (flow mode) and the impact area during the abrasion helps determine the workpiece's final surface roughness.

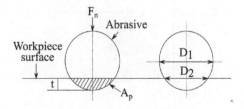

FIGURE 8.4 Abrasive particle indenting into workpiece surface and the sheared area.

8.2.1.2 Surface Roughness Model

During the analytical analysis of the surface roughness, it is assumed that the material's roughness peak is in the shape of a triangle with an initial height of R_{max} as shown in Figure 8.5. After the 1st reciprocating motion (stroke) of the MR fluid, the roughness peak height is decreased to R_1. After the 2nd stroke, the height is reduced to R_2 and so on. The total volume of material removed after the 1st stroke is calculated from Eq. (8.8).

$$\Delta V = \frac{1}{2} \times AD \times BC \times r \tag{8.8}$$

Now, the width of the material removed must be equal to the radius of the A_p (r) as shown earlier; similarly, the volume of the material removed can be calculated through Eqs. (8.9–8.11).

$$AD = R_{max} - R_1, \tag{8.9}$$

and

$$BC = 2\left(R_{max} - R_1\right)\tan\theta \tag{8.10}$$

Hence,

$$\Delta V = \left(R_{max} - R_1\right)^2 \tan\theta \times r \tag{8.11}$$

The total volume of material removed through a cell

$$\Delta V_c = \Delta V \times n_f \tag{8.12}$$

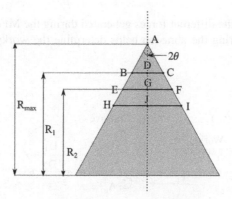

FIGURE 8.5 Height of roughness peaks after removal of material.

where n_f denotes the total number of triangular profiles on the workpiece. l_c represents workpiece length as shown in Figure 8.2. n_f can be calculated from Eq. (8.13).

$$n_f = \frac{l_c}{2R_{max} \tan \theta} \tag{8.13}$$

The material removed during the first stroke is calculated from Eq. (8.14).

$$\Delta V_{c1} = \frac{(R_{max} - R_1)^2 \times r \times l_c}{2R_{max}} \tag{8.14}$$

The height of the surface roughness peak (R_1) can be calculated in terms of the volume of material removed during the first stroke, as shown in Eq. (8.15).

$$R_1 = R_{max} - \left(\frac{2R_{max} \times V}{r \times l_c} \right)^{\frac{1}{2}} \tag{8.15}$$

Similarly, the workpiece material removed after the second stroke, in terms of the roughness peak height, can be determined through Eqs. (8.16) and (8.17).

$$BCFE = \left[(R_{max} - R_2)^2 - (R_{max} - R_1)^2 \right] \tan \theta \tag{8.16}$$

$$\Delta V_{c2} = \frac{\left[(R_{max} - R_2)^2 - (R_{max} - R_1)^2 \right] r \times l_c}{2R_{max}} \tag{8.17}$$

Hence, the height of the surface roughness peak (R_2) can be calculated in terms of the volume of material removed during the 2nd stroke, as shown in Eq. (8.18).

$$R_2 = R_{max} - \left[(R_{max} - R_1)^2 + \left(\frac{2VR_{max}}{r \times l_c} \right) \right]^{\frac{1}{2}} \tag{8.18}$$

Similarly, after the nth stroke, the height of the surface roughness peak (R_n) can be calculated from Eq. (8.19).

$$R_n = R_{max} - \left[(R_{max} - R_{n-1})^2 + \left(\frac{2VR_{max}}{r \times l_c} \right) \right]^{\frac{1}{2}} \tag{8.19}$$

8.2.2 SQUEEZE MODE

A concentrated and focused magnetic field is used to squeeze the MR fluid between the magnet and workpiece used as a finishing tool, e.g., BEMRF. Herein the magnet is placed inside a magnet holder, which rotates with certain rpm along the magnet's axis. The rotational motion generates the centrifugal action on the MR fluid placed at the polishing tool's tip. The combined action of magnetic force and centrifugal force assists the finishing operation. The forces generated during this operation are normal force and shear force [18].

8.2.2.1 Analysis of Forces

Normal force

The MR fluid consists of CIPs, under the influence of the external magnetic field producing a normal force on the workpiece surface, as calculated from Eq. (8.20).

$$F_n(z) = \frac{m * M}{B} B(z) \frac{dB(z)}{dz} \tag{8.20}$$

Shear Force

The normal force helps to indent the abrasive particles embedded in between the CIPs into the workpiece. The centrifugal force generated on the abrasives helps develop shear force, which can be calculated from Eq. (8.21).

$$F_s = (A - A') * \tau_y \tag{8.21}$$

Where the total area of impact of the indenting abrasive particle is represented by A, the area of impact of the indented particle is defined as A'. The yield stress (τ_y) of the MR fluid is a function of magnetic field strength B. With an increase in the magnetic field strength, the MR fluid's yield stress increases, as shown in Eq. (8.22). The abrasive particle's indentation depth and diameter can be calculated using the same principle applied in the flow mode of the MFAF process. As shown in Figure 8.6(a), the area of impact of an abrasive particle can be calculated as shown in Eq. (8.23):

$$\tau_y = -0.017B^2 + 50.182B - 2064 \tag{8.22}$$

$$A_p = \frac{(D_1)^2}{4} \left[\cos^{-1}\left(1 - \frac{2t}{D_1}\right) \right] - \left(\left(\frac{D_1}{2} - t\right)\sqrt{t(D_1 - t)}\right) \tag{8.23}$$

8.2.2.2 Surface Roughness Model

$$V_{abr} = A'\left(1 - \frac{R_a^i}{R_a^0}\right) * l_w \tag{8.24}$$

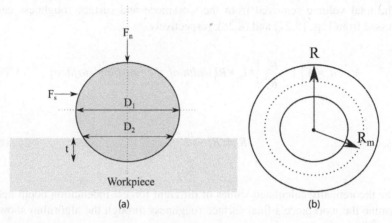

FIGURE 8.6 (a) Forces acting on the abrasive during squeeze mode and (b) circular path by abrasive on the workpiece during MFAF process.

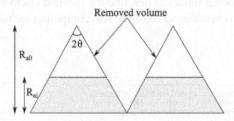

FIGURE 8.7 Shape of roughness peak.

The peak height of the initial roughness is represented as R_a^0 as shown in Figure 8.7. The volume of material removed from the workpiece can be determined from Eq. (8.24).

The length (l_w) covered by an abrasive during a single rotation of the polishing tool is calculated from Eq. (8.25). The mean diameter of an abrasive (R_m) is shown in Figure 8.6(b).

$$l_w = 2 * \pi * R_m = \pi * R \tag{8.25}$$

The MR fluid's microstructural study shows that the abrasives and CIP form a body-centered cube (BCC) structure. Here, the abrasive at the center is surrounded by CIP particles from the corners. The total number of active abrasive particles (n_s) is calculated from Eq. (8.26).

$$n_s = 0.4 * \left(\frac{l_w}{length\ of\ the\ one\ BCC\ unit\ cell} \right) \tag{8.26}$$

The total volume removed from the workpiece and surface roughness can be expressed from Eqs. (8.27) and (8.28), respectively.

$$V = n_s * A' \left(1 - \frac{R_a^i}{R_a^0} \right) * l_w * R \left(width\ of\ the\ finishing\ surface \right) \qquad (8.27)$$

$$R_a^i = R_a^0 - \frac{n_s * A'}{l_w} \qquad (8.28)$$

The theoretically calculated values of different forces' indentation depth help to determine the workpiece's final surface roughness through the algorithm shown in Figure 8.8.

The quality of surface produces during the MFAF mainly depends upon various input parameters like finishing time, working distance, magnetic field, MR fluid composition, and concentration. Optimizing such critical process parameters requires an explicit computational model to describe the finished component's surface characteristics. In the next section, some of the soft computing techniques are described briefly.

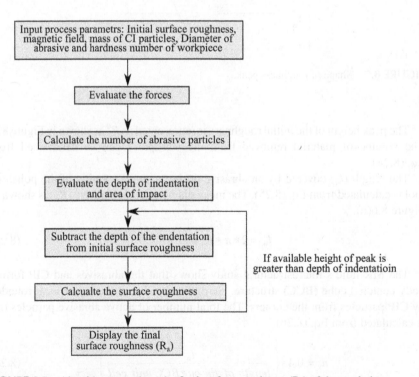

FIGURE 8.8   Algorithm to analyze the final surface roughness (R_a) of the workpiece.

8.3   AN OVERVIEW OF SOFT COMPUTING TECHNIQUES USED IN THE MFAF PROCESS

Soft computing is a computational method established based on AI, where a human decision-making system is analyzed, and based on that, different algorithms are generated. Zadeh [19] introduced soft computing in 1992; it is a well-designed algorithm capable of solving nonlinear, complex, multidimensional problems. Some of the impactful soft computing techniques are NN, GA, FL, JAYA, RSM, etc. A brief overview of these techniques is discussed in the following subsections.

8.3.1   NEURAL NETWORK

Neurons are processing units in human being used to transmit information to the nerve cell; approximately there are 10^{11} neurons available in the human brain. These neurons are parallelly connected. Different algorithms are developed based on these neurons' working principles known as a NN to mimic a human's decision-making system. Macculloch and Pitts [20] developed NN in 1943. These NNs are divided into different sub-layers, as shown in Figure 8.9(a).

The initial layer is nourished with the initial information regarding the problem; then, this information is transferred to the hidden layer. The connecting lines between these two layers provide weightage to each input data arbitrarily. The summation of these weighted data is passed through the activation function. This activation function decides which node it will pass to the next layer. Finally, through the application function, the output values are delivered to the output layer. The error between the obtained and the actual values are calculated, and the weights are adjusted to minimize them. Parallel computing can be possible through NN, which is computationally inexpensive [13].

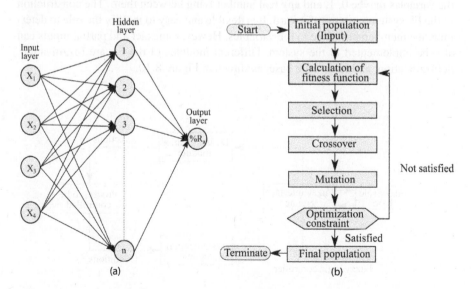

(a)   (b)

FIGURE 8.9   (a) Flow chart for the neural network algorithm (b) genetic algorithm.

8.3.2 GENETIC ALGORITHM

GA is based on Darwin's theory of natural evolution, "survival of fittest". This algorithm mimics the process of natural selection. The different GA phases used to achieve the optimal solutions are initial population, fitness function, selection, cross-over, and mutation.

Initial population: A set of random individuals termed as the population is selected; each individual contains a set of genes termed as a chromosome.

Fitness Function: This function defines each chromosome's fitness, and based on these fitness values, the probability of survival is determined.

Selection: This phase defines the selection of the individuals on the basis of fitness.

Cross-over: The new individuals are made from the cross-over of the parent populations.

Mutation: The population's variation is made by flipping a chromosome gene with a low random probability.

The evolved population goes through the same process again until an optimal solution is achieved, as shown in Figure 8.9(b). The benefit of GA is that it is a population-based search algorithm; hence convergence rate is high [13]. However, the result obtained through the GA is much closer to the experimental output as compared with the simulated annealing (SA).

8.3.3 FUZZY LOGIC

Zadeh [21] developed FL in 1965; as opposed to Boolean logic, the value assigned to the variables maybe 0, 1, and any real number lying between them. The construction of the FL system is straightforward. It is flexible and easy to modify the rule to determine the membership value to the variables. However, inexact and partial inputs can also be implemented in the system. Different modules of the FL are fuzzification, defuzzification, and fuzzy rule base, as shown in Figure 8.10.

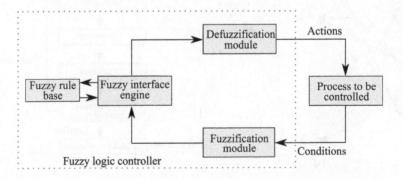

FIGURE 8.10 Flow chart for the fuzzy logic algorithm.

Fuzzification module: The input values of the variables are converted to the linguistic value. Crisp sets get transformed into the fuzzy subset.

Fuzzy rule base: Fuzzy sets and FL are used to model the relationship between the variables.

Defuzzification module: This is the reverse of fuzzification, where the fuzzy sets are converted into typical numerical values.

Nguyen et al. [22] performed the surface finishing operation with the assistance of MFAF on the Ni–P-coated SKD_{11} steel and achieved a surface roughness (R_a) of 0.561 nm. The process parameters were optimized with FL, and it was noticed that the influence of the abrasive size over surface quality is very high. The weightage of the working gap, abrasive size, CIP size, and magnetizing current over the surface roughness during MFAF (squeeze mode) is found to be 0.7461, 0.7797, 0.4914, and 0.6686 through FL.

8.3.4 JAYA

The Jaya computational technique is utilized to solve constraints as well as unconstraint problems. It is a global optimization algorithm that doesn't require a learning phase while solving problems. In the case of *teaching and learning-based optimization (TLBO)* algorithms, two phases are necessary to achieve the solution, i.e., teaching and learning. Still, in Jaya, only one phase is required. The algorithm of Jaya is based on adopting the best solution and ignoring the worst one while moving toward the optimized point, as shown in Figure 8.11(a). However, a few parameters are

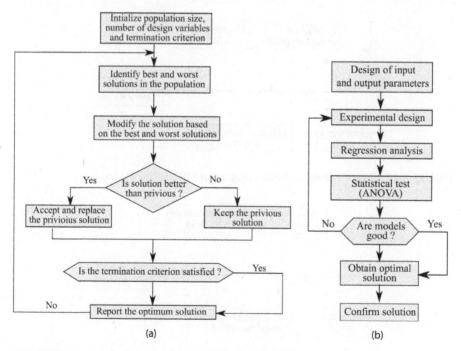

(a)   (b)

FIGURE 8.11   Flow chart for the (a) Jaya algorithm (b) RSM.

required to determine the solution through this algorithm. Anjanyulu et al. [23] optimized the process parameters of the MFAF process. The error obtained while choosing the optimum solution through the Jaya algorithm is 0.3%, much lower than GA.

8.3.5 RESPONSE SURFACE METHODOLOGY (RSM)

Box and Draper [24] developed the RSM algorithm in 1987, a multi-objective optimization technique. The optimized output variables (responses), which are the input parameters' function, can be determined through this algorithm. RSM helps to decrease the number of trials and provides the sensitivity of input variables over the response [25]. Its working pattern is elaborated in Figure 8.11(b). The alteration in the output variables is monitored when changes are made in the input parameters, and accordingly, the most crucial input parameters are obtained. In surface finishing, Chawla et al. [5] utilized RSM to optimize the process parameters during the MFAF process. A good relationship is observed between the experimental data and predicted values obtained through the RSM. However, the influence of extrusion pressure and magnetic field strength on the surface roughness during MFAF (flow mode) is 47.45% and 15.30%, respectively.

8.3.6 OPTIMIZATION OF PROCESS PARAMETERS AFFECTING SURFACE ROUGHNESS

Optimizing the process parameters during the MFAF process aims to achieve the best surface quality during surface finishing. The crucial process parameters during the flow mode and squeeze mode of the MFAF processes are shown in Figs. 8.12 and 8.13, respectively.

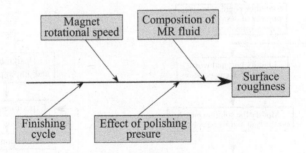

FIGURE 8.12   Ishikawa cause and effect diagram of MFAF (flow mode).

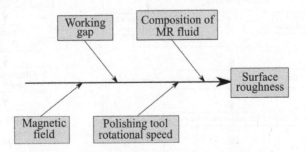

FIGURE 8.13   Ishikawa cause and effect diagram of MFAF (squeeze mode).

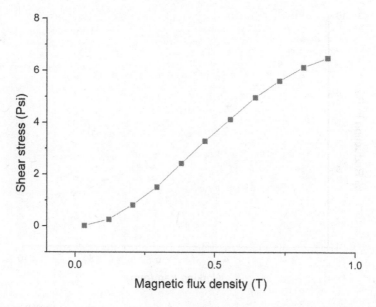

FIGURE 8.14 Variation in shear stress of MR fluid with a change in abrasive particles' size [27].

Composition of MR fluid

The composition and constituents of the MR fluid play a vital role in achieving a highly polished surface. The determining parameters that affect MR fluid behavior during surface finishing operations are the abrasive particles' size, CIP's size, concentration, hardness, etc. However, it is noticed that MR fluid having an identical size of abrasive and CIP particles shows higher yield stress than other compositions. When the abrasive size is greater than the CIPs, the MR fluid's yield stress decreases. This is because the number of the abrasive particle is less in the MR fluid than the CIP; still, the larger size of abrasive particles can break the CIP chain, which leads to a decrease in the strength of the MR fluid. However, when abrasive particles are smaller than CIP, a higher number of abrasive particles leads to breaking the CIP chain, which reduces the MR fluid strength [26]. With particular variation in the abrasive or CIP size, MR fluid's physical properties (viscosity and shear stress) change drastically, which affects the surface quality during the finishing operations. However, MR fluid's rheological properties can be controlled by varying the magnetic field during the surface finishing process, as shown in Figure 8.14.

8.3.6.1 Squeeze Mode

Working gap

The magnetic field's strength is the prime factor affecting the MR fluid's behavior during the surface finishing operation; the magnetic field influences the CIP in MR fluid. An increase in the magnetic field enhances the stability of the CIP

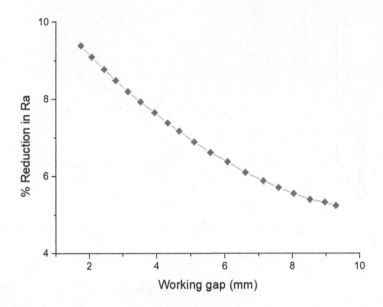

FIGURE 8.15 Variation in R_a with the change in working gap (*Source* [29] Reused with permission).

chain. However, it is observed that the magnetic field strength is inversely proportional to the axial distance from the magnet. Now, when the working gap decreases, the magnetic field at the point of impact increases. The high magnetic field leads to stable MR fluid; hence, the reduction in the surface roughness increases with the decreased working gap between the workpiece and the tool, as shown in Figure 8.15 [28].

The rotational speed of the polishing tool

The polishing tool's rotational speed generates the centrifugal force on the MR fluid; with an increase in the polishing tool's rpm above a specific limit, the holding capacity of the CIP chain breaks, and the reduction in the change in the R_a value is noticed as shown in Fig. 8.16. As the centrifugal force is directly proportional to the square of the polishing tool's rotational speed, the centrifugal force on the CIP is much higher than the normal force at a higher rotational speed of the polishing tool. Hence, the % reduction in R_a reduces with an increased rpm of the polishing tool [30].

Magnetizing current

The electromagnet is used in BEMRF to produce the magnetic field during the finishing operation. The magnetic field's strength mainly depends upon the current supplied to the electromagnet; with increased magnetic current, the magnetic field strength increases, leading to enhanced MR fluid properties. Hence, superior surface quality is achieved with better shear stress and the MR fluid's viscosity, as shown in Figure 8.17 [32].

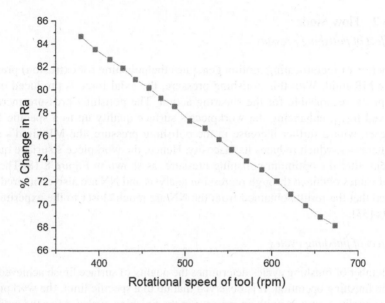

FIGURE 8.16 Variation in R_a with the change in rotational speed of the tool (*Source* [31] Reused with permission).

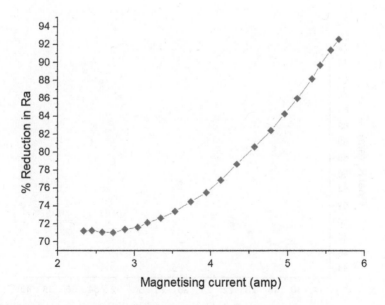

FIGURE 8.17 Variation in R_a with the change in magnetizing current to the electromagnet (*Source* [31] Reused with permission).

8.3.6.2   Flow Mode

Effect of polishing pressure

The action of reciprocating motion generates the polishing (or extrusion) pressure on the MR fluid. With this polishing pressure, the axial force is produced on the workpiece, responsible for the shearing action. The polishing pressure increases the axial force, enhancing the workpiece's surface quality up to a specific limit. However, with a further increase in the polishing pressure, the MR fluid's shear rate increases, which reduces its viscosity. Hence, the workpiece's surface quality worsens after the optimum polishing pressure, as shown in Figure 8.18. The predicted values obtained through regression analysis and NN are also compared. It is noticed that the results obtained from the NN are much closer to the experimental results [33].

Effect of finishing cycles

The number of finishing cycles determines the quality of surface finish achieved during the finishing operation. It is observed that up to a specific limit, the workpiece's surface quality improves with an increase in the finishing cycles. After the optimum number of finishing cycles, the surface roughness peak gets reduced. Later, a higher finishing force is required to remove the increased plateau of the workpiece's roughness peaks. Hence, no further improvement is noticed on the workpiece surface because of this, as shown in Fig. 8.19 [35].

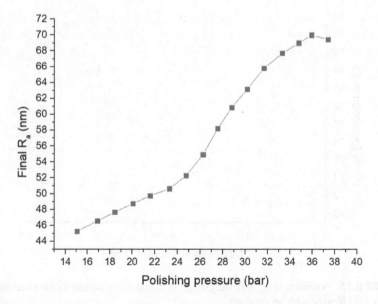

FIGURE 8.18 Variation in R_a with the change in polishing pressure (*Source* [34] Reused with permission).

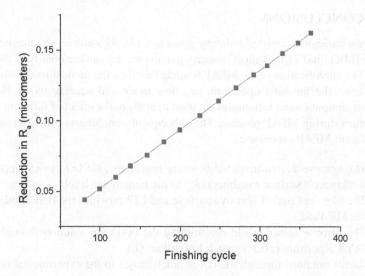

FIGURE 8.19 Variation in R_a with the change in finishing the cycle (*Source* [36] Reused with permission).

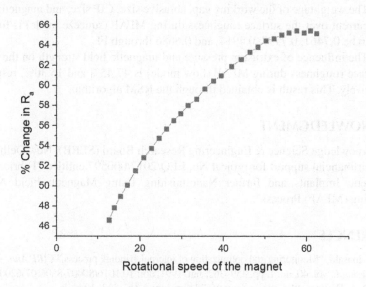

FIGURE 8.20 Variation in R_a with the change in magnet rpm.

Effect of magnet rpm

The magnet rpm increases the abrasive particles' effectivity, as it allows the abrasive particles to move in a helical path rather than the straight line one. With increased magnet rpm, the centrifugal force on the abrasive particles increases. Hence, an increased value of the centrifugal force enhances the quality of the workpiece surface, as shown in Figure 8.20 [37].

8.4 CONCLUSIONS

Magnetorheological-assisted finishing processes (MFAF) utilizes the magnetorheological (MR) fluid's rheological property to enhance the surface quality of the workpiece. The classification of the MFAF is made based on the mode through which MR fluid assists the finishing operation, i.e., flow mode and squeeze mode. However, different computational techniques are used to analyze the effect of different process parameters during MFAF process. The subsequent conclusions can be drawn from the different MFAF processes:

- *Magnetorheological-assisted finishing processes (MFAF)* can enhance the workpiece's surface roughness (R_a) to the nanometer level.
- The identical size of abrasive particle and CIP provides the best stability to the MR fluid.
- The error obtained while determining the optimum solution through the JAYA algorithm is 0.3%, much lower than GA.
- Result obtained through the *GA* is much closer to the experimental output than the *SA*.
- During the MFAF (squeeze mode), the most crucial parameter is the working gap among all other process parameters.
- The weightage of the working gap, abrasive size, CIP size, and magnetizing current over the surface roughness during MFAF (squeeze mode) is found to be 0.7461, 0.7797, 0.4914, and 0.6686 through FL.
- The influence of extrusion pressure and magnetic field strength on the surface roughness during MFAF (flow mode) is 47.45% and 15.30%, respectively. This result is obtained through the RSM algorithm.

ACKNOWLEDGMENT

We acknowledge Science & Engineering Research Board (SERB), New Delhi, India for theirfinancial support for project No. EEQ/2017/000597 entitled "Fabrication of Prosthetic Implants and further Nanofinishing Using Magnetic Field Assisted Finishing (MFAF) Process".

REFERENCES

1. I. Inasaki, "Monitoring and optimization of internal grinding process," *CIRP Ann. – Manuf. Technol.*, vol. 40, no. 1, pp. 359–362, Jan. 1991, doi: 10.1016/S0007-8506(07)62006-X.
2. D. A. Wallace, "Lapping apparatus," US patent 5,367,833, 1944.
3. M. S. Niranjan and S. Jha, "Experimental investigation into tool aging effect in ball end magnetorheological finishing," *Int. J. Adv. Manuf. Technol.*, vol. 80, no. 9–12, pp. 1895–1902, 2015, doi: 10.1007/s00170-015-6996-3.
4. L. Nagdeve, V. K. Jain, and J. Ramkumar, "Experimental investigations into nano-finishing of freeform surfaces using negative replica of the knee joint," *Procedia CIRP*, vol. 42, pp. 793–798, 2016, doi: 10.1016/j.procir.2016.02.321.
5. S. V. Jha, and K. Jain Ranga Komanduri, "Effect of extrusion pressure and number of finishing cycles on surface roughness in magnetorheological abrasive flow finishing (MRAFF) process," *Int J Adv Manuf Technol*, vol. 33, pp. 725–729, 2007, doi: 10.1007/s00170-006-0502-x.

6. A. Sadiq, and M. S. Shunmugam, "A novel method to improve finish on non-magnetic surfaces in magnetorheological abrasive honing process," *Tribol. Int.*, vol. 43, no. 5, pp. 1122–1126, 2010, doi: https://doi.org/10.1016/j.triboint.2009.12.038.

7. W. Kordonski, and A. Shorey, "Magnetorheological (MR) jet finishing technology," *J. Intell. Mater. Syst. Struct.*, vol. 18, no. 12, pp. 1127–1130, 2007, doi: 10.1177/1045389X07083139.

8. Z. Alam, F. Iqbal, S. Ganesan, and S. Jha, "Nanofinishing of 3D surfaces by automated five-axis CNC ball end magnetorheological finishing machine using customized controller," *Int. J. Adv. Manuf. Technol.*, vol. 100, no. 5–8, pp. 1031–1042, Feb. 2019, doi: 10.1007/s00170-017-1518-0.

9. L. Nagdeve, V. K. Jain, and J. Ramkumar, "Optimization of process parameters in nanofinishing of Co-Cr-Mo alloy knee joint," *Mater. Manuf. Process.*, vol. 00, no. 00, pp. 1–8, 2020, doi: 10.1080/10426914.2020.1750633.

10. S. Jha, and V. K. Jain, "Modeling and simulation of surface roughness in magnetorheological abrasive flow finishing (MRAFF) process," *Wear*, vol. 261, no. 7–8, pp. 856–866, Oct. 2006, doi: 10.1016/j.wear.2006.01.043.

11. Z. Alam, and S. Jha, "Modeling of surface roughness in ball end magnetorheological finishing (BEMRF) process," *Wear*, vol. 374, pp. 54–62, 2017.

12. M. Das, V. K. Jain, and P. S. Ghoshdastidar, "*Parametric Study of Process Parameters and Characterization of Surface Texture Using Rotational-Magnetorheological Abrasive Flow Finishing (R-MRAFF) Process*," in *ASME 2009 International Manufacturing Science and Engineering Conference*, 2009, pp. 251–260.

13. D. Ibrahim, "An overview of soft computing," in *Procedia Computer Science*, vol. 102, pp. 34–38, 2016, doi: 10.1016/j.procs.2016.09.366.

14. H. M. Pandey, "Jaya a novel optimization algorithm: What, how and why?," in *Proceedings of the 2016 6th International Conference – Cloud System and Big Data Engineering, Confluence 2016*, pp. 728–730, 2016, doi: 10.1109/CONFLUENCE.2016.7508215.

15. L. A. Zadeh, "Fuzzy logic," *Computer (Long. Beach. Calif).*, vol. 21, no. 4, pp. 83–93, 1988, doi: 10.1109/2.53.

16. A. K. Mondal, A. S. Rajput, D. Prasad, and D. Bose, "*Magnetic Field Assisted Finishing Processes*," Springer, Cham, pp. 211–234, 2020.

17. S. Kumar, V. K. Jain, and A. Sidpara, "Nanofinishing of freeform surfaces (knee joint implant) by rotational-magnetorheological abrasive flow finishing (R-MRAFF) process," *Precis. Eng.*, vol. 42, pp. 165–178, 2015.

18. D. A. Khan, Z. Alam, F. Iqbal, and S. Jha, "Design and development of improved ball end magnetorheological finishing tool with efficacious cooling system," in M. Shunmugam, and M. Kanthababu (eds.) *Advances in Simulation, Product Design and Development*, Springer, pp. 557–569, 2020.

19. Z. La, "*The calculus of fuzzy if-then rules*," in *In Fuzzy Engineering toward Human Friendly Systems – Proceedings of the International Fuzzy Engineering Symposium'91*, pp. 11–12, 1992.

20. W. S. McCulloch and W. Pitts, "A logical calculus of the ideas immanent in nervous activity," *Bull. Math. Biophys.*, vol. 5, no. 4, pp. 115–133, Dec. 1943, doi: 10.1007/BF02478259.

21. L. A. Zadeh, "Fuzzy sets," *Inf. Control*, vol. 8, no. 3, pp. 338–353, Jun. 1965, doi: 10.1016/S0019-9958(65)90241-X.

22. D. Nguyen, J. Wu, N. M. Quang, L. A. Duc, and P. X. Son, "Applying fuzzy grey relationship analysis and Taguchi method in polishing surfaces of magnetic materials by using magnetorheological fluid," *Int. J. Adv. Manuf. Technol.*, vol. 112, no. 5–6, pp. 1675–1689, Jan. 2021, doi: 10.1007/s00170-020-06567-1.

23. K. Anjaneyulu, and G. Venkatesh, "Optimization of process parameters of magnetic abrasive finishing using jaya algorithm," *Mater. Today Proc.*, Jul. 2020, doi: 10.1016/j. matpr.2020.06.568.

24. G. E. P Box Norman, R. Draper John Wüey and S. New York Glichester Brisbane Toronto Singapore, "Empirical Model-Building and Response Surfaces," 1987.

25. R. H. Myers, D. C. Montgomery, G. Geoffrey Vining, C. M. Borror, and S. M. Kowalski, "Response surface methodology: A retrospective and literature survey," *J. Quality Tech.*, vol. 36, no. 1. American Society for Quality, pp. 53–78, 2004, doi: 10.1080/00224065.2004.11980252.

26. L. Nagdeve, A. Sidpara, V. K. Jain, and J. Ramkumar, "On the effect of relative size of magnetic particles and abrasive particles in MR fluid-based finishing process," *Mach. Sci. Technol.*, vol. 22, no. 3, pp. 493–506, May 2018, doi: 10.1080/10910344.2017.1365899.

27. K. Özsoy, and M. R. Usal, "A mathematical model for the magnetorheological materials and magneto reheological devices," *Eng. Sci. Technol. an Int. J.*, vol. 21, no. 6, pp. 1143–1151, Dec. 2018, doi: 10.1016/j.jestch.2018.07.019.

28. P. Kala, V. Sharma, and P. M. Pandey, "Surface roughness modelling for double disk magnetic abrasive finishing process," *J. Manuf. Process.*, vol. 25, pp. 37–48, Jan. 2017, doi: 10.1016/j.jmapro.2016.10.007.

29. G. Ghosh, A. Sidpara, and P. P. Bandyopadhyay, "Experimental and theoretical investigation into surface roughness and residual stress in magnetorheological finishing of OFHC copper," *J. Mater. Process. Technol.*, vol. 288, p. 116899, Feb. 2021, doi: 10.1016/j.jmatprotec.2020.116899.

30. D. A. Khan, and S. Jha, "Selection of optimum polishing fluid composition for ball end magnetorheological finishing (BEMRF) of copper," *Int. J. Adv. Manuf. Technol.*, vol. 100, no. 5–8, pp. 1093–1103, 2019, doi: 10.1007/s00170-017-1056-9.

31. K. Saraswathamma, S. Jha, and P. V. Rao, "Experimental investigation into Ball end Magnetorheological Finishing of silicon," *Precis. Eng.*, vol. 42, pp. 218–223, Oct. 2015, doi: 10.1016/j.precisioneng.2015.05.003.

32. D. A. Khan, Z. Alam, and S. Jha, "*Nanofinishing of Copper Using Ball End Magnetorheological Finishing (BEMRF) Process*," in *Conference: ASME International Mechanical Engineering Congress and Exposition (IMECE) 2016*, Phoenix, Arizona, USA, 2016, doi: 10.1115/imece2016-65974.

33. S. Jha, V. K. Jain, and R. Komanduri, "Effect of extrusion pressure and number of finishing cycles on surface roughness in magnetorheological abrasive flow finishing (MRAFF) process," *Int. J. Adv. Manuf. Technol.*, vol. 33, no. 7–8, pp. 725–729, Jul. 2007, doi: 10.1007/s00170-006-0502-x.

34. S. Kumar, V. K. Jain, and A. Sidpara, "Nanofinishing of freeform surfaces (knee joint implant) by rotational-magnetorheological abrasive flow finishing (R-MRAFF) process," *Precis. Eng.*, vol. 42, pp. 165–178, Oct. 2015, doi: 10.1016/j.precisioneng. 2015.04.014.

35. M. Das, V. K. Jain, and P. S. Ghoshdastidar, "Nanofinishing of flat workpieces using rotational-magnetorheological abrasive flow finishing (R-MRAFF) process," *Int J Adv Manuf Technol*, vol. 62, no. 1, pp. 405–420, 2012, doi: 10.1007/s00170-011-3808-2.

36. M. Das, V. K. Jain, and P. S. Ghoshdastidar, "Fluid flow analysis of magnetorheological abrasive flow finishing (MRAFF) process," *Int. J. Mach. Tools Manuf.*, vol. 48, no. 3, pp. 415–426, 2008, doi: 10.1016/j.ijmachtools.2007.09.004.

37. A. Barman, and M. Das, "Soft computing techniques to model and optimize magnetic field-assisted finishing process and characterization of the finished surface," *Proc. Inst. Mech. Eng. Part C J. Mech. Eng. Sci.*, vol. 232, no. 17, pp. 3156–3168, Sep. 2018, doi: 10.1177/0954406217731116.

9 Numerical Analysis of Limited LOCA Event Involving Deflection of Pressure Tube

Ankit R. Singh
Indian Institute of Technology Bombay, India

Nitesh Dutt
College of Engineering Roorkee, India

Pradeep K. Sahoo
Botswana International University of Science and
Technology, Botswana

CONTENTS

9.1 INTRODUCTION

Indian Pressurized Heavy Water Reactor (PHWR) 220 MWe core has 306 horizontal channels, and each channel consists of two tubes in a concentric configuration. The inside tube is the pressure tube (PT), and the outer tube is the calandria tube (CT). Fuel bundles reside in the PT with 12 fuel bundles per channel. The core resides in a large cylindrical calandria, a pool of heavy water moderators at ambient pressure (Bajaj & Gore, 2006; Nayak & Banerjee, 2017). This heavy water moderator is a source of primary heat sink during an accident condition (Gillespie, Moyer, & Hadaller, 1985; Gupta, Dutt, Raj, & Kakokar, 1996; Nandan et al., 2012; Snell et al., 1988).

Calandria is surrounded by a large shield Calandria vault that contains light water. Calandria vault is a source of ultimate heat sink during the late phase of an accident (AERB, 2000; Blahnik, Luxat, Nijhawan, & Thuralsingham, 1993; Brown, Blahnik, & Muzumdar, 1984; Mathew, Kupferschmidt, Snell, & Bonechi, 2001). Pressurized heavy water (coolant) carries heat from fuel bundles. Both ends of the CT are fixed to Calandria tube sheets, while one end of the PT is fixed and the other end is free to expand. CO_2 gas is used as annular gas, which acts as a thermal insulator during regular operation.

Nuclear coolant channels are placed as a tube-in-tube type configuration, as depicted in Figure 9.1. The primary heat transfer mechanism is the convective heat transfer carried away by heavy water coolant under regular operation. In the postulated loss of coolant accident (LOCA) conditions, the heat transfer mechanism is coolant convection, natural convection through the PT and CT annulus, and convection over the outer surface of CT.

Beyond Design Basis Accident scenario includes LOCA and loss of Emergency Core Cooling System (ECCS). However, the moderator cooling supports the decay heat removal during such events. The severity of such events is restricted to a single channel; hence, these events are called limited core damage accidents (LCDA). A severe situation arises if the moderator heat-removal system fails along with other safety features such as firefighting water, such situations lead to Severe Core Damage Accidents (SCDA) (Dutt, Singh, & Sahoo, 2020; Gupta et al., 1996; Majumdar, Chatterjee, Lele, Guillard, & Fichot, 2014; Negi, Kumar, Majumdar, & Mukopadhyay, 2017). The blackout (SBO) situation is also classified as an SCDA event. Depending upon the internal pressure during limited accidents, PT will either balloon (in a high-pressure case) or sag/deflect downward (in a low-pressure case). These situations are illustrated in Figure 9.2.

The continuous moderator boil-off during SCDA leads to a dry-out and overheating of the channel. The overheated channels sag due to fuel bundles' weight and self-weight. The upper elevation channels will contact and rest over the below channel, and lastly below channel fails under high stress. This event is called channel disassembly – the disassembled channels are collected at Calandria bottom forming terminal debris.

Deflection of the channel typically occurs, as its internal pressure becomes equal to or less than one MPa. The stress-induced bending moment is insufficient to hold

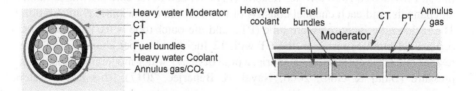

FIGURE 9.1 PT-CT position during regular operation.

(a)   (b)

FIGURE 9.2   PT-CT position during accident conditions: (a) high internal pressure; (b) low internal pressure.

the channel's plastic flow on its own at high temperatures. Therefore, strain or PT deflection could take place solely due to $\alpha \rightarrow (\alpha + \beta)$ phase-transformation under bending load, and this typically happens in the temperature range of 610°C to 925°C (Gillespie, Moyer, & Hadaller, 1984; Mani Mathew, 2004).

It becomes crucial to understand the behavior of the reactor channel following the postulated LOCA events or SCDA events concerning safety. Researchers studied the LOCA effect with or without ECCS failure. The study has further been extended to the situation with malfunctioning in the moderator cooling circuitry. One such analysis has been performed by Nandan et al. (2007) at IIT Roorkee to experimentally investigate the limited LOCA condition, which leads to deflection of the PT, where the accident is restricted to a single channel. In the present study, we are exploring the experimental outcomes numerically using numerical software.

The study's main objective is to model the effect of a fuel bundle or fuel simulator on the deflection of PT. In essence, we will explore the applicability of numerical techniques with a 3D model of the experimental channel and predict the deflection of the channel.

9.2   EXPERIMENTAL WORK ON THE PRESSURE TUBE DEFLECTION IN LIMITED CORE DAMAGE CONDITION

The schematic of the set-up is shown in Figure 9.3. The test section is a rectangular tank with a two-meter long channel placed horizontally. Channel comprises PT, CT, and weight simulator. The channel dimensions are mentioned in Figure 9.4. A PT is placed concentrically and symmetrically inside the CT through two end supports. One of the supports allows the expansion of the PT during temperature escalation. The CT is fixed to the tank's side, and the outer surface of CT is wrapped with ceramic wool to reduce heat loss to the atmosphere and prevent high-temperature oxidation. Fuel bundle weight is simulated through eight structural steel solid cylinders of 70 mm diameter and 250 mm length.

The heat was generated in the PT through resistive heating with a DC supply from a 42 kW rectifier. Bus bars from the rectifier were connected to both the ends of the PT with copper clamps. Argon is purged through PT and CT annulus intermittently during the experiment. The temperatures of PT and CT are measured in axial and circumferential directions. The displacement measurement of the PT is measured with potentiometers or linear variable differential transformers (LVDTs). An LVDT

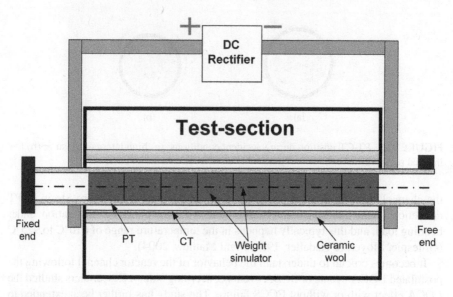

FIGURE 9.3   Schematic of the experimental test (Nandan et al., 2007).

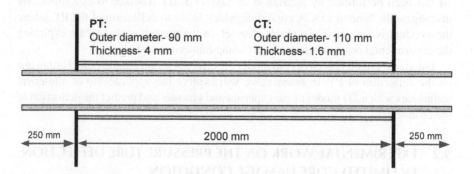

FIGURE 9.4   Dimension of a pressure tube and CT (Nandan et al., 2007).

Ceramic rod is inserted through a 2 mm diameter hole at the bottom of the CT to make contact with PT.

The PT was heated slowly to a temperature of 150°C, and then, a power ramp of 35.6 kW was applied. The power was maintained through the rest of the experiment and was put off once PT was observed to contact CT at one station. The crucial observations from the experiment are summarized in the following paragraphs.

The average temperature variation of the PT with time is shown in Figure 9.5. The initial temperature rise is identical at all axial locations. The temperature rise rate of the PT and CT and the PT's deflection are presented in Figure 9.6. The initiation of deflection is reported at 125s. At the initiation, the average temperature of the PT is 515°C. After about 175s, the deflection is said to increase significantly.

It is noted from Figure 9.6 that at 300s PT touches the CT, as it covers the annulus gap between the PT and CT. The corresponding average temperature of the PT is

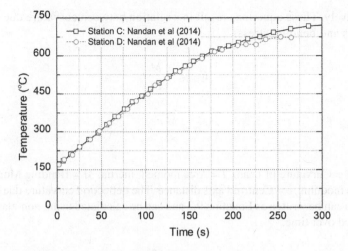

FIGURE 9.5 Temperature variation of PT (Nandan et al., 2007).

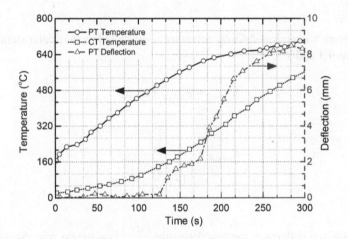

FIGURE 9.6 Deflection of PT (Nandan et al., 2007).

680°C. It is also reported that the rate of temperature rise of PT is much higher than the CT, which is evident because the PT was directly heated in the experiment. As the temperature of the PT increases, the radiative heat flux from the PT results in a significant rise in CT's temperature.

9.3 DEFLECTION MECHANISM AND MODEL

The deflection of PT could be calculated from simple beam theory with both ends fixed. High-temperature deflection is an overall effect of elastic and inelastic/plastic bending. The deflection curvature of a beam under elastic strain and longitudinal stress across a section can be given as Equations 9.1 and 9.2, respectively. These equations result in a relation between curvature due to elastic modulus and bending stress:

$$K_{elastic} = \frac{M}{EI}$$ (9.1)

$$\sigma = \frac{M.c}{I}$$ (9.2)

where K = Curvature of beam, I = Area moment inertia, M = Bending Moment, E = Young's modulus, c = Centroid axis distance. The deflection curvature due to inelastic strain can be written as Equation 9.3 and can be calculated using non-elastic strain integrated over time.

$$K_{plastic} = \frac{\varepsilon}{c} = \frac{\int_0^t \dot{\varepsilon}\,dt}{c}$$ (9.3)

The beam's overall deflection curvature is estimated as the scalar summation of Equations 9.1 and 9.3.

$$K_{overall} = K_{elastic} + K_{plastic}$$ (9.4)

$$K_{overall} = \frac{\dfrac{d^2 y}{dx^2}}{\left[1 + \left(\dfrac{dy}{dx}\right)^2\right]^{3/2}}$$ (9.5)

For minimal deflection, the first derivative $\left(\dfrac{dy}{dx}\right)$ is assumed zero. So, the overall deflection curvature equation is simplified to Equation 9.6. However, this assumption is valid only for deflection within the elastic regime or minimal deflection:

$$K_{overall} \approx \frac{d^2 y}{dx^2}$$ (9.6)

Deflection (y) of the PT at a particular section (x) is thus evaluated by double integration of Equation 9.7.

$$y = \iint \left\{ \frac{M}{EI} + \frac{\int_0^t \dot{\varepsilon}\,dt}{c} \right\} dx\,dx$$ (9.7)

9.3.1 A NON-ELASTIC MATERIAL FLOW MODEL

The literature shows that deflection of pressure tube will result as the temperature of channel shoots up. Above 30% of the melting point temperature of zirconium alloys, creep dominants for the material flow. Hence, thermally activated rate-dependent plasticity (creep) causes the deflection of the PT (Mathew et al., 2001; Nandan et al., 2010; Singh, Tariq, & Majumdar, 2020; Singh, Tariq, Sahoo, Majumdar, & Mukhopadhyay, 2021).

Material creep is classified into three distinct regions: primary, secondary, and tertiary creep. Most engineering studies are concerned with the primary and secondary creeps. A creep strain rate ($\dot{\varepsilon}$) depends mainly on stress (σ), strain (ε), time (t), and temperature (T), which is mathematically represented in Equation 9.8. Empirical correlations exist, representing the dependency of variables on strain rate. However, a generalized strain rate in Norton format is used here for illustration, and it is mentioned in Equation 9.9 with a temperature-dependent term.

$$\dot{\varepsilon}_{cr} = f\left(\sigma, \varepsilon, t, T\right) \tag{9.8}$$

$$\dot{\varepsilon}_{cr} = a.\sigma^b.e^{-c/T} \tag{9.9}$$

Furthermore, as shown by the longitudinal strain rate correlation given in Equation 9.10, the creep strain is developed by Shewfelt and Lyall (1985). k_2, mentioned in Table 9.1, is a temperature-changing factor; its value can be assumed zero below 750°C.

$$\dot{\varepsilon} = 8 \times 10^{10}.\sigma.\left(k_2 - \frac{\varepsilon}{\sigma}\right)^{2.4} e^{\left(\frac{-26670}{T}\right)} \tag{9.10}$$

Equation 9.7 for total deflection and Equation 9.10 for creep strain rate in a non-elastic regime can be used to calculate the deflection of PT analytically. However, we would want to explore the effect of the weight simulator on the deflection estimate in the present study, which is tiresome and challenging work to perform analytically. Hence, we will explore finite element software (ANSYS, 2020) in this study.

TABLE 9.1
Temperature dependency of parameter k_2

Temperature (°C)	k_2 (1/MPa)
800	0.0119
825	0.0114
850	0.0167
875	0.0134
900	0.01

Equation 9.10 is not precisely in the format necessary for use in ANSYS mechanical solver. Hence, this equation is written in the Usercreep subroutine (usercreep.F), written in the FORTRAN language. Temperature-dependent creep coefficients can be incorporated in the subroutine. Once this subroutine is ready, it can be compiled and linked to the solver during the start of the solver run by calling/upf command.

Three essential parameters are required in the subroutine: (1) creep rate equation; (2) derivative creep strain rate concerning stress; and (3) derivative creep strain rate with respect to strain. The derivatives help to estimate the global tangent matrix at element integration points. In essence, the subroutine uses solver variables such as temperature and stress and returns incremental strain as output to the solver for deflection calculation. Moreover, the solver works on an implicit algorithm, which is unconditionally stable.

The output of the subroutine needs to validate before its use in the study. However, Singh and Sahoo (2018) have already reported the subroutine's working using a single mesh element model. Hence, we are using the subroutine without further validation for its output.

9.4 NUMERICAL FORMULATION FOR LIMITED PT DEFORMATION

The following observations are made from the experiment that helped in simplifying the numerical modeling:

(1) Variations in the axial and circumferential temperatures for the PT are insignificant; hence, it is assumed that the channel length is exposed to the same temperature transient,
(2) Experiment was terminated as PT makes contact with the CT; hence, the post-contact condition is not essential for this study.

These observations alleviate the modeling of thermal aspects of the channel to predict the temperature, i.e., the experimental temperature of PT and CT (See Figure 9.5 and 9.6) may be directly used as a thermal load. We thus used only a structural model with explicitly applied thermal loads (Experimental temperature) for PT and CT. The model is formulated in two separate cases: (1) Model-1 without the weight simulators, (2) Model-2 with weight simulators.

In model-1, the uniformly distributed load is used as a structural load on the PT, while in model-2, all eight-weight simulators are modeled explicitly. These two models are finally compared to predict the effect of weight simulators on the deflection of the PT. Each model is described in the following paragraph in detail.

9.4.1 MODEL-1: WITH UNIFORMLY DISTRIBUTED LOAD

Figure 9.7 shows the schematic of the numerical model formulated for the numerical study of the limited deformation of the PT. For the weight calculation of the simulators, the density of structural steel is used, which equals 7850 kg/m³. The volume of each simulator is calculated as 9.62e-4 m³ from the diameter and length data.

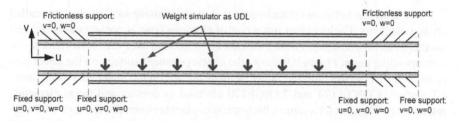

FIGURE 9.7   Schematic of a numerical model with boundary conditions.

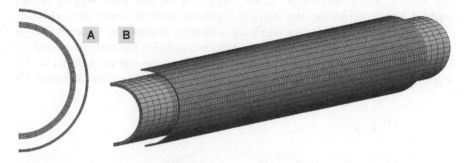

FIGURE 9.8   Body parts mesh elements: (a) circumferential mesh elements; (b) longitudinal mesh elements.

Therefore, the calculated weight of each simulator is 7.55 kg, and the uniformly distributed load is 296.3 N/m.

The extruded part of the PT on both sides is assumed as sliding support, i.e., the surface has only u-displacement. The fixed end of the PT has zero displacements in all directions, while the free end has u-displacement. Both ends of the CT are considered as fixed support with zero displacements. The gravity acceleration (9.8066 m/s²) is imposed to account for PT and CT's self-weight. Temperatures of PT and CT (See Figure 9.6) are applied as thermal load. It is assumed that the deflection will proceed in a vertical plane; thus, the concept of symmetry is used to reduce the total element counts.

The longitudinal creep, given in Equation 9.10, is used. The temperature dependency of the parameter mentioned in Table 9.1 is also included in the modeling. The creep is implemented with the help of Usercreep subroutine, as mentioned in the last section. The mesh elements used in the model are rendered in Figure 9.8; hex (SOLID186 is 3D 20 nodes) elements are used to meshing the PT and CT. It can model geometric non-linearity (viz. such as large deformation), material non-linearity (viz. creep, plasticity), and surface to surface contact.

Three and two-element layers in the thickness direction of PT and CT are used, respectively. Since the PT has more thickness than CT; therefore, an additional layer of the element is used for the PT. Twenty element divisions are used in the circumferential direction, while 150 element divisions are used along the length of PT and CT. The five longitudinal element divisions are used for the extruded length of the PT. With this, the total mesh elements become 15,600.

The interaction between contacting parts is modeled using surface elements called contact elements. Each contact has a pair of elements: one is contact elements, and another is target elements. These elements are generated over the mesh elements of corresponding parts. Figure 9.9 shows the contact element surface on the lower part of the outer PT surface and the target element surface on the lower part of the inner CT surface. CONTA174 and TARGE170 are used as contact and target elements, respectively. The contact assumes frictionless contact between the surfaces with augmented Lagrange formulation and gauss point as a detection method. The initial stiffness factor is selected as one; however, it is set to be updated by the solver during the solution. The simulation is run for 300 s with the auto time stepping feature enabled, which decides the time step size depending upon the convergence criteria.

The numerical result for the maximum deflection is shown in Figure 9.10. It is clear from the figure that the longitudinal creep correlation has under-predicted the deflection of PT. A possible explanation for this outcome can be deduced as: (1) Correlation (Equation 9.7) has been developed for the Canadian fabricated PT

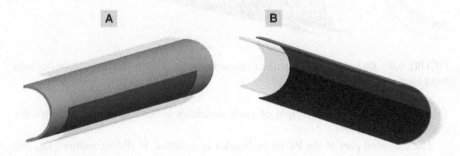

FIGURE 9.9   Contacting surface: (a) on the outer surface of PT; (b) on the inner surface of CT.

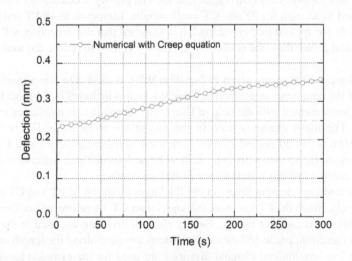

FIGURE 9.10   Deflection using longitudinal creep correlation (Eq. 9.7).

material, where the experiment (Nandan et al., 2007) has been performed for the Indian fabricated PT material. There is some difference in both fabrication techniques, which could alter the material configuration slightly among them. (2) Creep parameter k_2 is reported from 800°C onwards, while the maximum temperature in the experiment is reported below this creep initiation temperature. Thus, it could be postulated that the Indian PT has a much lower deflection initiation temperature for the thermal creep, contrary to its Canadian counterpart.

Currently, there is a scarcity of creep correlation for the Indian PT material. Hence, we have used a technique of accelerated creep equation, which is just an acceleration of Equation 9.7 with a scalar factor multiplication. Singh and Sahoo (2018) have reported this multiplication factor as six in their study on severe accident analysis. We have therefore used the same factor and simulated another case with this accelerated creep. Furthermore, we have modified the initiation temperature to 300°C, i.e., k_2 is set to 0.0119 between 300°C and 800°C.

Figure 9.11 shows the numerical result with the accelerated creep correlation; for the comparison, experimental deflection is also provided in the same figure. It is observed that the deflection has increased significantly with the application of the accelerated creep equation, but there is still a significant difference in both deflection curves. However, the contact time (280 s) estimated with the accelerated creep is nearly close to that of the experimentally reported time (281s). Our present focus is to study the effect of weight simulators on the deflection of PT. Hence, we have restricted our scope of exploration to this concern and used the accelerated creep for further study.

Before proceeding with the model-2 simulation, which includes the explicit modeling of weight simulators, it is needed to reduce the overall element counts intuitively. Hence, a mesh efficiency approach is adopted to evaluate the optimum meshing of PT and CT, maintaining accuracy. The effect of mesh elements across the thickness of the PT and the circumferential of PT and CT, along the longitudinal length of CT, is evaluated and presented in Figure 9.12.

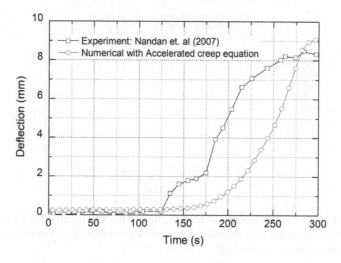

FIGURE 9.11   Deflection using accelerated creep correlation.

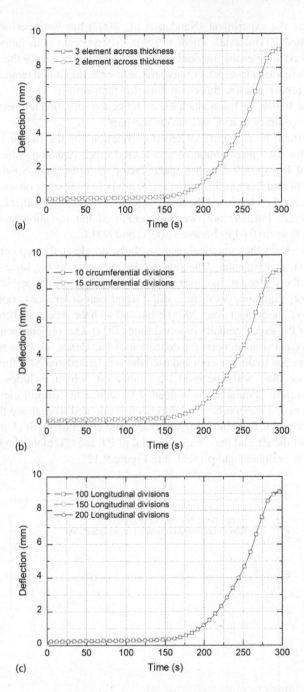

FIGURE 9.12 Effect of meshing on the deflection of the PT: (a) effect of element layers across the thickness of PT; (b) effect of elements in the circumferential direction; (c) effect of elements in the longitudinal direction.

It is found that the two elements across the thickness of PT (has an error of 0.013 % at end time) and 20 circumferential divisions (has an error of 0.17 % at end time) are sufficient to produce accurate results. Simultaneously, 150 longitudinal divisions (has of error of 0.4% at end time) give acceptable deflection. Therefore, for the meshing of model-2, we use two-element layers in the thickness direction and 20 and 150 divisions in circumferential and longitudinal directions, respectively.

9.4.2 MODEL-2: WITH WEIGHT SIMULATORS

The numerical model with the meshing is shown in Figure 9.13, whereas contacting surfaces are shown in Figure 9.14. Each Weight simulator has meshed with 20 circumferential divisions and ten longitudinal divisions – the meshing results in 21200 hexahedral elements. Different pairs of contact elements are generated over different contacting surfaces. CONTA174 and corresponding TARGE170 are used as contact and target elements, respectively. The gravity acceleration is activated to account for the weights.

As explained in the previous section, the relation between parts in the finite element domain is expressed in terms of contact elements. The parts are constrained

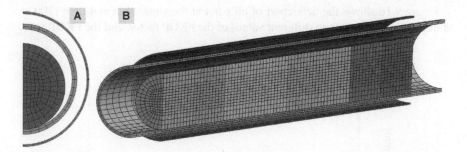

FIGURE 9.13 Body Parts Mesh elements: (a) circumferential mesh elements; (b) longitudinal mesh elements.

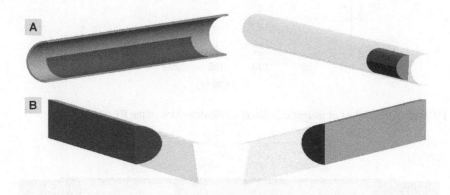

FIGURE 9.14 Contacting surfaces: (a) on PT and weight simulators; (b) among weight simulators.

using boundary conditions such as displacement. However, it is not easy to perform the simulation successfully if the parts will be contained mainly due to contact elements. In such a situation, a small perturbation during simulation will result in a rigid body motion (i.e., unconstrained motion), leading to numerical simulation divergence.

The model with weight simulators exhibits a similar problem since they are too constrained to contact the PT. Such a model can be successfully simulated using the bonded contact technology, where the contact and target elements are glued together during the simulation time. This avoids the rigid body movement of unconstrained parts. Such formulation is used by Singh, Sahoo, Majumdar, and Mukhopadhyay (2017) to predict channel deformation. However, bonded surfaces add extra stiffness to the deforming characters, which results in an error in overall deflection estimation.

Therefore, we proposed another concept in our formulation, which uses the contact opening factor (FKOP). The FKOP imposes a damping force on the relative motion between contact surfaces during contact opening and prevents sizable rigid body motions. Moreover, frictional contact formulation is used among the contacting surface with a friction coefficient of 0.2 and penetration tolerance factor of 0.1. The simulation is then performed with this formulation, and the deflection of the PT is given in Figure 9.15.

Figure 9.16 shows the deflection of all parts at the simulation end time (300 s). Multiple trials are run with different values of the FKOP factor, and the FKOP value

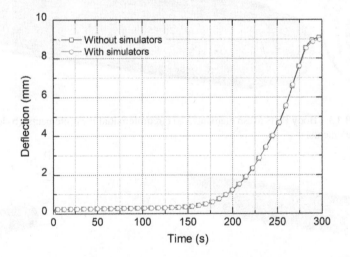

FIGURE 9.15 Effect of weight simulators on the deflection of the PT.

FIGURE 9.16 Deflection of all parts at the simulation end time (300 s).

of 100 has been found to help in the successful simulation with the weight simulators. It is found that the effect of weight simulators on the deflection of the PT is not prominent; however, an insignificant deviation in the deflection curves is seen after the contact.

Figures 9.17 and 9.18 show the Von-Mises equivalent stress along the bottom length of PT during the simulation for model-1 and model-2, respectively. The model-1, which is considered the uniform load for weight simulators, shows decreased stress with time until the contact. After that, the stress again shows increasing behavior. Since stress is the response of material against the applied load, initially, PT has high stress values. As the temperature rises, the PT undergoes

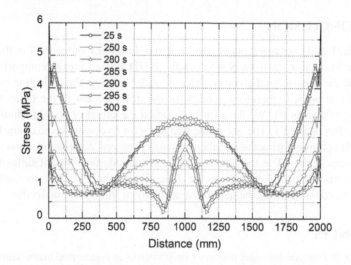

FIGURE 9.17 Equivalent Von-Mises stress along the PT length (Model-1).

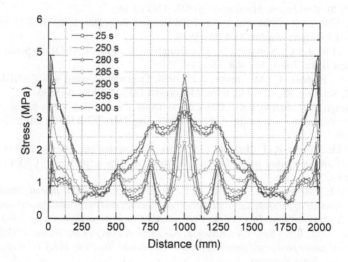

FIGURE 9.18 Equivalent Von-Mises stress along the PT length (Model-2).

creep flow that reduces the resistance response of the PT. However, an increase in stress after contact at 1000 mm is due to contact stress from the CT's stiffness. High-stress values at 0 mm and 2000 mm are contributed from the applied boundary conditions.

Model-2 shows a rise in equivalent stress at multiple locations (500 mm, 750 mm, 1000 mm, 1250 mm, and 1500 mm). These are the locations of weight simulators junctions. In the end time, the stress at 1000 mm goes beyond the initial stress in the PT. This shows the impact of simulators on the surface of PT, i.e., significant stress accumulation occurs in the PT at the junction of weight simulators. These stress accumulated locations can lead to local failure of PT during deformation.

9.5 CONCLUSION

A numerical study is carried out for the limited deformation of the PT. For this objective, experimental work from Nandan et al. (2007) is used for the comparison. It is found that the longitudinal creep correlation under-predicts the deflection of the PT. Thus, critical work is therefore needed for the development of the creep equation pertaining to Indian PT. We have further studied the effect of weight simulators on the deflection of the PT. The use of the FKOP factor to constrain the contacting surfaces from opening during the simulation is also explored. Significant stress accumulation occurs in the PT at the junction of weight simulators. Weight simulators will contribute an additional stiffness to the deflection of the PT. Thus, its contribution needs to be considered for the qualitative analysis of limited PT deflection.

REFERENCES

AERB, 2000. *Ultimate heat sink and associated systems in pressurised heavy water reactor (AERB SAFETY GUIDE No. AERB/SG/D-15)*. Atomic Energy Regulatory Board, Mumbai.

ANSYS, 2020. *Workbench: Mechanical APDL*. ANSYS Inc.

Bajaj, S.S., Gore, A.R., 2006. The Indian PHWR. *Nucl. Eng. Des.* 236, 701–722. https://doi.org/10.1016/j.nucengdes.2005.09.028

Blahnik, C., Luxat, J.C., Nijhawan, S., Thuralsingham, R., 1993. CANDU response to loss of all heat sinks. *Trans. Am. Nucl. Soc.* 69, 510.

Brown, R.A., Blahnik, C., Muzumdar, A.P., 1984. Degraded cooling in a CANDU reactor. *Nucl. Sci. Eng.* 88, 425–435. https://doi.org/10.13182/NSE84-A18596

Dutt, N., Singh, A.R., Sahoo, P.K., 2020. CFD analysis of suspended debris during postulated severe core damage accident of PHWR. *Nucl. Eng. Des.* 357, 110390. https://doi.org/10/ggrbfb

Gillespie, G.E., Moyer, R.G., Hadaller, G.I., 1984. *An experimental investigation of the creep sag of pressure tubes under LOCA conditions*, in: *Proc. 5th Annual CNS Conference.* pp. 68–72.

Gillespie, G.E., Moyer, R.G., Hadaller, G.I., 1985. *An experimental investigation into the development of pressure tube/calandria tube contact and associated heat transfer under LOCA conditions*, in: *Proc. of 6th CNS Annual Conference, Canada.* pp. 24–31.

Gupta, S.K., Dutt, B.K., Raj, V.V., Kakokar, A., 1996. *A study of the Indian PHWR reactor channel under prolonged deteriorated flow conditions*, in: *Proc. IAEA TCM on Advances in Heavy Water Reactors.*

IAEA, 2008. *Analysis of Severe Accidents in Pressurized Heavy Water Reactors (No. IAEA-TECDOC-1594)*. International Atomic Energy Agency, Vienna, Austria.

Lili, T., Kai, Y., Jingtian, Y., 2010. Analysis of severe core damage accident progression for the heavy water reactor. *Nucl. Sci. Tech.* 21(21), 251–256.

Majumdar, P., Chatterjee, B., Lele, H.G., Guillard, G., Fichot, F., 2014. ASTEC adaptation for PHWR limited core damage accident analysis. *Nucl. Eng. Des.* 272, 273–286. https://doi.org/10.1016/j.nucengdes.2013.10.011

Mani Mathew, P., 2004. *Severe core damage accident progression within a CANDU calandria vessel*, in: Presented at the *Workshops, MASCA Seminar*, Aix-en-Provence, France.

Mathew, P.M., Kupferschmidt, W.C.H., Snell, V.G., Bonechi, M., 2001. *CANDU-Specific Severe Core Damage Accident Experiments in Support of Level 2 PSA*. Washington DC, USA, pp. 1–8.

Nandan, G., Lele, H.G., Sahoo, P.K., Chatterjee, B., Kumar, R., Mukhopadhyay, D., 2007. *Experimental investigation of heat transfer during LOCA with failure of Emergency cooling system*, in: *5th International Conference on Heat Transfer, Fluid Mechanics and Thermodynamics, Presented at the HEFAT2007*, South Africa, p. 5.

Nandan, G., Majumdar, P., Sahoo, P.K., Kumar, R., Chatterjee, B., Mukhopadhyay, D., Lele, H.G., 2012. Study of ballooning of a completely voided pressure tube of Indian PHWR under heat up condition. *Nucl. Eng. Des.* 243, 301–310. https://doi.org/10.1016/j.nucengdes.2011.11.007

Nandan, G., Sahoo, P.K., Kumar, R., Chatterjee, B., Mukhopadhyay, D., Lele, H.G., 2010. Experimental investigation of sagging of a completely voided pressure tube of Indian PHWR under heatup condition. *Nucl. Eng. Des.* 240, 3504–3512. https://doi.org/10.1016/j.nucengdes.2010.05.042

Nayak, A.K., Banerjee, S., 2017. Pressurized heavy water reactor technology: Its relevance today. *J. Nucl. Eng. Radiat. Sci.* 3, 1–9. https://doi.org/10.1115/1.4035856

Negi, S., Kumar, R., Majumdar, P., Mukopadhyay, D., 2017. Full length channel pressure tube sagging study under postulated LOCA with un-availability of ECCS in an Indian PHWR. *Nucl. Eng. Des.* 320, 361–373. https://doi.org/10.1016/j.nucengdes.2017.06.017

Shewfelt, R.S.W., Lyall, L.W., 1985. A high temp longitudinal strain rate equation for Zr-2.5wt% Nb pressure tubes. *J. Nucl. Mater.* 132, 41–46. https://doi.org/10.1016/0022-3115(85)90391-5

Singh, A.R., Sahoo, P.K., 2018. Investigation of the channel disassembly behaviour of Indian 200MWe PHWR – A numerical approach. *Nucl. Eng. Des.* 339, 137–149. https://doi.org/10.1016/j.nucengdes.2018.09.008

Singh, A.R., Sahoo, P.K., Majumdar, P., Mukhopadhyay, D., 2017. *Study of thermo-mechanical deformation of scaled down fuel channel under Heatup*, in: *IHMTC-2017. Presented at the Proceedings of the 24th National and 2nd International ISHMT-ASTFE Heat and Mass Transfer Conference (IHMTC-2017)*, Begel House Inc., BITS Pilani-Hyderabad Campus, India, pp. 571–576. https://doi.org/10.1615/IHMTC-2017.790

Singh, A.R., Tariq, A., Majumdar, P., 2020. Experimental study on thermo-mechanical deformation of PHWR channel at elevated temperature. *Nucl. Eng. Des.* 364. https://doi.org/10.1016/j.nucengdes.2020.110634

Singh, A.R., Tariq, A., Sahoo, P.K., Majumdar, P., Mukhopadhyay, D., 2021. Longitudinal deformation study of pressure tube of Indian PHWR under high temperature transient. *Ann. Nucl. Energy* 8. https://doi.org/10.1016/j.anucene.2021.108160

Snell, V.G., Alikhan, S., Frescura, G.M., Howieson, J.Q., King, F., Rodgers, J.T., Tamm, H., 1988. *CANDU safety under severe accidents*, in: *International Symposium on Severe Accidents in Nuclear Power Plants*. p. 31.

10 Application of Configurational Force Concept to Calculate the Crack Driving Force in Presence of an Interface at Various Orientations

Abhishek Tiwari
Indian Institute of Technology Ropar, India

CONTENTS

10.1 INTRODUCTION

For structural integrity of components, resistance against fracture is a crucial design criterion. Since the theory of Griffith in 1921, there have been many approaches to understand and improve fracture resistance and, more importantly, optimize it to ensure safety and other properties. In this chapter, after a brief review of fracture mechanics, a recent concept of configurational force application in the field of fracture mechanics is discussed. The effect of placing a differently deforming material is explored by numerically modeling the interface. Further, the effect of change in inclination angle of the interface has been modeled, post-processed using the configurational force concept, and discussed.

In a fail-safe design approach, a pragmatic way of looking at structural integrity is to start with a flaw in the component. Making a component absolutely flawless is a very costly affair, if not impossible. Moreover, even if we start with a flawless component at some point in its application the damage will creep in, and a crack will initiate. Therefore, the safety depends on calculation and, if possible, predicting how long a component will survive after a crack has initiated. This field of study is known as fracture mechanics. A macroscopic crack has been studied using linear elasticity theory (Griffith 1921; Irwin 1956, 1960; Westergaard 1937; Williams 1957), and gradually more realistic plastic deformation was included in the theory of fracture mechanics, which led to the definition of J-integral (Rice 1968). This has been the most successful parameter in the field of fracture mechanics, and despite its limitations, it is still used in industries as well as in research.

There have been never-ending efforts from the researchers to improve the fracture resistance of materials, which has resulted in various techniques by which the crack resistance is improved. For instance, the crack in a material can be deflected by delamination, arrested by transformation toughening and delayed by fiber bridging. However, a recent application of configurational forces, which is explained in the next section in detail, has shown more clearly that a spatial change of mechanical properties such as elastic modulus or yield strength, can attract or deflect a crack and hence can decrease or increase the crack driving force (CDF). This effect of influencing the CDF by the presence of a region that comprises a spatial change in the mechanical property is called the *"material inhomogeneity effect"*. This effect, without using configurational forces was examined by Lee et al. (2007) on stress intensity factor. The effect of material inhomogeneity on CDF is because of the configuration forces that appear on the mechanical property transition boundary. This can be visualized by imagining an interface where a sudden transition of mechanical property occurs. In a real-world example, a dissimilar weld interface or an interface between two polymers of different strengths can be imagined. However, the material inhomogeneity effect will be true for any shape of inhomogeneity such as a void, a hole, a slit, an inclusion, or a second phase.

10.2 CONFIGURATIONAL FORCES-BASED J-INTEGRALS

Configurational forces (CFs) are thermodynamic forces arising from material defects (Maugin, 1937; Gurtin 1999; Kienzler and Herrmann 2000). A CF-vector $\mathbf{f}$ at a defect appears if the total potential energy of the system varies for change of one position of the defect to another. As a result, the defect feels a driving force to move in such a direction that the dissipation becomes a maximum (or the total energy becomes a minimum). This can be better understood by imagining an interstitial atom in a lattice. This atom shall have a lower energy state if it comes to the surface, as in the lattice the atom will be compressed by the surrounding atoms (Pauling 1929) due to the smaller interstitial size. Therefore, a force must be acting on the atom which will compel the atom to go to the surface. Whether or not the interstitial goes to the surface depends on the energy barrier provided by the path of the interstitial to reach the surface. The concept of CFs has been applied in various other fields of physics.

The $\mathbf{f}$ can be evaluated by the relation,

$$\mathbf{f} = -\nabla \cdot \mathbf{C} = -\nabla \cdot \left(\phi_e \mathbf{I} - \mathbf{F}^T \mathbf{S} \right) \tag{10.1}$$

In Eq. (10.1), $\nabla \cdot \mathbf{C}$ denotes the divergence of the configurational stress tensor (in the reference configuration), ϕ_e the elastic strain energy per unit volume, $\boldsymbol{I}$ the identity tensor, $\boldsymbol{F}^T$ the transposed of the deformation gradient, and $\boldsymbol{S}$ the 1st Piola-Kirchhoff stress tensor.

Divergence in physicality describes the outgoingness (if positive) of a field. Therefore, when the divergence of the configurational stress is zero, it can be physically interpreted as there is no net flow of the field, i.e., the outgoingness of the configurational stress is balanced by its inwardness. A non-zero value of the divergence means that the point in continuum space has either a source or a sink of the field of which the divergence is taken, which in this case is configurational stress (energy-momentum tensor).

Under small strain theory, Eq. (10.1) can be written as (Simha et al. 2008),

$$\mathbf{f} = -\nabla \cdot \mathbf{C} = -\nabla \cdot \left(\phi_e \mathbf{I} - \nabla \mathbf{u}^T \sigma \right) \tag{10.2}$$

where ∇ is the gradient operator, $\boldsymbol{u}$ the displacement vector, and σ the Cauchy stress tensor. Note that $\boldsymbol{C}$ has been denoted in literature also as energy-momentum tensor (Eshelby 1975).

If the body contains a sharp crack as shown in Figure 10.1, a single CF-vector, $\mathbf{f}_{tip}$, is generated at the crack tip, which is for infinitesimal strain setting given by

$$\mathbf{f}_{tip} = -\lim_{r \to 0} \int_{\Gamma_r} \left(\phi_e \mathbf{I} - \nabla \mathbf{u}^T \sigma \right) \mathbf{m} \; dl \tag{10.3}$$

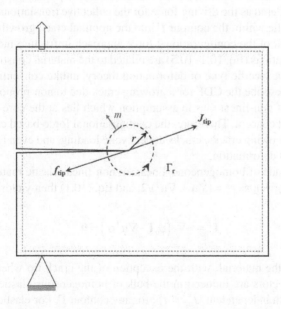

FIGURE 10.1 Schematic of configurational force at the crack tip and J-integral vector in cracked specimen loaded in Mode-I.

Γ_r designates a contour drawn from the lower to the upper crack flank at a distance r from the crack tip, m denotes the unit normal vector to the contour. By considering the crack tip dissipation, it follows that the negative CF-vector at the tip is the CDF, which can also be expressed in form of the near-tip J-integral vector, $\mathbf{J}_{tip} = -\mathbf{f}_{tip}$ (Simha et al. 2005).

The commonly used scalar CDF term, the near-tip J-integral J_{tip}, is the projection of the vector $\mathbf{J}_{tip}$ into the crack growth direction,

$$J_{tip} = \mathbf{e} \cdot \left(-\mathbf{f}_{tip}\right) = \mathbf{e} \cdot \mathbf{J}_{tip} = \mathbf{e} \cdot \lim_{r \to 0} \int_{\Gamma_r} \left(\phi_e \mathbf{I} - \nabla \mathbf{u}^T \sigma\right) \mathbf{m} \; dl \tag{10.4}$$

where $\mathbf{e}$ denotes the unit vector in the nominal crack growth direction. Eq. (10.4) is similar in nature to the conventional J-integral (Rice 1968). The scalar J-integral for an arbitrary contour Γ is obtained by the projection of the vector in the expected crack growth direction or basis vector directions as,

$$J_\Gamma = \mathbf{e} \cdot \mathbf{J}_\Gamma = \mathbf{e} \cdot \int_\Gamma \left(\phi_e \mathbf{I} - \nabla \mathbf{u}^T \sigma\right) \mathbf{m} \; dl = -\mathbf{e} \cdot \int_{A_\Gamma} \mathbf{f} \, dA = -\mathbf{e} \cdot \left(\mathbf{f}_{tip} + \int_{A_\Gamma \cap A_r} \mathbf{f} \, dA\right) \tag{10.5}$$

The right side of Eq. (10.5) means that J_Γ is equal to the negative of the summation of all CF-vectors that originate in the area inside the contour Γ, projected into the nominal crack growth direction (Simha et al. 2008). The parameter $A\Gamma$ denotes the area inside the contour Γ, and A_r the area inside the near-tip contour Γ_r. The J-integral J_Γ can be considered as the driving force for the collective translational movement of all defects that lie within the contour Γ into the nominal crack growth direction $\mathbf{e}$.

The highlight of the configurational force approach in fracture mechanics is that none of the equations (Eq. 10.1–10.5) are related to the material constitutive relations and neither on a specific type of deformation theory, unlike conventional J-integral which cannot describe the CDF for a growing crack due to non-proportional loading and violation of non-linear elastic assumption which lies at the core of the conventional J-integral concept. Therefore, the configurational force-based calculations can be applied for growing cracks, cracks under cyclic loading, and even for cracks under time-dependent deformation.

In a body made of homogeneous, linear, or non-linear elastic material, the linear strain tensor is given as $\boldsymbol{\varepsilon}^e = (\nabla \mathbf{u} + \nabla \mathbf{u}^T)/2$, and Eq. (10.1) then yields

$$\mathbf{f}^e = -\nabla \cdot \left(\phi_e \mathbf{I} - \nabla \mathbf{u}^T \sigma\right) = 0 \tag{10.6}$$

everywhere in the material, with the exception of the crack tip where $\mathbf{f}_{tip}$ emerges. Since no CF-vectors are induced in the bulk of homogeneous elastic materials, the J-integral is path independent, $J_{tip}^e = J_\Gamma^e$, for any contour Γ. For elastic materials, the CF-based J-integral of Eq. (10.5) is identical to the conventional J-integral introduced by Rice (1968). If the elastic body contains a material inhomogeneity, the

J-integral becomes path dependent, since the material inhomogeneity induces additional CF-vectors (Simha et al. 2005, 2008). In such inhomogeneous elastic materials, the near-tip J-integral, J_{tip}^{e} of Eq. (10.4), gives the magnitude of the CDF.

In a homogeneous elastic–plastic material and small strain assumption, the total strain tensor comprises elastic and plastic parts, $\varepsilon = \varepsilon^{e} + \varepsilon^{p}$. The strain energy density is given by $\phi = \phi_{e} + \phi_{p}$, where only the elastic part of the strain energy density, ϕ_{e}, is recoverable; ϕ_{p} is the plastic strain energy density. The CF in the bulk of the elastic–plastic body, $\mathbf{f}^{ep}$, can be evaluated by Eq. (10.2). It has been shown that the bulk CF depends on the stress and the gradient of the plastic strain in the form (Simha et al. 2008),

$$\mathbf{f}^{ep} = \sigma \cdot \frac{\partial \varepsilon^{p}}{\partial \mathbf{x}} \tag{10.7}$$

It follows that CFs appear in plastically deformed regions of the body and that the J-integral becomes path dependent according to Eq. (10.7).

The CDF is given by the so-called incremental plasticity J-integral evaluated for a contour Γ_{PZ} drawn around the crack-tip plastic zone (Kolednik et al. 2014),

$$J_{PZ}^{ep} = \mathbf{e} \cdot \mathbf{J}_{PZ}^{ep} = \mathbf{e} \cdot \int_{\Gamma_{PZ}} \left(\phi_{e} \mathbf{I} - \nabla \mathbf{u}^{T} \sigma \right) \mathbf{m} \; dl \tag{10.8}$$

J_{PZ}^{ep} can be easily evaluated for contained yielding conditions where the crack-tip plastic zone is completely surrounded by elastically deformed material. Only an approximate evaluation of J_{PZ}^{ep} is possible for general yielding conditions where plasticity extends through the whole ligament of the body.

The application of the conventional J-integral in non-linear fracture mechanics is based on the assumption of deformation plasticity, i.e., an elastic–plastic material is treated as if it was non-linear elastic. This means that the total strain energy density, $\phi = \phi_{el} + \phi_{pl} = \phi_{nlel}$, is inserted instead of ϕ_{el} into Eqs. (10.1–10.6). The conventional J-integral, J_{nlel}, is path independent because of the non-linear elastic assumption, although it cannot be related to the true CDF as the real material deforms in elastic–plastic manner. The relation between the conventional and the incremental plasticity-based J-integral calculations are discussed in Kolednik et al. (2014) in great detail. It is important to note that, for a contour Γ_{PZ} around the crack-tip plastic zone, Eq. (10.6) yields identical results for the non-linear elastic and the elastic–plastic material, if the whole contour passes through regions where only elastic deformation occurs. Therefore, both conventional and configurational force-based J-integrals are path independent on and beyond this contour.

The CF-concept has enabled also the determination of the driving force for growing cracks under either constant or cyclic loading conditions. In the first case, the CDF is given by the incremental plasticity J-integral, J_{actPZ}^{ep}, for a contour around the active crack-tip plastic zone, Γ_{actPZ}, (Kolednik et al. 2014). In the second case, the CDF is given by the cyclic J-integral, $\Delta J_{\text{actPZ}}^{ep}$, for a contour Γ_{actPZ} (Ochensberger and Kolednik 2015, 2016). CDF differs for a cyclically loaded crack; its purpose is to allow the prediction of the crack propagation rate of a fatigue crack (Paris et al. 1961; Paris & Erdogan, 1963). The active plastic zone is the plastically deformed region at the current crack tip position, in contrast to the plastic wake.

It should be noted that CFs appear also in regions where eigen- or thermal strains are present in a body, since $\nabla \cdot \mathbf{C}$ becomes non-zero. Eigenstrains can be treated in the same way as the plastic strains (Simha et al. 2005), and the magnitude of the bulk CFs follows a relation analogous to Eq. (10.7).

The CF concept has enabled the evaluation of the driving force of stationary and growing cracks in elastic–plastic materials and materials with eigenstrains.

10.3 MATERIAL INHOMOGENEITY: INFLUENCE OF CHANGE IN MATERIAL'S PROPERTY

As explained in Section 10.2, the CFs arise in a continuum space in presence of a defect. The defect here can be anything which has a different mechanical property in the reference configuration. A change in material property can be created deliberately to use the effect of the material inhomogeneity for better crack resistance of materials. Defects in a material exist at every scale and their effect also changes with the scale or magnitude of the observation. For instance, as explained earlier an interstitial can also result in a configurational force, however, the scale we are looking at is of the concern to fracture mechanics and at this scale defects like a hole, or a gradual or sudden transition of material's mechanical properties can results in a configurational force which will influence the CDF. This can be visualized in Figure 10.2, where Figure 10.2(a) shows a homogeneous cracked specimen in which the far-field J-integral is same as the one at the tip, i.e. J_{tip}, as long as there is proportional loading and non-linear elastic assumption holds. In Figure 10.2(b), a cracked specimen with an interface is shown where across interface the mechanical property takes a sharp jump. Due to this change in mechanical property additional CFs arise on the interface and hence the path independence of the J-integral is lost.

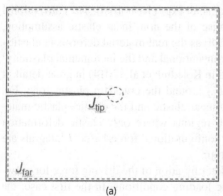
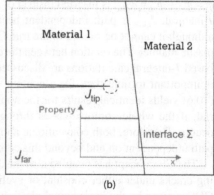

(a)   (b)

FIGURE 10.2 Crack driving force in (a) a homogeneous cracked body showing far-field J-integral and J-integral at crack tip and (b) in a heterogeneous body where a sharp transition of material property takes place along an interface.

A sudden transition of mechanical properties is realized when two different materials are joined together. For engineering interest, these events can happen when two differently deforming materials are welded or joined in the form of a composite. Dissimilar metal welds are extensively used in Pressurized Water Reactors (Brayshaw et al. 2019; Blouin et al. 2014; Wang et al. 2013; Jang et al. 2008) and in aerospace and aeronautical applications (Zhang et al. 2016; Vaidya et al. 2010; Khan et al. 2017; Abidi et al. 2020).

In presence of an interface, Σ, where a sudden change in the mechanical property occurs, CFs arise at the interface which can be expressed as,

$$f_\Sigma = \left(\llbracket \phi \rrbracket \mathbf{I} - \llbracket \mathbf{F}^T \rrbracket \langle \mathbf{S} \rangle \right) n \mathrm{d}l \tag{10.9}$$

where, $\llbracket \ \rrbracket$, represents the jump in the parameter from left to right of the interface and $\langle \ \rangle$ represents the average value of the left- and right-side values of the parameter. If the coordinate system in the reference configuration are expressed with X then in presence of an interface bulk CFs can arise in a contour ξ, which can be described as,

$$f = -\nabla_x \phi(\mathrm{F}, \mathrm{X}) \tag{10.10}$$

where, ∇_x is the explicit gradient operator. The detailed explanation of the bulk CFs are provided in Kolednik et al. 2014.

Due to these additional CFs arising in presence of an interface, the J-integral becomes path dependent and the difference between CDF at the tip, J_{tip} and J_{far} is given by,

$$J_{tip} - J_{far} = -\mathbf{e} \cdot \left(\int_\xi -\nabla_x \phi(\mathbf{F}, \mathbf{X}) + \int_\Sigma \left(\llbracket \phi \rrbracket \mathbf{I} - \llbracket \mathbf{F}^T \rrbracket \langle \mathbf{S} \rangle \right) n \mathrm{d}l \right) = C_{inh} \tag{10.11}$$

where, C_{inh} is the term which defines material inhomogeneity effect on the CDF. If C_{inh} is positive, the crack tip is attracted by the interface and if it is negative the CDF is suppressed. Usually, in a completely immiscible composites such as polymer glass composites the interface between glass and polymer will have a sudden change in elastic modulus or plastic properties, whereas in a metallic composite where diffusion is possible between the two different material a gradual change in material property can exist. In the latter case, instead of interface as a surface a thicker zone will act as material inhomogeneity and C_{inh} will have to be obtained for this zone.

From the previous studies on effect of interfaces on the CDFs by Simha et al. 2005 and Simha et al. 2008, the behavior of CDF is known in presence of a sharp interface at 90° from the crack plane. The J_{tip}/J_{far} ratio increases as a crack tip moves closer to the interface between a stiff and a compliant material resulting in a crack tip anti-shielding effect or in simple words it can be said that the crack is attracted toward this interface and opposite is true for a compliant to stiff transition of crack tip.

10.4   EFFECT OF ORIENTATION OF MATERIAL INHOMOGENEITY ON CRACK DRIVING FORCE

Effect of interfaces and interlayers has been studied in great detail and the same has been discussed in Section 10.3. Kolednik et al. 2010 has studied the effect of inclined interface on the CDF of a fatigue crack. The study showed that the inclination causes the crack to deflect from the nominal expected path. The study was performed for angle of orientation in the range of −15° to 45°. The study included elastic–plastic material with residual stresses contribution in the CDF. The effect of inclination was investigated in with irregular intervals of angle of orientation and it was found the J_{tip} increased with increasing angle of orientation. For a better understanding and possible application in dissimilar metal welds in nuclear and aerospace applications as well as for polymer matrix composites, it is required to understand the effect of orientation in greater detail on the crack driving force. In this study the effect of orientation of a sharp interface is studied at five different orientation angles for linear elastic material using finite element analysis performed on Abaqus commercial software. The CFs are calculated using the method described in Kolednik et al. 2019.

To understand the effect of orientation on CDF due to the interface between two differently deforming material, simple numerical situation of two materials with different elastic moduli are modeled. Material 1 has an elastic modulus of 200 GPa and Material 2 has an elastic modulus of 70 GPa. The CDF calculation is performed by modeling compact tension half symmetric specimen geometry in two dimension using plane strain quadrilateral elements with reduced integration. The initial ligament length is of 50 mm. The orientation angle direction and schematic of interface at 85° is shown in Figure 10.3(a). The regular mesh scheme close to the crack tip is shown in Figure 10.3(b). The effect of interface on the CDF is calculated by modeling each orientation at nine different locations along the ligament. This is to simulate the effect of interface on the crack tip as it moves under the application of load. Orientations of 120°, 90°, 85°, 60° and 45° are modeled. For each orientation, the initial crack tip is located at 50mm from the load line and the interface always stands at 10 mm distance from the initial crack tip location. L is the distance between crack tip and the interface intersection with the ligament line. At the crack tip location, the ratio of L with ligament length b_o is −0.2. As the crack tip moves closer to the interface, it acquires L/b_o of −0.16, −0.12, −0.08, −0.04, −0.02, 0.02, 0.04, 0.08, 0.12. At $L/b_o = 0$, the crack tip would be at the intersection of the interface with the ligament. The J-integral is calculated using a post-processing code which calculates the J_{nlel} by summing the bulk CFs in the contour region. The values of J_{nlel} is calculated for contours of sizes 1 mm, 2 mm, 5 mm, 10 mm and for a larger contour which represents the far-field J-integral, J_{far}.

It has been discussed in detail by Kolednik et al. (2014) that the contour should be so selected that it avoids the region at the crack tip which usually has discretization error. The discretization error appears because of crack-tip stress singularity. Therefore, the element size in these models is of 0.1×0.1 mm². For a contour of 1 mm length, there are ten elements along the length. The contours are depicted in Figure 10.3(a). The CFs, which appear due to discretization error of FE analysis, are shown in

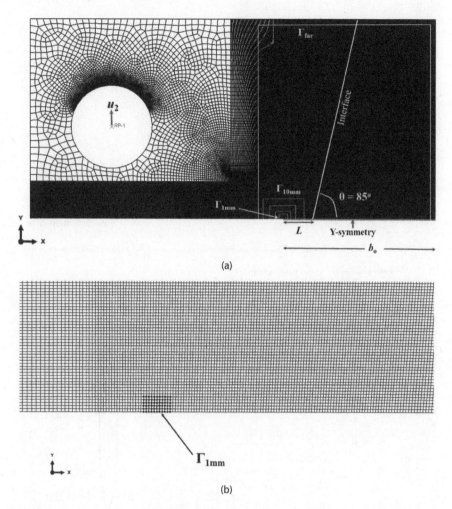

FIGURE 10.3 (a) Finite element (FE) model geometry depicting parameters, L, b_o with boundary conditions and contours (Γ_{1mm}, Γ_{2mm}, Γ_{5mm}, Γ_{10mm} and Γ_{far}) and (b) the elements of CPE4R type with $0.1 \times 0.1 mm^2$ size at the crack tip with the node-set used for calculation of J-integral for Γ_{1mm} contour.

Figure 10.4(a), which is a result from the same compact tension (CT) geometry modeled for a homogeneous linear elastic material with an elastic modulus of 200 GPa.

At J_{far} of 42.2 kJ/m², the CFs appear close to crack tip in a region of a radius of five elements. The smallest contour used in the analyses presented is 10×10 elements in size. To avoid the interference of CFs arising at the interface due to change in the mechanical property i.e. elastic modulus in this case, with those arising due to discretization error at the crack tip, the closest calculation is made when the crack-tip is ten elements away from the interface (see Figure 10.4(b)). Calculations closer than this is expected not to be accurate. Therefore, only a trend at the interface can be expected but exact values cannot be obtained at crack-tip locations closer than 1mm to the interface.

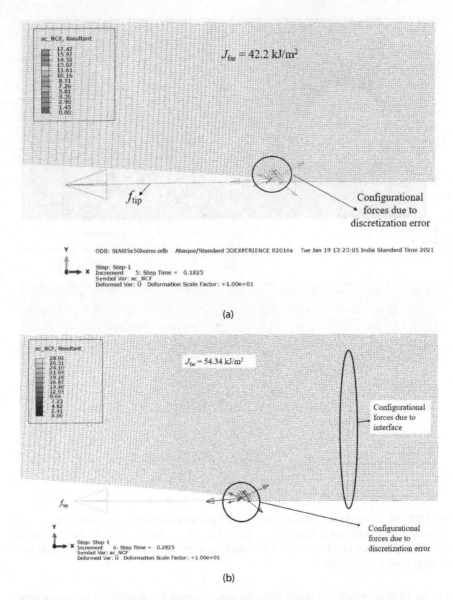

FIGURE 10.4 (a) Configurational forces appearing due to discretization error of crack-tip singularity in a homogeneous linear elastic material and (b) configurational forces appearing at the interface at a sharp interface between linear elastic materials with different elastic moduli.

In Figure 10.5, half of J_{nlel} values for contours of 1 mm, 2 mm, 5 mm, 10 mm, and remote J_{far} is plotted against half of the load line displacement, $v_{LLD}/2$ for model with an interface at 90° to the crack plane for $L/b_o = -0.2$. As all the contours except Γ_{far} do not contain any part of the interface and because the material is linear elastic the non-linear J, i.e., J_{nlel} is path independent. However, because J_{far} contains the

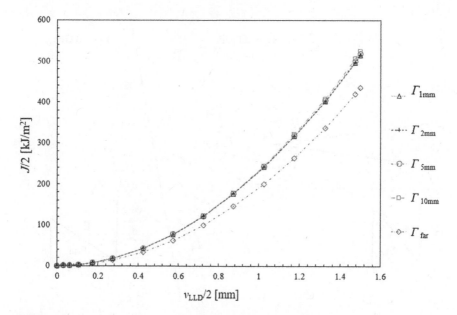

FIGURE 10.5 J_{nlel} for different contour sizes for an interface oriented at 90° with the crack plane.

interface, it has additional CFs arising from the jump in elastic modulus across the interface. Due to the additional CFs on the interface as per Eq.(10.9) and Eq.(10.10), J_{far} has a different value. The difference in what is supplied (J_{far}) and what is being felt at the crack tip (J_{tip}), is the inhomogeneity term, C_{inh} in Eq.(10.11).

The physical significance of this behavior can be understood by the schematic representation of strain energy density in Figure 10.6. For contours, Γ_1, Γ_2, and Γ_3, as there are inhomogeneities contained in the contour region the strain energy density ϕ can be represented by the area under the stress-strain curve shown in the left-hand lower side in Figure 10.6. Because J_{nlel} assumes non-linear elastic behavior and there is no unloading because the crack does not grow, the values of J_{nlel} for Γ_1, Γ_2, and Γ_3 must be same, similar to the Γ_{1mm}, Γ_{2mm}, Γ_{5mm} and Γ_{10mm} in Figure 10.5. However, for Γ_4 in Figure 10.6 which is similar to Γ_{far} in Figure 10.5, the scenario is different as this contour contains the interface.

For this contour Γ_4, the part in the yellow region will have ϕ different from that in the uncolored region of Γ_4. The strain energy density for yellow part of Γ_4 is represented by the area under stress-strain curve shown in the right-hand side lower part of Figure 10.6. The part of contour which falls on the right side of the interface absorbs less energy because of lower elastic modulus in comparison to the part on the left-hand side of the interface. Therefore, for the same amount of supplied, J_{far}, the crack tip will have more of it available because less is being consumed in the yellow part which is not at the crack tip. This makes J_{tip} to be higher than J_{far}, meaning that more CDF is available. This means that the crack is attracted toward an interface between a stiff and a compliant material (from the direction of crack). The opposite is true for compliant to stiff transition of crack. As long as the material is

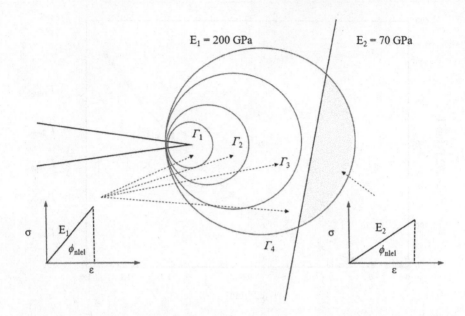

FIGURE 10.6 Schematic representation of effect of interface and asymmetric energy distribution due to difference in elastic moduli of the materials across the interface.

elastic, the non-linear assumption does not create any abnormality or difference between J calculated by conventional method or configurational force method. In the elastic–plastic material, however, ϕ will comprise of ϕ_{el} and ϕ_{pl} and ϕ_{pl} parts, which is irreversible and will not contribute to the true CDF. This is where the conventional J and configurational force J differ.

10.5 RESULTS AND DISCUSSIONS

The normalized CDF in the form of ratio of J_{tip} and J_{far} represents to what extent the CDF has increased or decrease at constant J_{far}. A higher value of J_{tip}/J_{far} means that a crack possesses more potential to extend for the same J_{far}. A lower value of J_{tip}/J_{far} means that sufficient amount of energy is not available for the crack to move. For homogeneous specimen under non-linear elastic assumption this ratio should have a value of 10. A lower value of J_{tip}/J_{far} is benefitial as it means to drive the crack furthermore load has to be applied. The lower J_{tip}/J_{far} therefore indicates the improvement in fracture resistance. However, this lower value of J_{tip} should be smaller than the material resistance R for the crack to be arrested.

The normalized CDF is plotted for CT specimen results with an interface at 90° in Figure 10.7 at different crack-tip locations. The crack-tip locations are presented in the normalized form as a ratio of the distance between the current crack-tip location L and the initial ligament length b_o. For crack-tip locations left to the interface, this ratio is negative and for post-interface locations it's positive.

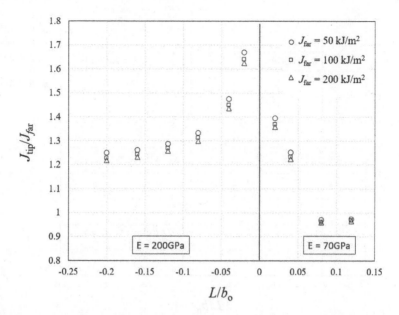

FIGURE 10.7 Normalized crack driving force at different crack-tip locations (L/b_o) for and interface at $\theta = 90°$ at three different values of J_{far} (50, 100, and 200 kJ/m²).

As expected and previously investigated by Tiwari et al. 2020, the CDF increases as the crack approaches to the interface. As soon as the crack is beyond the interface the J_{tip}/J_{far} ratio drops and reaches close to one with crack tip moving further from the interface. The influences of interface on the crack-tip location pre-interface and post-interface are different as the interface is unloaded when crack has crossed through it. This is why J_{tip}/J_{far} drops fast, whereas it even starts from a value >1 for pre-interface locations (i.e., for L/b_o <0). The influence of interface on CDF is shown for three different J_{far} values of 50, 100, and 200 kJ/m². For all three J_{far} values, the trend of the CDF remains same with higher CDF for higher J_{far} values.

With the established trend of the CDF with crack-tip locations for a simple case of interface at 90° to the crack plane, we proceed toward the effect of orientation on the same. The normalized CDF for 120°, 90°, 85°, 60°, and 45° are shown in Figure 10.8 for different crack-tip locations. The plot in Figure 10.8 and further results are at a constant J_{far} of 50 kJ/m². It becomes clear from Figure 10.8 that the pre-interface CDF for 90° and 85° are approximately same. As the angle of orientation decreases, the CDF also decreases. Opposite trend is seen for post-interface crack-tip locations. The CDF is highest for 45° and decreases with increasing angle except for 120° for which the CDF is lowest.

The reason for such behavior can be realized with the help of Figure 10.3(a). As the angle decreases from 90° for post-interface crack-tip location, the upper part of the interface will be closer to the crack tip in comparison to a higher angle. Hence, the effect of interface for a smaller angle for post-interface location will be stronger and will results in much lower CDF in comparison to a higher angle or for reference with the interface at 90°. As the angle increases from 90°, opposite will be true.

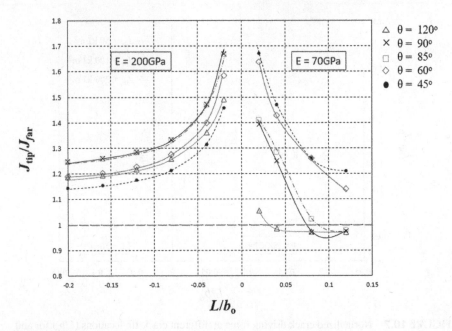

FIGURE 10.8   Normalized crack driving force at different crack-tip locations (L/b_o) for and interface at different orientations at a J_{far} of 50 kJ/m^2.

For the pre-interface locations, the upper part of the interface will be closer to the crack tip for $\theta > 90°$ and will be away for $\theta < 90°$. The upper part of the interface will be further away as θ will decrease from $90°$. This will result in a lighter influence of interface on CDF in comparison to $90°$ interface. This behavior can be clearly seen in Figure 10.9, where the normalized CDF is plotted against angle of orientation for different crack-tip locations. For negative values of L/b_o with increase in angle of orientation, the CDF increases and the opposite can be seen in Figure 10.9 for $L/b_o > 0$, i.e. for post-interface locations.

It is to be noted that the finite elements calculation is performed on a half symmetric model. Therefore, an opposite effect is expected due to the lower part of the specimen geometry modeled due to the inclination of the interface, the lower part will not be exactly symmetric to the upper part. Let us discuss the effect of lower part of the CT specimen with inclined interface at an angle of $45°$. In this case, for the upper part as shown in Figure 10.3(a), the interface will cause the J_{tip}/J_{far} ratio to be smaller in comparison to a $90°$ interface (see Figure 10.8 and 10.9) for pre-interface crack-tip locations. However, in the lower symmetric part the interface will be making an angle of $-135°$ (see Figure 10.10). This would mean that the lower part of the interface will be closer to the crack tip causing higher value of J_{tip}/J_{far}. This asymmetric distribution of configuration forces will cause a net J-vector to be away from the crack plane, causing the crack to deviate. The probable path of crack due to the angle of inclination is not under the scope of this work; however, it is noteworthy that the orientation of the interfaces can be used as a tool to optimize the crack resistance as well as it can also be used to make the crack follow a predetermined path.

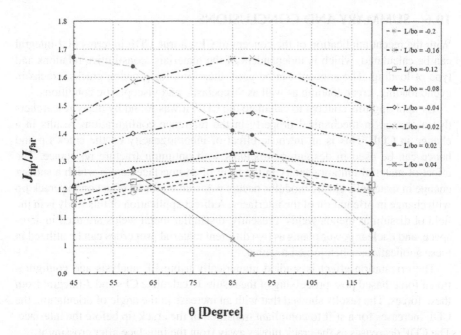

FIGURE 10.9 Normalized crack driving force at $L/b_o = -0.2$ with change in orientation of the interfaces.

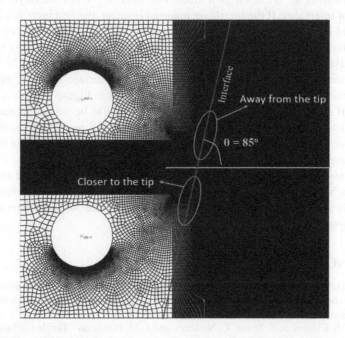

FIGURE 10.10 Effect of the lower part of the modeled geometry due to the inclined interface on the crack tip.

10.6 SUMMARY AND CONCLUSIONS

With the recent application of the concept of CFs, a true CDF in terms of J-integral can be calculated, which is independent of the materials' constitutive relations and type of loading. The concept can be used, therefore, for growing cracks, cracks in cyclic loading, creep loading as well as viscoelastic and viscoplastic conditions.

By applying this concept to calculate the CDF, it was found by many researchers that a change in mechanical property in the reference configurations results in a change in CDF. This is an intrinsic effect of inhomogeneity on the crack tip and hence can be used in designing structures with optimum fracture resistance. The effect of material inhomogeneity in the form of an interface across which a sudden change in material's elastic modulus occurs was studied for its effect on the crack tip with change in orientation of the interfaces. A direct application of this study is in the field of dissimilar metals welds. Presently, dissimilar metal welds are used in aerospace, and nuclear power plants as two different material properties can be utilized in these applications when joined together.

The orientation effect is studied numerically using FE analysis and configurational force-based post-processing of the results to calculate CFs and J-integral from these forces. The results showed that with an increase in the angle of orientation, the CDF increases for a stiff to compliant transition of the crack tip before the interface. The CDF decreases as the crack moves away from the interface after crossing it.

The opposite is expected to be true for a compliant to stiff transition. The opposite nature has been confirmed in previous studies by Kolednik et al. 2014 and by Tiwari et al. 2020, on their studies on interlayers. An interlayer is a case of a pair of interfaces. If the interlayer material is softer, the first interface between matrix and interlayer (stiff to compliant) will be exactly the opposite of the second interface (compliant to stiff) between interlayer material and matrix material. An interlayer can, in most cases, stop a crack (only if the reduced CDF is lower than the material's resistance R); however, a single interface may not, especially when it is a stiff to compliant transition, as evident from Figure 10.8. In a compliant to stiff transition, the curves in Figure 10.8 would be mirror images along $J_{tip}/J_{far} = 1$ line. This means that a reduction in CDF will be realized as the crack goes closer to the interface, and if the J_{tip} decreases to a value $<R$, then the crack will get arrested.

The study also helps in those cases where interfaces are needed to have an orientation. In such cases, a preferred orientation can be chosen, based on a configurational force study similar to this work with detailed material behavior simulation to have a safer component in terms of fracture.

REFERENCES

Abidi, M. H., N. Ali, H. Ibrahimi, S. Anjum, D. Bajaj, A. N. Siddiquee, M. Alkahtani, and A. Ur Rehman. "T-FSW of dissimilar aerospace grade aluminium alloys: Influence of second pass on weld defects." *Metals* 10, no. 4 (2020): 525.
Blouin, A., S. Chapuliot, S. Marie, C. Niclaeys, and J.-M. Bergheau. "Brittle fracture analysis of dissimilar metal welds." *Engineering Fracture Mechanics* 131 (2014): 58–73.

Brayshaw, W. J., A. J. Cooper, and A. H. Sherry. "Assessment of the micro-mechanical fracture processes within dissimilar metal welds." *Engineering Failure Analysis* 97 (2019): 820–835.

Eshelby, J. D. "The elastic energy-momentum tensor." *Journal of Elasticity* 5, no. 3 (1975): 321–335.

Griffith, A. A. "VI. The phenomena of rupture and flow in solids." *Philosophical Transactions of the Royal Society of London. Series A, Containing Papers of a Mathematical or Physical Character* 221, no. 582–593 (1921): 163–198.

Gurtin, M. E. "The nature of configurational forces." In J. M. Ball, D. Kinderlehrer, P. Podio-Guidugli, M. Slemrod (Eds.) *Fundamental Contributions to the Continuum Theory of Evolving Phase Interfaces in Solids*, pp. 281–314. Springer, Berlin, Heidelberg, 1999.

Irwin, G. R. "Fracture mode transition for a crack traversing a plate." *Journal of Basic Engineering*, 82, (1960): 417–423.

Irwin, G. R. *Onset of fast crack propagation in high strength steel and aluminum alloys*. Naval Research Lab, Washington DC, 1956.

Jang, C., J. Lee, J. S. Kim, and T. E. Jin. "Mechanical property variation within Inconel 82/182 dissimilar metal weld between low alloy steel and 316 stainless steel." *International Journal of Pressure Vessels and Piping* 85, no. 9 (2008): 635–646.

Khan, N. Z., A. N. Siddiquee, Z. A. Khan and A. K. Mukhopadhyay. "Mechanical and microstructural behavior of friction stir welded similar and dissimilar sheets of AA2219 and AA7475 aluminium alloys." *Journal of Alloys and Compounds*, 695, 2017: 2902–2908.

Kienzler, R., and G. Herrmann. *Mechanics in material space: With applications to defect and fracture mechanics*. Springer Science & Business Media, 2000.

Kolednik, O., J. Predan, and F. D. Fischer. "Cracks in inhomogeneous materials: Comprehensive assessment using the configurational forces concept." *Engineering Fracture Mechanics* 77, no. 14, (2010): 2698–27110.

Kolednik, O., J. Predan, F. D. Fischer, and P. Fratzl. "Improvements of strength and fracture resistance by spatial material property variations." *Acta Materialia* 68, (2014): 279–294.

Kolednik, O., R. Kasberger, M. Sistaninia, J. Predan, and M. Kegl. "Development of damage-tolerant and fracture-resistant materials by utilizing the material inhomogeneity effect." *Journal of Applied Mechanics* 86, no. 11, (2019): 111004.

Lee, J. Jin-Wu, I. K. Lloyd, H. Chai, Y.-G. Jung, and B. R. Lawn. "Arrest, deflection, penetration and reinitiation of cracks in brittle layers across adhesive interlayers." *Acta Materialia* 55, no. 17 (2007): 5859–5866.

Maugin, G. A. "Eshelby stress in elastoplasticity and ductile fracture." *Int. J. Plast.* 118, no. 1 (1937): 26–29.

Ochensberger, W., and O. Kolednik. "Overload effect revisited– Investigation by use of configurational forces." *International journal of fatigue* 83, (2016): 161–173.

Ochensberger, W., and O. Kolednik. "Physically appropriate characterization of fatigue crack propagation rate in elastic–plastic materials using the *J*-integral concept." *International Journal of Fracture* 192, no. 1 (2015): 25–45.

Paris, P. C., M. P. Gomez, and W. E. Anderson. "Paul C. Paris." *The Trend in Engineering at the University of Washington* 13, (1961): 9–14.

Paris, P., and F. Erdogan. "A critical analysis of crack propagation laws." *Journal of Basic Engineering* 85, (1963): 528–533.

Pauling, L. "The principles determining the structure of complex ionic crystals." *Journal of the American Chemical Society* 51, no. 4 (1929): 1010–1026.

Rice, J. R. "A path independent integral and the approximate analysis of strain concentration by notches and cracks." *Journal of Applied Mechanics* 35, (1968): 379–386.

Simha, N. K., F. D. Fischer, G. X. Shan, C. R. Chen, and O. Kolednik. "J-integral and crack driving force in elastic–plastic materials." *Journal of the Mechanics and Physics of Solids* 56, no. 9 (2008): 2876–2895.

Simha, N. K., F. D. Fischer, O. Kolednik, J. Predan, and G. X. Shan. "Crack tip shielding or anti-shielding due to smooth and discontinuous material inhomogeneities." *International Journal of Fracture* 135, no. 1, (2005): 73–93.

Tiwari, A., J. Wiener, F. Arbeiter, G. Pinter, and O. Kolednik. "Application of the material inhomogeneity effect for the improvement of fracture toughness of a brittle polymer." *Engineering Fracture Mechanics* 224, (2020): 106776.

Vaidya, W. V., M. Horstmann, V. Ventzke, B. Petrovski, M. Koçak, R. Kocik, and G. Tempus. "Improving interfacial properties of a laser beam welded dissimilar joint of aluminium AA6056 and titanium Ti₆Al4V for aeronautical applications." *Journal of Materials Science* 45, no. 22 (2010): 6242–6254.

Wang, H. T., G. Z. Wang, F. Z. Xuan, C. J. Liu, and S. T. Tu. "Local mechanical properties of a dissimilar metal welded joint in nuclear powersystems." *Materials Science and Engineering: A* 568 (2013): 108–117.

Westergaard, H. M. "What is known of stresses." *Engineering News Record*, 118 no. 1 (1937): 26–29.

Williams, M. L. "On the Stress at the Base of a Stationary Crack, Journal of Applied Mechanics." *Transactions ASME* 24 (1957): 109–114.

Zhang, C. Q., J. D. Robson, and P. B. Prangnell. "Dissimilar ultrasonic spot welding of aerospace aluminum alloy AA2139 to titanium alloy TiAl₆V₄." *Journal of Materials Processing Technology* 231 (2016): 382–388.

11 Thermal Contact Conductance Prediction Using FEM-Based Computational Techniques

Ashwani Kumar
Technical Education Department Uttar Pradesh Kanpur, India

Sachin Rana
ABES Institute of Technology Ghaziabad, India

Yatika Gori
Graphic Era University Dehradun, India

Neelesh Kumar Sharma
Indian Institute of Technology Patna, India

CONTENTS

DOI: 10.1201/9781003202233-13

11.1   INTRODUCTION

All engineering surfaces exhibit some level of microscopic roughness. The resistance
to heat flow through a contact interface occurs because only a small portion (usually
1%–2%) of the nominal surface area is actually in contact. Heat may pass through the
interface via three paths: conduction through the contact spots, conduction through
the gas present in the gap between the surfaces, and radiation across the gap.
Convection may be neglected due to the small length scales involved. Also, radiation
does not play a significant role at temperatures below 500°C.

Since the conductivity of the gas is much smaller than that of the substrate,
most of the heat is constrained to flow through the contact spots. This constriction
and subsequent spreading of heat flow lines in the two materials in contact mani-
fests as thermal resistance at the interface. The total resistance of the surface is
found by summing, in parallel, the constriction resistances of all of the contact
spots. This thermal resistance impedes the heat flow across the interface, when
two solid bodies at different temperatures are brought into mechanical contact,
and results in a temperature drop as shown in Figure 11.2 This resistance, commonly known as thermal contact resistance, is well-explained by the fact that the
real contact area is exceedingly small as compared to the apparent contact area,
due to the presence of roughness and waviness of the engaging surfaces. As the
interstitial material, such as air, is a poor heat conductor and the radiative heat
transfer is often insignificant, a large portion of the heat flow converges to the
discrete solid-solid contact spots, as illustrated in Figure 11.1. Hence, the increase

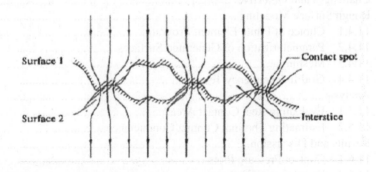

FIGURE 11.1   Convergence of heat flow through contact spots.

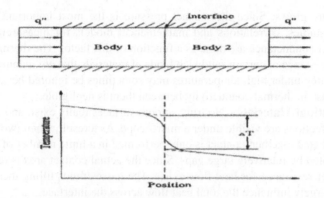

FIGURE 11.2 Illustration of thermal contact conductance.

in the heat-flow path length causes the thermal contact resistance. Its reciprocal is called thermal contact conductance, defined as

$$h_c = \frac{q}{A\Delta T} \qquad (11.1)$$

Where q is the heat flow rate, A is the apparent contact area, and ΔT is the temperature drop at the interface.

With assumptions of a particular statistical distribution of surface parameters, as well as a mode of deformation for contacting asperities, contact conductance is predicted for given material properties and applied load. Since the surface parameters, such as asperity height, density, slope, and radius of curvature, are not intrinsic properties of the surface, but instead can vary with the sampling frequency the contact conductance is dependent on the properties of the surfaces such as hardness, roughness and waviness, real contact area and the contact pressure. Thermal contact conductance plays an important role in all thermal systems where a mechanical contact is involved. Recently, such conductance has received special interest and attention in small-scale heat removal systems, such as microelectronics and in heat transfer between superconductor films.

Surface roughness is a measure of microscopic irregularity, whereas the macroscopic errors of form include flatness deviations, waviness and, for cylindrical surfaces, out of roundness. Two solid surfaces apparently in contact touch each other only at a few individual spots as shown in Figure 11.1. Even at relatively high contact pressure of the order of 10 Mpa, the real area of contact for most metallic surfaces is only about 1% to 2% of the nominal contact area [1].

11.1.1 FACTORS INFLUENCING THERMAL CONTACT CONDUCTANCE

A. **Contact Pressure** – The contact pressure is the factor of most influence on contact conductance. As contact pressure grows, contact conductance grows (and consequentially, contact resistance becomes smaller). This is attributed to the fact that the contact surface between the bodies grows as the contact pressure grows. Since the contact pressure is the most important factor, most studies, correlations and mathematical models for measurement of contact conductance are done as a function of this factor. The thermal contact resistance of certain sandwich kinds of materials that are manufactured by rolling under high temperatures may sometimes be ignored because the decrease in thermal conductivity between them is negligible.

B. **Interstitial Materials** – No truly smooth surfaces really exist, and surface imperfections are visible under a microscope. As a result, when two bodies are pressed together, contact is only performed in a finite number of points, separated by relatively large gaps. Since the actual contact area is reduced, another resistance for heat flow exists. The gasses/fluids filling these gaps may largely influence the total heat flow across the interface.

C. **Effect of Loading–Unloading Cycles** – As the interface between two rough surfaces is subjected to loading–unloading cycles, an increase in thermal contact conductance is observed for the second and subsequent loadings compared to the first loading, mainly due to the progressive nature of the deformation of asperities on the surfaces in contact. An increase in thermal contact conductance is observed for successive loadings, but only during the first few cycles. After the second loading cycle, the thermal contact conductance (TCC) comes to a steady value with respect to loading cycle.

D. **Effects of Material Properties** – Material properties of the two surfaces in contact have a significant effect on the observed TCC. The effects of material properties such as micro-hardness and yield strength are less easily discerned. Although aluminum has a much higher conductivity, aluminum and brass contacts have similar TCC values for pressures in the region of 0.5–1.5 MPa. This indicates that higher yield strength leads to a reduction in TCC. The higher yield strength of aluminum means that geometrically equivalent asperities on brass begin to deform plastically before those on aluminum.

Surface roughness, waviness and flatness, effect of temperature, surface deformations, surface cleanliness are important factor to be considered for TCC.

11.1.2 Importance and Applications

For the conservation and efficient use of thermal energy it is important to be able to control the heat transfer properties of the engineering constructions involved in its use. This may require on the one hand the minimization of heat transfer to reduce losses or, on the other, its maximization to avoid unacceptably high temperature gradients for a given thermal flux. TCC is an important factor in a variety of applications, largely because many physical systems contain a mechanical combination of two materials. Some of the fields where contact conductance has importance are as follows:

A. **Electronic Packaging** – The exposed surface area of many of today's high powered electronic packages is no longer sufficient for the removal of the heat generated during normal operation. Heat sinks are a commonly used,

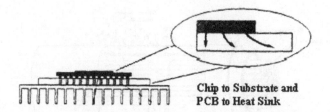

Chip to Substrate and
PCB to Heat Sink

FIGURE 11.3   Constriction and spreading resistances in electronics packaging.

low cost means of increasing the effective surface area for dissipating heat
by means of convective air cooling. While the use of a heat sink lowers
the fluid-side thermal resistance, it also introduces an interface resistance
across the contact formed between itself and the package case. Under some
circumstances, this contact resistance can be substantial, impeding heat
flow and reducing the overall effectiveness of the heat sink. Figure 11.3
depicts the constriction resistance from chip to substrate and from a printed
circuit board (PCB) to heat sink in an electronic packaging [3].

B. **Microelectronics** – The heat generation within microelectronic circuits,
printed circuit boards, and multichip modules has become the primary factor that limits the physical size of both the individual components and multichip modules. The thermal environment and associated high temperatures
could lead to overheating, significantly reducing component performance
and increasing the possibility of total failure. As a consequence, there is
an increasing interest in improving the thermal conductance at interfaces
within microelectronic systems [4].

C. **Biomedicine** – Heat transfer mechanisms in and between biomaterials have
received increasing attention with the development of synthetic organs, tissues, and other implantable devices. Knowledge of the thermal properties
of biomaterials, as well as in situ tissue, is essential to modeling and to
the development of new materials and organs. There are numerous circumstances in which the temperature gradient and the temperature difference
between body tissues or fluids and implantable materials affect the overall
performance or comfort of implantable devices. In a recent technique that
deals with the use of a laser catheter for removal of plaque in both normal
and diseased vessels by means of vaporization, the removal of plaque using
laser techniques requires accurate temperature control as well as knowledge
of the temperature gradients, tissue thermal conductivity, and TCC of the
plaque-arterial wall interface [5].

D. **Nuclear Reactor** – In systems designed with high heat flux components
such as nuclear power plants and fusion systems, it is important that the
thermal behavior of the solid-solid interface be precisely known. The fuel
channels of ATR (advanced thermal reactor) consist of pressure tubes (PTs)
separated from the surrounding concentric calandria tubes (CTs) by carbon
dioxide (CO_2) gas-filled gap. Cool heavy water moderator surrounding each
vertical CT acts as a sink for the decay heat in the postulated loss of coolant

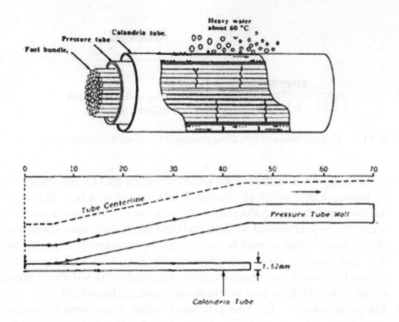

FIGURE 11.4 Configuration of PT and CT.

accidents (LOCAs) as shown in Figure 11.4. The postulated accident conditions will result in high temperatures both in fuel and PTs. As a result, fuel elements and PTs may deform such that the former comes in contact with the latter while PT comes in contact with the surrounding CT, respectively. The rate of heat transfer to the moderator will depend on the temperatures of fuel and PT at contact and the thermal conductance between the contacting fuel and PT, and that between PTs and CTs. In order to solve this heat transfer problem and to realistically assess the maximum fuel temperatures, it is required to determine the total thermal conductance [6].

E. **Space Applications** – Plans for the development of a space station, and the continuing desire for more efficient space thermal power systems, suggest the need for a better understanding of thermal contact resistance as it relates to space applications. The desirability of providing optimum spacecraft systems necessitates the improvement of thermal control systems. Further study of contact conductance utilizing selected materials in space environments is important [7].

F. **Heat Exchanger** – Generally, the fin-tube heat exchanger is manufactured through mechanical expansion of tube to tighten the contact between fins and tubes. The features of heat transfer through interfaces cannot be clarified because of the irregular contact of interface. Therefore, the thermal contact resistance between fins and tubes interfaces is investigated [9].

In field of advanced materials, I.C. engine, metal-forming and super conductor have applications of TCC.

11.2 LITERATURE REVIEW

Leung et al. [1] presents a theoretical investigation, directed toward the application of statistical mechanics to the prediction of contact conductance. This paper addresses contact conductance from macroscopic and microscopic viewpoints in order to demonstrate the promise of the statistical mechanics approach. Williamson et al. [2] have studied the surface deformation effect on thermal contact conductance. In similar manner Marotta et al. [8] have investigated TCC effect of coating one material on another. They have coated sintered copper coating on ferro alloy. Other researchers [10, 11] also studied the measurement technique of TCC using different methods. Cooper et al. [12] developed a TCC model (CMY model). They consider the resistance to the flow of heat between two thick solid bodies in contact in a vacuum. Existing analyses of single idealized contacts are summarized and compared, and then applied, together with results of recent electrolytic analogy tests, to predict the conductance of multiple contacts "appropriately" or "inappropriately" distributed at the interface. Reconsideration of the theory of interaction between randomly rough surfaces shows how the parameters required predicting heat transfer can be determined in principle by simple manipulation of typical profiles of the mating surface, together with an approximation from deformation theory. Mikic [13] presents the major theoretical contributions in the area of thermal contact resistance dealt with a simple contact, multiple contacts, directional effects, and others. In this work, Mikic eliminated the concept of the contour area and the macroscopic construction is related directly to non-uniformity of the macroscopic heat flux and specifically, to the pressure distribution over the interface. The result is incorporated into an expression which relates the overall contact resistance (microscopic and macroscopic) to the pressure distribution and surface properties. Mikic [14] considered the TCC of nominally flat surfaces in contact was considered. The emphasis of the work is on effect of the mode of deformation on the value of conductance. For rough nominally flat surfaces in contact thermal contact resistance was explicitly evaluated for assumed pure plastic deformation, plastic deformation of asperities and elastic deformation of the substrates and pure elastic deformation. In addition, a criterion determining mode of deformation for given surfaces in contact (given geometry, and materials) was given. The surfaces were described with Gaussian distribution of height and random distribution of the slope.

Sayles and Thomas [15] derived a completely general relationship for the conductance of the elastic contact of an isotropic Gaussian random surface with a flat in terms only of properties of the bulk material and quantifiable parameters of the surface topography. We have further shown that the average contact size is sensibly independent of load and that the increase in the area of real contact is almost entirely due to an increase in the number of contact spots. It is true that our expression for the conductance is an upper-bound one, but against that our derivation contains no disposable constants. It holds over a range of separations and surface wavelengths that correspond to normal engineering practice. Gibson [16] considered the flow through abutting cylinders having, in general, different conductivities. As Cooper has shown that the contact spot may be considered as an isotherm. Thus, the simpler problem of one cylinder with heat flux specified over part of the boundary and the remainder being an isotherm had been considered. This mixed boundary value problem had been solved analytically. This problem was reduced to a Fredholm integral equation of the second kind, which had been solved by a standard procedure. The temperature distribution and, finally, an expression for the thermal contact resistance was derived. Antonetti et al. [17] has studied experimentally to enhance the thermal contact conductance using metallic coating. McGee et al. [18] developed a line-contact model for the thermal resistance of a cylinder-flat contact. Resistance due to heat flow constriction across the solid-to-solid contact was calculated. The model was also accounted for the variation in resistance across the gas-filled gap, on either side of the contact, as a function of gas pressure. Experiments measurements were performed to compare with the theoretical models and also to enhance the understanding of the mechanisms controlling the resistance to heat transfer across a cylinder-flat joint, and good agreement was obtained over a limited range of experimental parameters. Thermal experiments were performed on several materials in a vacuum and in helium and argon atmospheres at pressures up to atmospheric. Two types of contact-resistance tests were performed, in the first, the pressure was maintained as low as possible, and measurements were made with various values of applied contact load. Huang et al. [19] present the inverse solution methodology based on the conjugate gradient method is developed for estimating the variation of air-gap resistance with time from the transient temperature measurements taken with thermocouples inside the casting region and at the outer mold surface. The conjugate gradient method, which utilizes the function estimation approach is used to solve the inverse solidification problem to determine the unknown time-wise variation of the contact conductance between the mold and casting region. The results show that the conjugate gradient method requires much less computer time than the least-squares method, is less sensitive to measurement errors, and does not require prior information for the functional form of the unknown quantity.

Tsai and Crane [20] presented an analytical temperature distribution solution to the one-dimensional symmetric system with heat flux on one outside surface and insulation on the other. The analysis provided a theoretical basis for transient measurement of TCC. Antonetti et al. [21] developed a correlation for an approximate TCC, which does not depend upon the surface asperity slope. In this study, a strong correlation exists between the surface roughness and the asperity slope, and regression analysis established an equation relating the two parameters. By using the equation relating the roughness and asperity slope, a new approximate contact conductance expression was developed, which does not contain the asperity slope. Madhusudana [22] demonstrated that in many applications involving contact heat transfer, including electronic components, the contact pressure is low or moderate. It is shown that at these pressures, conduction through the interstitial gas is the dominant mode of heat transfer. It is further proved that the heat transfer coefficient at the interface at low contact pressures depends on the heat flow direction, but the mechanism of this 'rectification' is different to that postulated for cases where the solid spot conduction is predominant. Nishino et al. [23] was studied the TCC in a vacuum environment under low applied load with square test plates made of aluminum alloy. Two kinds of contact geometries were examined: the contact between a practically flat rough surface and an approximately spherical one, and the contact of similarly flat rough surfaces. By using a pressure-measuring film that is capable of visualizing contact pressure distributions, a new technique for predicting the TCC was developed. The technique evaluates the microscopic and macroscopic thermal constriction resistances from the real contact pressure distribution, which is measured by means of digital image processing from the color density pattern appearing in the film. Sridhar and Yovanovich [24] made a measurement on TCC on ground-lapped interfaces of tool steel, and they were compared with the recently proposed elastoplastic model. The type of deformation associated with contact conductance measurements of ground-lapped interfaces of untreated tool steel was elastoplastic, whereas heat-treated tool steel underwent fully elastic deformation. It is found that the interface equivalent elastic modulus for tool steel and some previous data are higher than the equivalent bulk elastic modulus. Sridhar and Yovanovich [25] proposed a new thermal elastoplastic contact conductance model for isotropic conforming rough surfaces. This model is based on surface and thermal models used in the Cooper, Mikic, and Yovanovich plastic model but it differs in the deformation aspects of the TCC model. The model incorporates the recently developed simple elastoplastic model for sphere-flat contacts, and it covers the entire range of material behavior, i.e., elastic, elastoplastic, and fully plastic deformation. Previously data were either compared with the elastic model or the plastic model assuming a type of deformation a priori. The model is used to reduce previously obtained isotropic contact conductance data, which cover a wide range of surface characteristics and material properties. For the first time, data can be compared with both the elastic and plastic models on the same plot. This model explains the observed discrepancies noted by previous workers between data and the predictions of the elastic or plastic models. For the first time, TCC data have been reduced using both the elastic and elastoplastic models. This new procedure for data reduction eliminates the dilemma of assuming a type of deformation a priori.

Ohsone et al. [26] present a non-contact optical technique for measuring the TCC between wafer-like thin solid samples. The technique is based on heating one solid surface by a modulated laser beam and monitoring the corresponding temperature modulation of the other solid surface across the interface using the reflectance of a probe laser beam. Tseng [27] reviewed the modeling approaches and correlations used to study the interface heat transfer phenomena of the roll-strip contact region in rolling processes. The TCC approach was recommended for modeling the interface phenomena. Zhao et al. [28] present an elastic-plastic asperity microcontact model for contact between two nominally flat surfaces. The transition from elastic deformation to fully plastic flow of the contacting asperity is modeled based on contact-mechanics theories in conjunction with the continuity and smoothness of variables across different modes of deformation. The relations of the mean contact pressure and contact area of the asperity to its contact interference in the elastoplastic regime of deformation are respectively modeled by logarithmic and fourth-order polynomial functions.

Benigni et al. [29] carried out an investigation which deals with the possibility and the first measurements of TCC of cylindrical joints by the periodic method at high temperature. The mathematical principle is presented. The calibration of the method is performed on a composite nickel/nickel cylinder, the interface of which has a controlled macro roughness. The experimental results are compared with the results of a geometric conductance model. Lee et al. [30] developed an analytical model for predicting constriction and spreading resistances associated with heat transfer from various electronic components under different modes of cooling. The model assumes a heat source in contact with a larger cold plate which is in turn cooled with a convective heat transfer coefficient specified over the sink surface. Huang et al. [31] determined temporally and circumferentially varying TCC of a plate finned tube heat exchanger by reading the simulated transient temperature measurement data from the thermocouples located on the plate. The thermal properties of the fin and tube were assumed to be functions of temperature, making the problem non-linear. For the non-linear inverse problem, the conjugate gradient method (CGM) was used for minimization. Sunil Kumar and Ramamurthi [32] determined the influence of the roughness, waviness, and flatness of surfaces on contact conductance. The surface waviness is observed to influence surface contact conductance more strongly compared to flatness deviations. Jeng et al. [33] investigated contact heat conduction from the viewpoint of interface contact and developed a conductance model considering elastic, elastoplastic and fully plastic deformation. The results obtained from experiments were compared with those calculated from the theoretical model. Increasing the load tends to increase the real contact area, thus increasing the surface contact heat conduction rate.

Kim et al. [34] developed a new tool including an experiment and a numerical calculation for the estimation of the TCC between the fin collar and tube surface, and pursues the evaluation of the factors affecting the TCC in a fin-tube heat exchanger. A finite-difference numerical scheme has been used for the data reduction of the experimental data to evaluate the TCC. Sunil Kumar et al. [35] developed a theoretical model to predict the TCC based on randomly occurring contacts between asperities on surfaces of cylindrical and spherical bodies. Sunil Kumar and Ramamurthi [36] determined the influence of variations of interface temperature in the range 50–300 K on the TCC between aluminum and stainless-steel joints was determined. Predictions were done by modeling the deformation at the interface for different values of surface finish and contact pressure over the range of interface temperatures. Both elastic and plastic deformation was considered. Singhal et al. [37] developed a numerical model for the prediction of TCC at metal-to-metal contact interfaces. In parallel, an experimental facility had been constructed to measure TCC across an interface, especially at the low pressures (encountered in electronics cooling applications) to validate the model developed. Zhao et al. [38] measured the roughness profiles of some common machined surfaces were measured. Four different criteria for determining contact peaks are presented. Jeong et al. [39] investigated new factors such as fin types and manufacturing types of the tube affecting the TCC and to find a correlation between the TCC and the effective factors in fin-tube heat exchangers with 7 mm tube. Jeng et al. [40] conducted an experimental and theoretical study to investigate the effects of diamond film coatings on the TCC. A model based on statistical elastoplastic surface contact mechanics was developed to study the TCC of coating material. Zou et al. [41] developed a random number model based on fractal geometry theory to calculate the TCC of two rough surfaces in contact. Study is carried out by geometrical and mechanical investigations. Shojaeefard et al. [42]

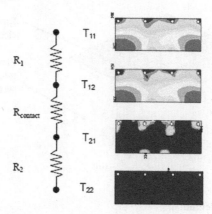

FIGURE 11.5   Thermal circuit and temperature plots for a bolted plate system.

investigated an inverse heat conduction problem for estimating the periodic TCC between one-dimensional, constant property contacting solids with CGM of function estimation. This method converges very rapidly and is not so sensitive to measurement errors. M. K. Thompson and J. M. Thompson [43] studied that in many finite element models, thermal contact resistance is either neglected or included only as an intrinsic characteristic of the system. This work presents a discussion of the issues associated with predicting thermal contact resistance and methods for overcoming some of the difficulties associated with doing so in ANSYS. In a thermal system, the analogy to Ohm's Law has a similar form and is stated as: $\Delta T = QR$ (Figure 11.5).

Where ΔT is the average temperature drop across the thermal body, Q is the heat flowing through that body, and R is the thermal resistance of the body. When the temperature drop occurs at the interface between two bodies or materials, the phenomenon is referred to as thermal contact resistance (TCR).

In this study, several considerations for the prediction of thermal contact resistance, including the multi-scale and multiphysics nature of thermal contact resistance and the determination of a representative value of TCC were discussed.

Poroshin et al. [44] had studied which is devoted to the problem of transferring of the results of the real surface roughness measurements into universal finite element method analysis software ANSYS. The detailed methods for both 2D surface profile and 3D surface topography transferring are proposed. The software module for surface roughness transferring into ANSYS according to the proposed methods is designed.

Peyrou et al. [45] presents in this paper two methods for generating rough surfaces, one using the real shape with an original reverse engineering method and the other one by using a parametric design language to generate a normally distributed rough surface (Figure 11.6). As an application to demonstrate the power of these methods, we choose to predict by simulation the electrical contact resistance and the real contact area between rough surfaces as a function of the contact force. This application is a major concern in RF MEMS ohmic Switches and shows an original approach to extract a guideline in choosing a design, materials, and process flow to

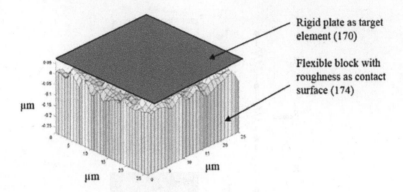

FIGURE 11.6   Model definition.

minimize the contact resistance. To perform the finite element analysis, we choose the combined method based on penalty and Lagrangian methods called the augmented Lagrange method.

This application shows a powerful method to analyze surface topography effects such as roughness. Due to its great interest, this application would be intensively studied in order to give a guideline for the material choice by studying the impact of different materials' properties and roughness distribution on the resistance.

Kathryn Thompson [47] studied four methods of generating normally distributed rough surfaces using the ANSYS Parametric Design Language (APDL): (1) moving all nodes, (2) moving selected nodes by location, (3) creating key points and (4) creating shell primitives. The resulting geometry, meshes, and solutions are then compared to demonstrate the strengths and weaknesses of the various methods and recommendations are made based on the results.

Very limited methods for creating rough surfaces in finite element models have been proposed. Thompson et al. [48] created three-dimensional surfaces with a normal Gaussian distribution by moving the surface nodes. Hyun et al. [49] created three-dimensional self-affine fractal surfaces by moving all of the nodes in the model. Bhushan, et al. [50] created single asperities (hemispherical point, hemispherical scratch, and groove (triangular) scratch) by modifying the solid model. However, there has been little or no discussion about the relative merit of the various methods based on the quality of solution that each produces or alternate methods for creating surface roughness.

11.3   CHALLENGES AND OBJECTIVE

The challenges in numerical analysis are as follows:

   A. Generation of rough surface profiles.
   B. Evaluation of real contact area.

For the analysis of structures, users have historically chosen to model the gross geometry of the system and ignore the details to simplify the model and reduce computation times. Surface effects are generally ignored in the analysis, partly because of the difficulty in creating a reasonable model of the surface and partly because of the assumption that surface effects are less important. However, surface effects caused by surface topography are of major concern in some engineering applications and may become critical when contact is involved.

Evaluation of real contact area by experimental methods is also a difficult task because no instrument can be put at the contact interface therefore no work has been done experimentally to evaluate real contact area and uses the nominal contact area in place of real contact area. The prediction of real contact area is needed in many applications such as electronic packaging, I.C. Engines, etc., to obtain the actual heat transfer rate and provide required cooling effect.

Some of the available commercial finite element programs have incorporated features that permit the user to customize or modify the program to increase the capabilities and flexibility of the program. This allows the programs to bridge the gap between commercial and custom codes. In ANSYS, these capabilities include a scripting language called the APDL and User Programmable Features which allow the user to write FORTRAN routines to create a custom version of ANSYS. Because of the difficulty in creating the model of rough surface in contact and evaluation of real contact area, I have chosen to create the rough surface geometry in a FEM ANSYS to predict the correct estimation of heat transfer between contacting surfaces.

The objective of this study is to perform theoretical modeling (FEM-based) in order to predict the real contact area, using ANSYS. Modeling and estimation of real contact area involve: generation of rough surface profiles, creating interface model of real surfaces, evaluation of real contact area under varying loading conditions, varying surface roughness and varying materials/physical properties of a specimen.

11.4   ROUGH SURFACE MODELING

There are number of challenges associated with modeling surfaces and systems where the behavior at the surface is critical. First, the geometry at the surface is often irregular and complex. Surface damage such as pitting or scratching can further increase the complexity of the surface topography, making it difficult to formulate a geometric model that accurately describes the surface topography. Surface chemistry may have a large effect on the behavior of surfaces in contact. Adhesion and molecular bonding between surfaces may contribute to the contact forces holding the surfaces together. Adsorption, absorption, oxidation, and corrosion may change the chemical composition and the material properties at the surface. Material properties are also temperature and length scale dependent which further complicates the model.

In addition to natural oxides, other surface layers both intentional and unintentional may be present at the surface. Protective oxides may be placed on the surface through anodizing operations. Surfaces may be coating with pure metals such as copper and gold to increase the electrical or thermal conductivity of the surface. Compounds such as titanium carbide may be added to increase the hardness. It is often difficult to judge the thickness and material properties of these surface layers and the resulting layers may not be uniform in thickness or in quality. Sub-surface imperfections such as voids or cracks may be present. These can have a significant impact on the behavior of the surface and lead to phenomena such as delamination wear, however they are difficult to measure and predict.

Materials such as air, lubricants, dirt, and wear particles may be present in the interface between two surfaces which can affect the mechanical, thermal, and electrical behavior at the interface. Materials can also transfer from one surface to another across an interface. Traditionally, surfaces were modeled analytically by ignoring the surface details, including the details by using experimentally derived coefficients, or simplifying the surface geometry. Surface asperities have been modeled using a variety of geometric shapes and the behavior of a single pair of interacting asperities was often assumed to describe the behavior of a pair of interacting surfaces covered in asperities. Surfaces have also been modeled using probability distributions and by using fractals. In most cases, these assumptions were not made because they were shown to accurately represent the system of interest, but because they made modeling possible.

More recently, numerical contact simulations have been used to model surface phenomena. Some researchers choose to write custom programs to perform the numerical analysis while others have chosen to use commercial finite element programs. However, the geometric assumptions used in the analytical models have followed into the numerical models primarily because the computational costs associated with solving increasingly complex geometry were considered to be prohibitive.

11.4.1 Choice of Finite Element Program

The finite element method can be applied either by writing a custom finite element program or by using a commercial finite element program. Some of the available commercial finite element programs have also incorporated features that permit the user to customize or modify the program to increase the capabilities and flexibility of the program. This allows the programs to bridge the gap between commercial and custom codes. In ANSYS, these capabilities include a scripting language called the APDL and User Programmable Features which allow the user to write FORTRAN routines to create a custom version of ANSYS. The cost and learning curve associated with using a commercial code with customization features are far outweighed by the benefit of adapting these types of programs for non- traditional contact and tribological applications. For this work, the ANSYS finite element program will be used although the methods should be applicable to a wide range of finite element programs.

11.4.2 Parameterization of Generated Surfaces

Rough surfaces profile is composed of a large number of surface asperities which can be thought of as local deviations from the nominal or average surface height. Real surfaces may exhibit roughness (high frequency asperities), waviness (medium frequency asperities) and surface form (the general shape of the surface neglecting roughness and waviness). Surfaces may also exhibit lay, which is a directional

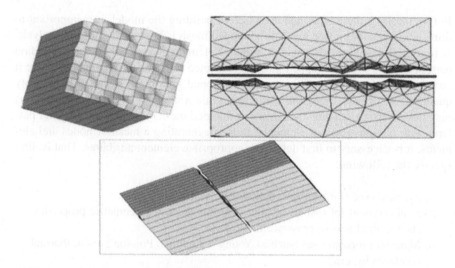

FIGURE 11.7   Rough surfaces in contact to create interface.

characteristic of the surface "such as a parallel, circular, radial or multidirectional pattern". Surface roughness and other surface topography may be a function of more than one roughness (waviness, form or lay) distribution. Consider a simple surface in Cartesian coordinates that is characterized by a single modal distribution of roughness and does not exhibit waviness, surface form, or lay. The rough surface can then be fully described by six parameters. Two parameters should describe the asperity density, or the amount of space between asperities: one in x and another y. Three parameters should describe the asperity size: length (x), width (y), and height or depth (+/− z). The last parameter should describe the asperity shape and define any coupling between the first five parameters if appropriate. These parameters may have values that are constants, functions, or statistical distributions.

Here we have assumed the asperity density to be equal in X and Y directions and set to a single parameter: **ASPDENS** = 5000 asperities/unit length. The asperity sizes in X and Y were assumed to be equal to each other and set to a value of 1/**ASPDENS**, thus each asperity is 200 μm × 200 μm square. The asperity size in z was assumed to be a normal Gaussian distribution with a mean of 0 and a standard deviation of 0.33 * **MAXZ** meters, where **MAXZ** is the maximum roughness of the surface (Figure 11.7). The length, width, and height of model are set to 5 mm.

11.4.3 Meshing

The process for generating a mesh of nodes and elements consists of three general steps:

1) Set the element attributes.
2) Set mesh controls (optional). ANSYS offers a large number of mesh controls from which we can choose as needs dictate.
3) Meshing the model.

Before meshing the model, and even before building the model, it is important to think about whether a free mesh or a mapped mesh is appropriate for the analysis. A free mesh has no restrictions in terms of element shapes, and has no specified pattern applied to it. A mapped mesh is restricted in terms of the element shape it contains and the pattern of the mesh. A mapped area mesh contains either only quadrilateral or only triangular elements, while a mapped volume mesh contains only hexahedron elements. In addition, a mapped mesh typically has a regular pattern, with obvious rows of elements. Before generating a mesh of nodes and elements, it is necessary to first define the appropriate element attributes. That is, first specify the following:

1) Element type (for example, SOLID185, SOLID 187, etc.).
2) Real constant set (usually comprising the element's geometric properties, such as thickness or cross-sectional area).
3) Material properties set (such as Young's modulus, Poisson's ratio, thermal conductivity, etc.)
4) Element coordinate system.

Smart element sizing (smart sizing) is a meshing feature that creates initial element sizes for free meshing operations. Smart sizing gives a better chance of creating reasonably shaped elements during automatic mesh generation. This feature, which is controlled by the **SMRTSIZE** command, provides a range of settings (from coarse to fine mesh) for meshing. Figure 11.8 shows the FE mesh created by smart sizing from 10 (coarse) to 1 (fine).

Refinement is an excellent tool in free meshing by which meshing at any specific region can be refine. Since, in this work, contact is the area of interest it is necessary to be fine mesh at contact and coarse mesh at other areas of model. In smart sizing the whole volume has fine mesh which increases the number of nodes and elements and thus increases the computational time. But by refinement the only area in contact can bi refine and thus it reduces the computational time. Refinement is far

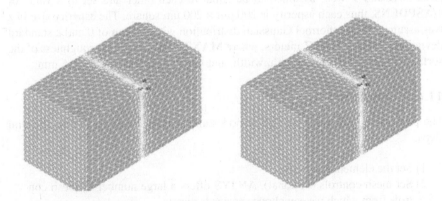

FIGURE 11.8   Mesh created by smart sizing from 10 to 1.

FIGURE 11.9 Mesh created by refinement from 0 to 5.

better tool than smart sizing in ANSYS for this work where a fine mesh is needed at a specific location. Figure 11.9 shows the area of contact with refinement from 0 (coarse) to 5 (fine).

In both the method (smart sizing and refinement) of free meshing we don't have control on element shape and element number. To calculate the area of contact after the analysis with different loading it is necessary to aware about the number of elements in contact and the element shape or element contact area. The main disadvantage of free meshing is to control the number and shape of elements.

11.4.4 GRID INDEPENDENCY TEST

To save computational time, the grid independency test has been performed (Table 11.1). We can use more and more fine mesh in the model which results increase in computational time. This test has been performed to determine which element

TABLE 11.1
Results of grid independency test

Element Division	No. of Elements in Contact	Contact Area (μm)	% Change in Contact Area
1	225	0.08	NA
2	900	0.68	88.2
3	2025	0.93	26.8
4	3600	1.08	13.8
5	5625	1.18	8.4
6	8100	1.22	3.2

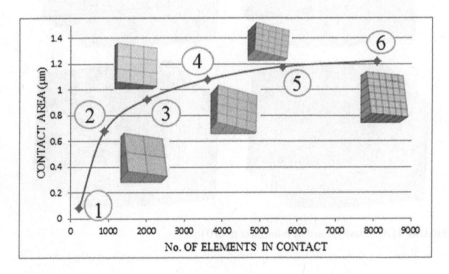

FIGURE 11.10 Contact area vs. number of elements.

shape the results do not have significant change. When performing the test, the model is meshed by varying element divisions from 1 to 6 and performs analysis to get the results in the form of a real contact area. The real contact area first changes a large amount with the increase in number of elements and becomes almost constant at 6 number of element divisions as shown in Figure 11.10.

As the number of element divisions increases from 1 to 6, the total number of elements as well as the number of elements in contact also increases. The real area of contact first changes and become constant at 6 number of element divisions as shown in the table. The change in real contact area remains less than 5% at 6 number of element divisions, so for this model to evaluate real contact area we freeze 6 number of element divisions to get FE mesh.

11.5 ANALYSIS

Analysis consists of modeling of real contact area and TCC based on real contact area. Real contact area has been evaluated using ANSYS and a MATLAB program and based on real contact area, TCC has been evaluated using available co-relation (33).

11.5.1 EVALUATING REAL CONTACT AREA

After modeling and meshing, the analysis was performed to estimate real contact areas under varying loading conditions. Mapped meshing has been found suitable for estimating each element's area at the interface to get the real contact area. In this work, real contact area has been evaluated by using ANSYS and MATLAB. A small program in MATLAB has been used to evaluate the real contact area. An element table has been created in ANSYS to evaluate real contact areas. The element table contains each and every element in contact with their corresponding value of contact gap. It is assumed that the element having zero gaps is in perfect contact with the target element. The data in the element table has been saved and imported into MATLAB and fined the elements having zero contact gap by using a program. MATLAB counts the no. of elements having zero contact gap and multiply to the area of one element to get the real contact area. The area of one element can be calculated by using the formula:

$$Area\ of\ one\ element = \left(\frac{Asperity\ size}{No.of\ element\ division}\right)^2$$

Real Contact Area = Area of one element × No. of elements having zero contact gap

11.5.2 ESTIMATING THERMAL CONTACT CONDUCTANCE

The governing equation

$$\frac{\partial^2 T}{\partial x^2} + \frac{1}{r}\frac{\partial T}{\partial r} + \frac{\partial^2 T}{\partial r^2} = 0 \tag{11.2}$$

The appropriate boundary conditions are

$$k_1 \frac{\partial T}{\partial z} = \frac{Q}{2\pi b\left(b^2 - r^2\right)^{1/2}}, z - 0, 0 < r < b$$

$$k_1 \frac{\partial T}{\partial z} = 0, z = 0, b < r < c$$

$$k_1 \frac{\partial T}{\partial z} = \frac{Q}{\pi c^2}, z = \infty$$

$$k_1 \frac{\partial T}{\partial r} = 0, r = c \tag{11.3}$$

Using the method of variable separation, we can write the temperature distribution at $z = 0$ as

$$T_0 = C_0 + \frac{8}{\pi} \frac{Q}{4k_1 b} \frac{c}{b} \sum_{n=1}^{\infty} \frac{\sin(\alpha_n b) J_1(\alpha_n b)}{(\alpha_n c)^3 J_o^2(\alpha_n c)} = C_0 + \frac{Q}{4k_1 b} \Psi\left(\frac{b}{c}\right)$$

$$\Psi\left(\frac{b}{c}\right) = \frac{8}{\pi}\left(\frac{c}{b}\right) \sum_{n=1}^{\infty} \frac{\sin(a_n b) J_1(a_n b)}{(a_n c)^3 J_o^2(a_n c)} \tag{11.4}$$

Where Ψ is the constriction parameter of contact area heat transfer.

When the two contact surfaces are of different heat transfer coefficients, the temperature difference can be given by

$$\Delta T_c = \frac{Q}{4k_s b} \Psi\left(\frac{b}{c}\right) \tag{11.5}$$

Where k_s is the mean thermal conductivity, b is the mean radius, b is the mean radius of the real contact area, and c is the nominal contact area.

The CMY Model showed that the relation between the contact area and the heat transfer coefficient can be written as

$$h_c = \frac{2k_s N b_c}{A_n} \sum_{i=1}^{N} \frac{1}{\Psi(\varepsilon)} = 2k_s n b_c \sum_{i=1}^{N} \frac{1}{\Psi(\varepsilon)} \tag{11.6}$$

Mikic proposed the constriction parameter of contact area heat transfer as

$$\Psi(\varepsilon) = (1-\varepsilon)^{1.5} = \left(1 - \sqrt{\frac{A_t}{A_n}}\right)^{1.5} \tag{11.7}$$

From Equations (11.6) and (11.7), the TCC can be expressed as

$$h_c = \frac{2k_s n b_c}{\left(1 - \sqrt{\frac{A_t}{A_n}}\right)^{1.5}} \tag{11.8}$$

where n is the density of asperities and b_c is the mean radius of spherical asperity.

11.6 RESULTS AND DISCUSSION

After meshing and creation of contact pairs, the analysis can be started with a properly constrained model. When performing analysis on the FE model, all of the key points on the bottom of the block were constrained in all three directions, i.e., x, y, and z and all the key points at the above side of the block were constrained in x and y direction to prevent the rigid body displacement and rotation, leaving the surface free to deform. A varying load is applied on all the key points above the block to move the contact surface toward the target surface and analysis starts. Results have been produced with varying loading conditions and varying roughness.

11.6.1 CONTACT PRESSURE PLOTS

Figure 11.11 shows the contact pressure plots of aluminum and mild steel models of roughness 10 μm with different loading condition. The real contact area in case of aluminum at 5N is about 95% and the real contact area in case of mild steel at 5N is about 80% at roughness 10 μm but as the roughness increases the percentage of real contact area decreases. Pressure plots for aluminum and mild steel having the roughness of 40 μm with different loading conditions has been compared in the Figure 11.12.

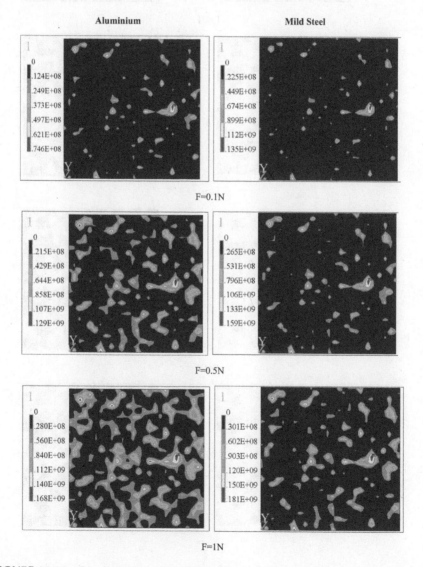

FIGURE 11.11 Comparison of pressure plots between aluminum and mild steel models of roughness 10 μm with different loading conditions. *(Continued)*

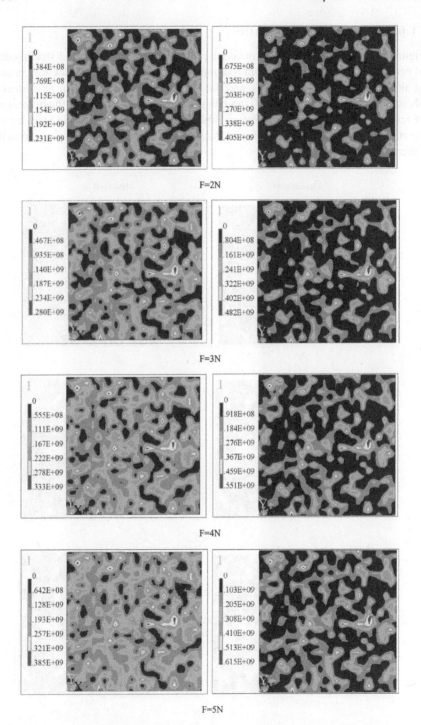

F=2N

F=3N

F=4N

F=5N

FIGURE 11.11 (CONTINUED) Comparison of pressure plots between aluminum and mild steel models of roughness 10 μm with different loading conditions.

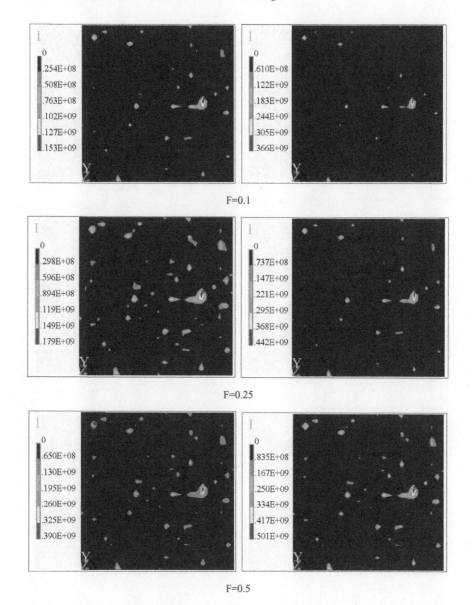

FIGURE 11.12 Comparison of pressure plots between aluminum and mild steel models of roughness 40 µm with different loading condition. (*Continued*)

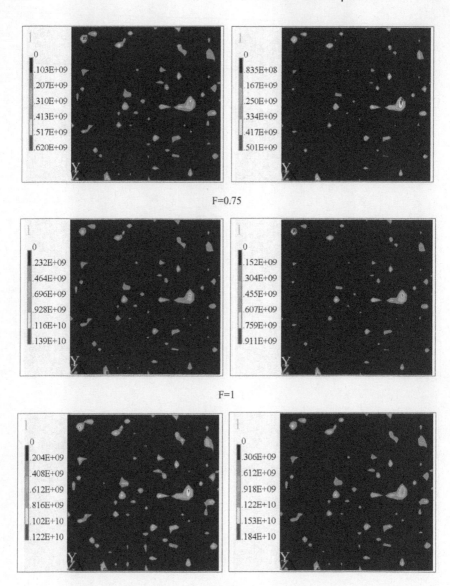

FIGURE 11.12 (CONTINUED)

In the contact pressure plots it is clear that elements having zero pressure are not in contact and the elements having some value of pressure are in contact hence the pressure plots also shows the area of contact. It is clear that for the same loading condition the area in contact for mild steel is less than as compared to area in contact for aluminum for both the roughness of 10 μm and 40 μm. The maximum contact pressure is located at the same place for both the materials because both materials have same roughness but the value of maximum contact pressure in case of mild steel is greater than as compared to aluminum. The percentage of real contact area in case of aluminum at 2N is about 21% and the percentage of real contact area in case of mild steel at 2N is about 19%.

11.6.2 Roughness Effect

11.6.2.1 Mild Steel Model

Figure 11.13 shows the effect of surface roughness on real contact area at different loading condition. The analysis has been performed on the rough surface model of the surface roughness of 10 μm, 25 μm, and 40 μm at 0.5 N, 1 N, 2 N, 3 N, 4 N, and 5 N to see the effect of roughness at different loading condition on real contact area. As the surface roughness increases the real contact area decreases for the same load due the increase in gap between the rough surfaces.

Figure 11.14 shows the relation between the surface roughness and percentage of real contact area at different loading condition. The analysis has been performed on the rough surface model of the surface roughness of 10 μm, 25 μm, and 40 μm at 0.5 N, 1 N, 2 N, 3 N, 4 N, and 5 N to see the effect of roughness at different loading condition on the percentage of real contact area. With the increase in surface roughness the real contact area decreases because of the increase in gap between the contacting surfaces and hence the percentage of real contact area decreases.

11.6.2.2 Aluminum Model

Figure 11.15 shows the effect of surface roughness on real contact area at different loading condition. The analysis has been performed on the rough surface model of the surface roughness of 10 μm, 25 μm, 40 μm, and 80 μm at 0.1 N, 0.25 N, 0.5 N, 0.75 N, 1 N, and 2N to see the effect of roughness at different loading condition on real contact area. As the roughness increases, real contact area decreases for all of the loading conditions.

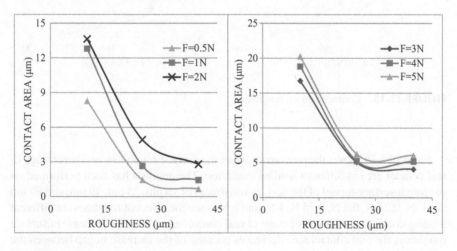

FIGURE 11.13 Contact area vs. roughness.

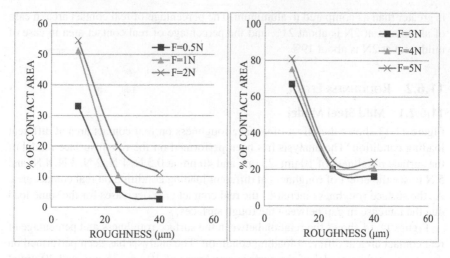

FIGURE 11.14   Percentage of contact area vs. roughness.

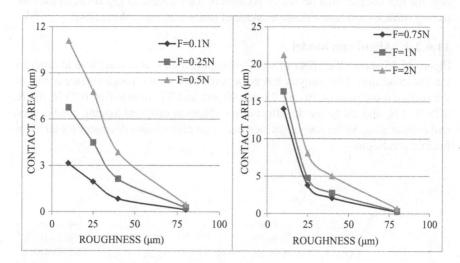

FIGURE 11.15   Contact area vs. roughness.

Figure 11.16 shows the relation between the surface roughness and percentage of real contact area at different loading condition. The analysis has been performed on the rough surface model of the surface roughness of 10 µm, 25 µm, 40 µm, and 80 µm at 01. N, 0.25 N, 0.5 N, 0.75 N, 1 N, and 2 N to see the effect of roughness at different loading condition on the percentage of real contact area. With the increase in surface roughness the real contact area decreases because of the increase in gap between the contacting surfaces and hence the percentage of real contact area decreases.

11.6.3 Loading Effect

In this work the solid model of rough surface is converted into FE model by meshing the model and analysis has been performed with varying roughness and varying loading condition with different materials. Results for various loading condition and different roughness has been saved and compared. The first comparison has been made to quantify the loading effect on different roughness for aluminum and mild steel.

11.6.3.1 Mild Steel Model

Figure 11.17 shows the effect of nominal pressure on contact pressure for the model having the surface roughness of 10 µm, 25 µm, and 40 µm. At fine surfaces, mean contact pressure increases with the load. At coarse surfaces (≥40 µm), no significant effect of load up to 200 kPa on contact pressure.

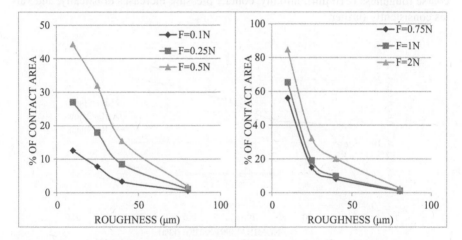

FIGURE 11.16 Percentage of contact area vs. roughness.

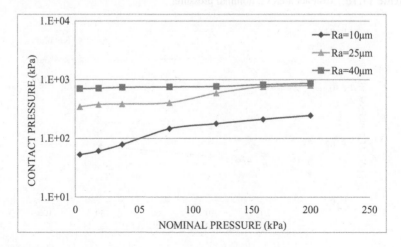

FIGURE 11.17 Contact pressure vs. nominal pressure.

Figure 11.18 shows the effect of nominal pressure on real contact area on the models having the surface roughness of 10 μm, 25 μm. and 40 μm. As the nominal pressure increases, the real area of contact increases for all of the roughness values. The real contact area of the model of surface roughness 10 μm has sharp increase up to the pressure of 40 kPa while the real contact area of the model having the roughness of 25 μm and 40 μm increases monotonically. As the roughness increases, the real contact area decreases and do not have much effect on small increase in pressure.

11.6.3.2 Aluminum Model

Figure 11.19 shows the effect of nominal pressure on contact pressure for different roughness values of 10 μm, 25 μm, 40 μm, and 80 μm. As the nominal pressure increases, the contact pressure also increases for all of the roughness values. At fine roughness (10 μm), increase in contact pressure varied consistently with load. At coarse roughness (≥10 μm), initially, contact pressure increases drastically, and varies consistently further.

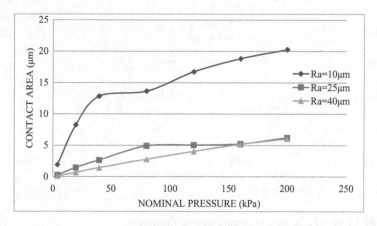

FIGURE 11.18   Contact area vs. nominal pressure.

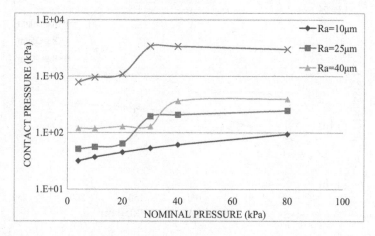

FIGURE 11.19   Contact pressure vs. nominal pressure.

Figure 11.20 shows the effect of nominal pressure on real contact area with the models having the surface roughness of 10 μm, 25 μm, 40 μm, and 80 μm. For fine surfaces (10 μm), real contact area varies with pressure. For course surfaces (10 μm ≤ 80 μm), contact area typically decreases after initial increase and then varies monotonically with pressure. The decrease in real area of contact is because of a change of elastic deformation to plastic deformation.

11.6.4 MATERIAL EFFECT

Figure 11.21 shows the effect of hardness on real contact area with the models having the surface roughness of 10 μm at different loading condition. Real contact area decreases with the increase in hardness at different loading conditions. At

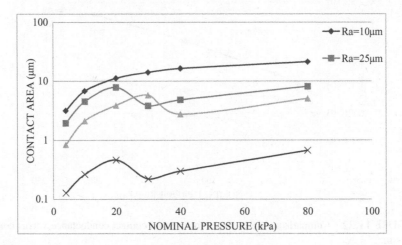

FIGURE 11.20   Contact area vs. nominal pressure.

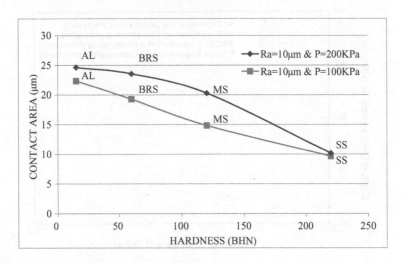

FIGURE 11.21   Contact area vs. hardness.

higher hardness (stainless steel), the load from 100 kPa to 200 kPa does not have significant change in real contact area.

11.6.5 THERMAL CONTACT CONDUCTANCE

Figure 11.22 and Figure 11.23 shows the variation of TCC with pressure for different materials. Figure 11.25 is the results from the present wok for three materials: aluminum, mild steel, and stainless steel at increasing pressure. Figure 11.23 is the result taken from Jeng et al. [33] for aluminum and stainless steel at both increasing and decreasing pressure. For the same pressure, TCC in the process of decreasing

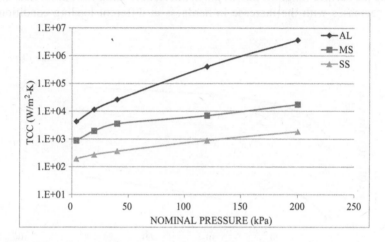

FIGURE 11.22   Comparison of the curves for thermal contact conductance versus contact pressure.

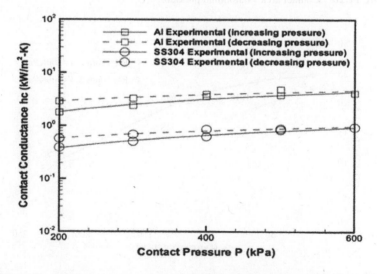

FIGURE 11.23   TCC vs. contact pressure for aluminum and stainless steel.

pressure is larger than that in the process of increasing pressure due to plastic deformation. The trend of the curves of TCC with pressure is same for the models of aluminum and stainless steel for both the study. It is clear that TCC increases of all the materials with the increase in pressure.

Figure 11.24 shows the variation of TCC with roughness for different loading condition with specimen made of aluminum. Increasing load resulted in terms of increase in TCC. At lower load, an increase in roughness resulted in increase in TCC. At load (≥1 N), increasing roughness resulted decrease in TCC.

Figure 11.25 shows the effect of hardness on TCC with the models having the surface roughness of 10μm at different loading condition. Increasing hardness decreases TCC for both of the loading values. At higher hardness (stainless steel), the effect of loading (<200 kPa) gets diminished.

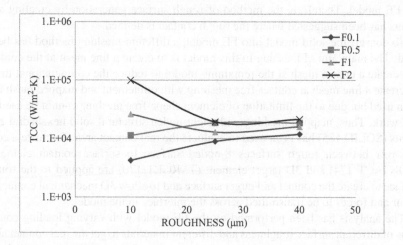

FIGURE 11.24   TCC vs. roughness.

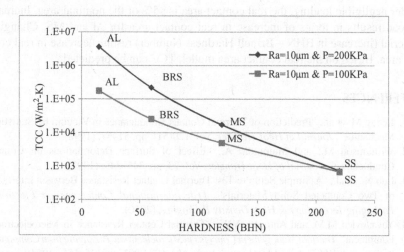

FIGURE 11.25   TCC vs. hardness.

11.7 CONCLUSION

Most of the users have chosen to model the gross geometry of the system and ignore the details of surface effect to simplify the model and reduce computation times. Surface effects are generally ignored in the analysis, partly because of the difficulty in creating a reasonable model of the surface and partly because of the assumption that surface effects are less important. However, surface effects caused by surface topography are of major concern in most of the engineering applications and may become critical when contact is involved.

In this work two method of rough surface generation has been adopted and a contact has been created between two models of rough surface. In the first method (by moving all nodes) of rough surface generation, the FE model is detached from the solid model and thus it is difficult to transfer the solid model boundary conditions to the FE model. Therefore, the method of rough surface generation by creating key points has been suggested where the rough contact is defined.

To convert the solid model into FE model, a different meshing method has been used. The main aim of meshing in this model is to create a fine mesh at the contact and create a coarse mesh at the remaining model to reduce the computational time. To create a fine mesh at contact-free meshing with refinement and mapped mesh has been used but due to the limitation of element shape-free meshing cannot be used in this work. Thus, mapped meshing with eight-noded structural solid hexahedral elements (SOLID 185) has been used to evaluate the real contact area. To create a contact pair between rough surfaces 8-noded surface to surface contact elements (CONTACT 174) and 3D target element (TARGET 170) are applied to the rough surface to create the contact and target surface and to allow 3D mechanical contact to occur and forces to be transmitted across the interface in the model.

The analysis has been performed on the FE model with varying loading conditions of different surface roughness and different materials to get the real contact area and thus TCC. Effect of varying parameters on real contact area has been obtained. Increase in roughness (10–40 μm) increases the real contact area for Al and MS. Under negligible loading, the real contact area is ≤5% of the nominal area. Increase in load results in terms of increase in real contact area for Al and MS. Changing material (increase in BHN – Brinell Hardness Number) results decrease in real contact area. Based on the real contact area model, TCC can be predicted.

REFERENCES

1. Leung M. et al., "Prediction of Thermal Contact Conductance in Vacuum by Statistical Mechanics", *Journal of Heat Transfer*, Vol. 120(51), pp. 51–57, (1998).
2. Williamson M., and Majumdar A., "Effect of Surface Deformations on Contact Conductance", *Journal of Heat Transfer*, Vol. 23, pp. 345–349, (1992).
3. Luo X., et al., "A Simple Setup to Test Thermal Contact Resistance Between Interfaces of Two Contacted Solid Materials", *11th International Conference on Electronic Packaging Technology & High Density Packaging*, (2010).
4. Yovanovich M.M. and Antonetti V.W., "Thermal Contact Resistance in Microelectronic Equipment", *Thermal Management Concepts in Microelectronic Packaging from Component to System, ISHM Technical Monograph Series* Vol. 6984-003, pp. 135–151, (1984).

5. Singhal V., et al., "An Experimentally Validated Thermo-Mechanical Model for Prediction of Thermal Contact Conductance", *International Journal of Heat and Mass Transfer*, Vol. 48, pp. 5446–5459, (2005).

6. Mochizuki H., and Quaiyum M.A., "Contact Conductance between Cladding/Pressure Tube and Pressure Tube/Calandria Tube of Advanced Thermal Reactor", *Journal of Nuclear Science and Technology*, 31, pp. 726–734, (1994).

7. Peterson G.P., and Fletcher L.S., "Measurement of Thermal Contact Conductance and Thermal Conductivity of Anodized Aluminum Coatings", *Journal of Heat Transfer*, 112/585, pp. 579–585, (1990).

8. Marotta E., et al., "Thermal Contact Conductance of Sintered Copper Coatings on Ferro Alloy", *ASME Journal of Heat Transfer*, Vol. 121, pp. 177–182, (1999).

9. Jeng Y., et al., Thermal Contact Conductance of Coated Surfaces, *Wear*, Vol. 260, pp. 159–167, (2006).

10. Rosochowska M., et al., "Measurement of Thermal Contact Conductance", *Journal of Materials Processing Technology*, Vol. 135, pp. 204–210, (2003).

11. Mantelli M., and Yovanovich M.M., *"Thermal Contact Resistance – Theory and Applications"*, Lecture Notes, University of Waterloo (1980).

12. Cooper M.G., et al., "Thermal Contact Conductance", *International Journal Heat Mass Transfer*, Vol. 12, pp. 279–300, (1969).

13. Mikic B., "Thermal Constriction Resistance Due to Non-Uniform Surface Conditions; Contact Resistance at Non-Uniform Intjzrface Pressure", *International Journal Heat Mass Transfer*, Vol. 13, pp. 1497–1500, (1970).

14. Mikic B., "Thermal Contact Conductance; Theoretical Considerations", *International Journal Heat Mass Transfer*, Vol. 17, pp. 205–214, (1974).

15. Sayles R.S., and Thomas T.R., "Thermal Conductance of a Rough Elastic Contact", *Applied Energy* Vol. 2, pp. 249–267, (1976).

16. Gibson R.D., "The Contact Resistance for a Semi- Infinite Cylinder in a Vacuum", *Applied Energy*, Vol. 2, pp. 57–65, (1976).

17. Antonetti V.W., and Yovanovich M.M., "Enhancement of Thermal Contact Conductance by Metallic Coatings: Theory and Experiment", *Journal of Heat Transfer*, Vol. 107, pp. 513–519, (1985).

18. McGee G.R., et al., "Thermal Resistance of Cylinder Flat Contacts: Theoretical Analysis and Experimental Verification of a Line Contact Model", *Nuclear Engineering and Design*, vol. 86, pp. 369–385, (1985).

19. Huang C.H., et al., "Conjugate Gradient Method for Determining Unknown Contact Conductance During Metal Casting", *International Journal Heat Mass Transfer*, Vol. 35, pp. 1779–1786, (1992).

20. Tsai Y.M., and Crane R.A., "An Analytical Solution of a One- Dimensional Thermal Contact Conductance Problem with One Heat Flux and One Insulated Boundary Condition", *Journal of Heat Transfer*, Vol. 114, pp. 503–505, (1992).

21. Antonetti V.W., et al., "An Approximate Thermal Contact Conductance Correlation", *Journal of Electronic Packaging*, Vol. 115, pp. 131–134, (1993).

22. Madhusudana C.V., "Thermal Contact Conductance and Rectification at Low Joint Pressure", *International Communications in Heat Mass Transfer*, Vol. 20, pp. 123–132, (1993).

23. Nishino K, et al., "Thermal Contact Conductance Under Low Applied Load in a Vacuum Environment", *Experimental Thermal and Fluid Science*, Vol. 10, pp. 258–271, (1995).

24. Sridhar M.R., and Yovanovich M.M., "Thermal Contact Conductance of Tool Steel and Comparision with Model", *International Journal of Heat and Mass Transfer*, Vol. 39, pp. 831–839, (1996).

25. Sridhar M.R., and Yovanovich M.M., "Elastoplastic Contact Conductance model for Isotropic Conforming Rough Surfaces and Comparison with Experiments", *Journal of Heat Transfer*, Vol. 118/1, pp. 3–9 (1996).

26. Ohsone Y., et al., "Optical Measurement Of Thermal Contact Conductance Between Wafer-like Thin Solid Samples", *Journal of Heat Transfer*, Vol. 121, pp. 954–963, (1999).

27. Tseng A.A., "Thermal Modeling of Roll and Strip Interface in Rolling Processes: Part 2D Simulation", *Numerical Heat Transfer, Part A*, Vol. 35, pp. 135–154, (1999).

28. Zhao Y., et al., "An Asperity Microcontact Model Incorporating the Transition from Elastic Deformation to Fully Plastic Flow", *ASME Journal of Heat Transfer*, Vol. 122, (2000).

29. Benigni P., et al., "High Temperature Contact Conductance Measurement across Cylindrical Joint by Periodic Method: Feasibility and Results", *International Journal of Heat and Mass Transfer*, Vol. 43, pp. 4217–4227, (2000).

30. Lee S., et al., "Constriction/Spreading Resistance Model For Electronics Packaging", *ASME Thermal Engineering Conference*, Vol. 4, (1995).

31. Huang C.H., et al., "A Non Liner Inverse Problem for the Prediction of Local Thermal Contact Conductance in Plate Finned Tube Heat Exchanger", *Heat and Mass Transfer*, Vol. 37, pp. 351–359, (2001).

32. Kumar S.S., and Ramamurthi K., "Influence of Flatness and Waviness of Rough Surfaces on Surface Contact Conductance", *Journal of Heat Transfer*, Vol. 125, pp. 394–402 (2003).

33. Jeng Y., et al., "Theoretical and Experimental Study of a Thermal Contact Conductance Model for Elastic, Elastoplastic and Plastic Deformation of Rough Surfaces", *Tribology Letters*, Vol. 14, pp. 251–259, (2003).

34. Kim C.N., et al., "Evaluation of Thermal Contact Conductance Using a New Experimental-Numerical Method in Fin-Tube Heat Exchangers", *International Journal of Refrigeration*, Vol. 26, pp. 900–908, (2003).

35. Kumar S.S., et al., "Thermal Contact Conductance for Cylindrical and Spherical Contacts", *Journal of Heat and Mass Transfer*, Vol. 40, pp. 679–688, (2004).

36. Kumar S.S., and Ramamurthi K., "Thermal Contact Conductance of Pressed Contacts at Low Temperatures", *Cryogenics*, Vol. 44, pp. 727–734, (2004).

37. Zhao J.F., et al., "Prediction of Thermal Contact Conductance based on the Statistics of the Roughness Profile Characteristics", *International Journal of Heat and Mass Transfer*, Vol. 48, pp. 974–985, (2005).

38. Jeong J., et al., "A Study on the Thermal Contact Conductance in Fin–Tube Heat Exchangers With 7 mm Tube", *International Journal of Heat and Mass Transfer*, Vol. 49, pp. 1547–1555, (2006).

39. Zou M., et al., "Fractal Model for Thermal Contact Conductance", *Journal of Heat Transfer*, Vol. 130, pp. 101301–101309 (2008).

40. Shojaeefard M.H., et al., "Inverse Heat Transfer Problem of Thermal Contact Conductance Estimation in Periodically Contacting Surfaces", *Journal of Thermal Science*, Vol. 18, pp. 150–159, (2009).

41. Thompson M.K., and Thompson J.M., "Considerations for Predicting Thermal Contact Resistance in Ansys", *Department of Civil and Environmental Engineering, KAIST*, Daejeon, Korea.

42. Poroshin P., et al., "Transfer of the Surface Roughness Geometry into the Universal Fem Software Ansys", *Advanced Engineering*, Vol. 32, pp. 1846–5900, (2009).

43. Peyrou D., et al., "Effect of Contact Force between Rough Surfaces on Real Contact Area and Electrical Contact Resistance", *MEMSWAVE Workshop 2007*, Barcelona, Spain (2007).

44. Thompson M.K., and Thompson J.M.," *Methods for Generating Probabilistic Rough Surfaces in Ansys*", Lecture Notes, *Department of Civil and Environmental Enginering, KAIST*, (2009).

45. Thompson M.K., "Methods for Generating Rough Surfaces in Ansys'" Lecture Notes, Mechanical Engineering Dept, MIT, (2009).

46. Thompson M.K., et al., "The Effect of Surface Roughness on the Pressure Required for Coupler Sealing." *International ANSYS Users Conference & Exhibition*, (2004).

47. Hyun S., et al., "Finite Element Analysis of Contact Between Elastic Self-Affine Surfaces", *Physical Review. E, Statistical, Nonlinear, and Soft Matter Physics*, 70, p. 026117, (2004).

Section C

Materials Engineering

12 Viscoelastic Composites for Passive Damping of Structural Vibration

Satyajit Panda, Abhay Gupta, and
Rajidi Shashidhar Reddy
Department of Mechanical Engineering, Indian Institute of
Technology Guwahati, India

CONTENTS

12.1 INTRODUCTION

Viscoelastic materials are homogeneous and isotropic materials that exhibit an important property of energy-dissipation during time-dependent deformation. These materials are composed of long intertwined and cross-linked molecular chains where every chain contains thousands or millions of atoms. During the time-dependent deformation, the internal molecular interaction occurs, and it gives rise to the macroscopic material properties like stiffness and damping (Baz, 2019). However, the energy-dissipation property of viscoelastic materials is exploited enormously for attenuation of vibration of thin-walled structures in many engineering systems like railway wheels (Jones and Thompson, 2000), outlet guide van (Tomlinson 1990), computer hardware (Rao, 2003), machine tools (Marsh and Hale, 1998; Shi et al., 2017), spinning disks (Seubert et al., 2000), compression blades (Sun and Kari, 2010), blast protective military equipment, wind energy systems, radiofrequency antenna, biomedical devices (Herrmann et al., 2005; Birman and Kardomatea, 2018), door/floor panels, fuselage section covers of aircraft (Rao, 2003; Cunha-Filho et al.,

DOI: 10.1201/9781003202233-15

2016), etc. In these structural applications, the commonly used viscoelastic materials are styrene-butadiene rubber (SBR), Paracril-BJ, Polymer Blend, butyl rubber, Viton-B, LD-400, Soundcoat N5, 3 M-467, etc. (Jones, 2001). However, the available literature shows different passive damping arrangements in the use of these viscoelastic materials like tuned damper, edge damping, free/unconstrained layer damping (UCLD), constrained layer damping (CLD), etc. (Grootenhuis, 1970; Nakra, 2000). Among these various passive damping arrangements, the UCLD and CLD treatments are popular ones to attenuate the bending mode of vibration of thin-walled structures. The damping mechanism in these viscoelastic damping treatments is delineated in the next section, along with a literature review.

12.1.1 UNCONSTRAINED/CONSTRAINED LAYER DAMPING TREATMENT

In the free/UCLD treatment, the viscoelastic damping layer is freely attached to the surface of a substrate layer/structure (Figure 12.1.(a)). However, under the time-dependent bending deformation of this layered structure, the passive damping mainly arises through the extensional/compressional deformation of the viscoelastic layer (Figure 12.1b)). The UCLD treatment was proposed by Oberst and Frankenfeld (1952).

Later, this concept of passive damping is utilized for attenuation of vibration of different structures (Ungar and Kerwin Jr, 1964; Markus, 1976; Okazaki et al., 1994; Mead, 2007; Cortes and Elejabarrieta, 2008; Sun et al., 2018; Zarraga et al., 2019). Further, this passive damping treatment is used in the form of the partial UCLD treatment, particularly for achieving improved damping with its reduced mass/weight. The partial UCLD treatment is devised by locating the viscoelastic material patches optimally over the host structure surface instead of the use of the same damping material on the layer form (Lunden, 1979; Parthasarathy et al., 1985; Cheng and Lapointe, 1995; Roy and Ganesan, 1996; El-Sabbagh and Baz, 2014).

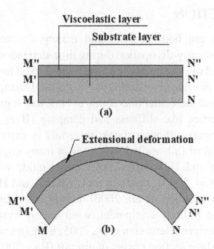

FIGURE 12.1 Schematic diagrams of a layered structure with the UCLD treatment; (a) undeformed and (b) deformed layered structures.

Besides, the stand-off layer has also been introduced between the viscoelastic layer and host structure surface to enhance the passive damping in the UCLD treatment (Liang et al., 2020). Further research on this topic addresses a 1–3 viscoelastic composite layer that can be used in place of the pure viscoelastic layer to achieve better passive damping in the UCLD treatment (Kumar and Panda, 2016). However, parallel to this UCLD treatment, another damping arrangement emerged as the CLD treatment.

surface for achieving improvedIn the CLD treatment, a viscoelastic layer is constrained by a stiff constraining layer against the host structure surface (Figure 12.2(a)). Here, the damping in the layered structure arises mainly due to the transverse shear deformation of the constrained viscoelastic layer (Figure 12.2(b)) during the time-dependent bending deformation of the layered structure. This arrangement of passive damping is also called as passive CLD (PCLD) treatment since the viscoelastic damping layer is constrained by a passive constraining layer.

The PCLD/CLD treatment was introduced by Swallow (Swallow, 1939), while Kerwin (Kerwin Jr, 1959) reported its mathematical formulation. Later, it is used to suppress the vibration of different structural elements (Mead and Markus, 1969; Lu et al., 1979; Hamdaoui et al., 2015; Zhai et al., 2019; Ren et al., 2020). However, the available research in this context addresses the PCLD/CLD treatment in different configurations, in particular, to augment its passive damping capability. Among these available PCLD/CLD configurations, the most significant ones are the multi-layered CLD and stand-off layered configurations (Torvik and Strickland, 1972; Alam and Asnani, 1984; Garrison et al., 1994; Masti and Sainsbury, 2005; Zheng et al., 2014). Similar to the UCLD treatment, the CLD treatment is also used in the form of the patch that is located optimally over the host structure surface for achieving improved

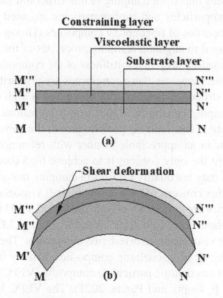

(a)

(b)

FIGURE 12.2 Schematic diagrams of a layered structure with CLD treatment; (a) undeformed and (b) deformed layered structure.

damping with less mass/weight of the damping arrangement (Marcelin et al., 1992; Zheng et al., 2005; Khalfi and Ross, 2016). In this line, different optimization techniques are also available in the literature for deciding the optimal size and locations of the CLD patches over the host structure surface (Madeira et al., 2015; Arikoglu, 2017; Wang et al., 2020; Fang et al., 2020). However, the damping capability of the UCLD treatment is much lesser than that of the CLD treatment (Grootenhuis, 1970), and thus CLD treatment is the most popular means of viscoelastic damping for passive vibration control of thin-walled flexible structures.

12.1.2 Viscoelastic Composites for UCLD/CLD Treatment

In most of the available studies on the UCLD/CLD treatment, monolithic viscoelastic materials are utilized. However, some available studies address viscoelastic composites in place of the monolithic viscoelastic material for having better damping capability of the UCLD/CLD treatment. In this line, the first one appeared as nanocomposite that is comprised of single-walled/multi-walled carbon nanotubes (CNTs) embedded in the epoxy matrix where the improved material damping appears due to the stick–slip mechanism (Rajoria and Jalili, 2005; Khan et al., 2011; DeValve and Pitchumani, 2014; de Borbon et al., 2014). The damping effectiveness of this kind of viscoelastic composite is demonstrated in some studies through the CLD arrangement, where it is revealed that the damping capability of the neat epoxy can be augmented by the inclusion of CNTs (Rajoria and Jalili, 2005; de Borbon et al., 2014). Another type of viscoelastic composite appeared by the inclusion of soft viscoelastic particles within the stiff elastic matrix (Attipou et al., 2013). The stiff elastic matrix is used for the material of load-carrying structural member, while viscoelastic particles' inclusion provides improved damping in that structural member.

Further, silica nanoparticles and rubber particles are used as the inclusions to improve damping properties of fiber/epoxy composites (Huang and Tsai, 2015). The rubber particles are used to augment damping properties of the fiber/epoxy composite, while the corresponding drop in the stiffness of the composite is compensated by the silica nanoparticles. However, these composites are primarily used as the materials for load-carrying structural members. So, they possess high stiffness that is a reason for their low damping capability, while various inclusions are used to improve their damping properties. The damping performance of the corresponding CLD treatment does not appear in an appreciable manner with reference to the conventional CLD treatment, where the only concern is to achieve high damping. Therefore, the aforesaid composites may not qualify for fruitful damping materials in the CLD treatment. However, for this concern of high damping, soft viscoelastic materials having high material loss factor are commonly used in the CLD/UCLD treatment. The damping performance of these soft viscoelastic materials in the CLD treatment is further enhanced by devising viscoelastic matrix-based composites. The available research in this line shows two kinds of viscoelastic composites, namely 0–3 viscoelastic composite (0–3 VEC) and viscoelastic particulate composite (VEPC) (Kumar et al., 2018; Gupta et al., 2020a, b; Gupta and Panda, 2021). The VEPC is comprised of solid graphite particles embedded in a soft viscoelastic matrix. The inclusion of graphite particles has not much contribution to the material loss factor of the particulate

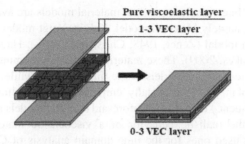

FIGURE 12.3 Schematic diagram of the 0–3 VEC along with its component layers.

composite; however, the stiffness of the VEPC increases significantly due to the inclusion resulting in augmented stored energy as well as energy-dissipation in the CLD treatment (Gupta et al., 2020a, b; Gupta and Panda, 2021). In another type, i.e., 0–3 VEC, rectangular graphite wafers are used to compose a 1–3 VEC layer with a soft viscoelastic matrix having high material loss factor. This 1–3 VEC is further sandwiched between two identical monolithic viscoelastic layers to form the 0–3 VEC layer (Figure 12.3). Here, the graphite wafers provide additional constraint to the deformation of the viscoelastic phase in the 0–3 VEC layer so that the transverse shear deformation of the viscoelastic phase enhances during bending deformation of the composite layer resulting in significantly improved damping capability of CLD treatment (Kumar et al., 2018). However, the aforesaid available studies reveal that both the 0–3 VEC and VEPC damping materials perform well in the CLD treatment compared to the use of the conventional monolithic viscoelastic materials. But the selection of one of these damping composites is an important concern for achieving effective damping performance of the CLD treatment in an application. A corresponding comparison study is presented in the following sections through the finite element (FE) modeling of the CLD treatment comprising 0–3 VEC or VEPC damping layer.

12.2   MATHEMATICAL MODELING OF CLD TREATMENT

The arrangement of CLD treatment commonly appears in the form of a layered structure (Figure 12.2). For mathematical modeling of this layered structure, several analytical and numerical approaches are now available in the literature (Huang et al., 2001; Hu et al., 2008; Ferreira et al., 2013). Among these different modeling approaches, the appropriate one seems to appear by the use of layer-wise theory because of the significant difference in the stiffness between the viscoelastic layer and substrate/constraining layer. In the layer-wise theory, the definition of the kinematic of deformation is confined to the domain of individual layer, and it facilitates to account the structural behavior of all layers separately while the layers are made of materials with different stiffness properties (Reddy, 2003). The implementation of this layer-wise theory in the modeling of the CLD treatment having viscoelastic composite damping layer is demonstrated in the next section.

However, another important aspect in the modeling of the CLD treatment is to select a fruitful viscoelastic material model for the damping layer. As it is now concerned with the linear analysis, several material models are available for the viscoelastic materials, namely Maxwell model, Kelvin–Voigt model, Zener model and Poynting–Thomson model (Zener, 1948; Christensen, 2012; Haddad, 1995; Lakes, 1998; Lakes and Lakes, 2009). These material models are commonly known as the classical viscoelastic material models in the time domain, while the complex stiffness method (Scanlan, 1970) is usually employed for modeling the viscoelastic materials in the frequency domain. The aforesaid material models in the time domain may not provide the realistic behavior of a viscoelastic material (Baz, 2019). However, the most used ones for the time-domain analysis of CLD treatment are Anelastic Displacement Field (ADF) model (Lesieutre and Bianchini,1995) and Golla-Hughes-McTavish (GHM) model (Golla and Hughes, 1985; McTavish and Hughes, 1993). These viscoelastic material models work well in the modeling of the viscoelastic layer of CLD treatment; however, a disadvantage in these material models is the appearance of additional coordinates resulting in an increase in the system degrees of freedom. So, the computational cost increases significantly, particularly for the computation in the FE framework.

In contrast, there is no requirement of additional degrees of freedom for another type known as fractional order derivative (FOD) viscoelastic model (Bagley and Torvik, 1983; Bagley and Torvik, 1985). The FOD viscoelastic model is capable of capturing realistic behavior of a viscoelastic material (Pritz, 1996) while it provides other advantage of low computational cost in the FE framework (Schmidt and Gaul, 2001; Galucio et al., 2004; Datta and Ray, 2015; Sahoo and Ray, 2019). However, a mathematical derivation for FE modeling of the CLD treatment using this FOD viscoelastic material model and the layer-wise theory is demonstrated in the following section.

12.3 FINITE ELEMENT FORMULATION

The damping effectiveness of the CLD treatment using 0–3 VEC or VEPC damping layer is demonstrated here for the vibration control of a plate element. Figure 12.4 shows the schematic diagram of this plate element integrated with a CLD layer at its top surface, where the damping layer is made of either 0–3 VEC or VEPC. The VEPC is usually characterized as an isotropic material, having effective material properties on a macroscopic scale (Gupta et al., 2020b). In contrast, the 0–3 VEC

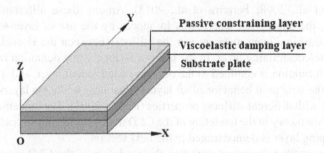

FIGURE 12.4   Schematic diagram of the plate integrated with a CLD layer.

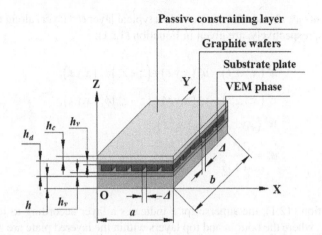

FIGURE 12.5 Schematic diagram of the plate with passively constrained 0–3 VEC layer.

cannot be taken with its effective properties since the dimensions of graphite wafers are on a macro scale. Therefore, the 0–3 VEC would be modeled explicitly where its phases' material properties are to be accounted directly in the mathematical model. This consideration of every phase's material properties in the 0–3 VEC poses complexity in the mathematical modeling of the corresponding CLD treatment, and it would be somewhat manageable in the FE framework. So, the CLD treatment of the plate is modeled here in the FE framework. However, the FE modeling of the plate with CLD treatment is presented in this section considering the 0–3 VEC damping layer, since the material configuration of the corresponding plate (Figure 12.5) appears in a complex manner. This FE model can also be used for the use of the VEPC layer instead of the 0–3 VEC layer, as it is described at the end of this section.

(Figure 12.5) shows the material configuration of a substrate plate with the constrained 0–3 VEC layer within the Cartesian coordinate system (xyz) having the origin (O) at one corner of the overall plate. The geometrical dimensions of the substrate plate, constraining layer, and 0–3 VEC damping layer are indicated in the figure (Figure 12.5). However, the gap (Δ) between two consecutive rectangular graphite wafers in the 0–3 VEC layer is considered to be equal in the x and y directions. Also, the number (n_g) of wafers along the x direction is considered to be equal to that in the y direction since the overall plate is taken with the square shape $a = b$.

The thicknesses of the 0–3 VEC damping layer and its monolithic viscoelastic layers (Figure 12.5) are denoted by h_d and h_v, respectively, while the thickness of the 1–3 VEC layer within the 0–3 VEC layer (Figure 12.4) is indicated by the thickness ratio as, $r_v = h_v/h_d$. The layer-wise theory is utilized here to define the kinematics of deformation of the layers where the displacement field in every layer is defined according to the Taylor series expansion with respect to the thickness (z) coordinate. Further, thin layers are considered within the overall layered plate, and thus the Taylor series expansion is taken here up to the first-order derivative term as it appears in the first-order shear deformation theory (FSDT) (Reddy, 2003). Accordingly, the displacements u^k, v^k, and w^k at any point in a typical layer (k^{th} layer) along the x, y, and z directions, respectively, are given in Equation (12.1):

$$u^k\left(x,y,z,t\right) = u_0^k\left(x,y,t\right) + \left(z - z_o^k\right)\phi_x^k\left(x,y,t\right),$$

$$v^k\left(x,y,z,t\right) = v_0^k\left(x,y,t\right) + \left(z - z_o^k\right)\phi_y^k\left(x,y,t\right),$$

$$w^k\left(x,y,z,t\right) = w_0\left(x,y,t\right), \tag{12.1}$$

$$\phi_x^k = \left.\frac{\partial u^k}{\partial z}\right|_{z_o^k}, \quad \phi_y^k = \left.\frac{\partial v^k}{\partial z}\right|_{z_o^k}$$

In Equation (12.1), the superscript k indicates a layer according to its values as $1,2,3,\ldots,N_L$, where the bottom and top layers within the layered plate are specified by the values of k as 1 and N_L, respectively; z_o^k is the thickness coordinate at the bottom surface of the kth layer; u_o^k and v_o^k are the displacements along the x and y directions, respectively at any point over the bottom surface ($z = z_o^k$) of the kth layer. Usually, the thickness of a thin plate remains almost the same under the bending deformation (Reddy, 2003). So, the deformation of the layered plate along the vertical direction is assumed as constant over its thickness, and this vertical deformation is represented by the symbol w_o. Now, implementing the continuity of displacements (Equation (12.2)) at the interlayer surfaces, the in-plane displacements (u^k, v^k) at a point in the kth layer can be expressed in terms of the generalized displacements ($u_o, v_o, \phi_x^k, \phi_y^k$, $k = 1, 2,\ldots, N_L$) as given in Equation (12.3).

$$u_o^{k+1} = u^k\Big|_{z=z_o^{k+1}}, \quad v_o^{k+1} = v^k\Big|_{z=z_o^{k+1}} \tag{12.2}$$

$$u^k\left(x,y,z,t\right) = u_0\left(x,y,t\right) + z^k\phi_x\left(x,y,t\right),$$

$$v^k\left(x,y,z,t\right) = v_0\left(x,y,t\right) + z^k\phi_y\left(x,y,t\right),$$

$$\phi_x = \left\{\phi_x^1 \quad \phi_x^2 \quad \ldots \ldots \quad \phi_x^{(N_L-1)} \quad \phi_x^{N_L}\right\}^{\mathrm{T}}, \quad \phi_y = \left\{\phi_y^1 \quad \phi_y^2 \quad \ldots \quad \ldots \quad \phi_y^{(N_L-1)} \quad \phi_y^{N_L}\right\}^{\mathrm{T}}$$

$$z^k = \left\{z_1^k \quad z_2^k \quad \ldots \quad \ldots \quad z_{(N_L-1)}^k \quad z_{N_L}^k\right\} \tag{12.3}$$

In Equation (12.3), u_o and v_o are the in-plane displacements at any point on the bottom surface ($z = 0$) of the overall plate along the x and y directions, respectively. The other thickness coordinates for different layers in the layered plate are given in Equation (12.4), where h_k represents the thickness of the kth layer:

$$z_1^k = z\left(\text{for } k = 1\right) \text{ or } h_1\left(\text{for } k > 1\right)$$

$$z_2^k = 0\left(\text{for } k < 2\right) \text{ or } \left(z - h_1\right)\left(\text{for } k = 2\right) \text{ or } h_2\left(\text{for } k > 2\right)$$

$$z_{(N_L-1)}^k = 0 \left(\text{for } k < (N_L - 1)\right) \text{ or } \left(z - h_1 - h_2 - \dots - h_{(N_L-2)}\right)$$
$$\left(\text{for } k = (N_L - 1)\right) \text{ or } h_{(N_L-1)} \left(\text{for } k > (N_L - 1)\right)$$

$$z_{N_L}^k = 0 \ \left(\text{for } k < N_L\right) \text{ or } \left(z - h_1 - h_2 - \dots - h_{(N_L-1)}\right) \ \left(\text{for } k = N_L\right) \qquad (12.4)$$

The displacement components (u^k, v^k and w^k) at any point within the kth layer can also be expressed in a generalized form, as presented in Equation (12.5) where O is the null vector of size ($1 \times N_L$):

$$d^k = \left(d_t + Z_r^k d_r\right), \quad d^k = \{u^k \quad v^k \quad w^k\}^T,$$

$$d_t = \{u_0 \quad v_0 \quad w_0\}^T, \ d_r = \{\phi_x^T \quad \phi_y^T\}^T \ Z_r^k = \begin{bmatrix} z^k & O_{(1 \times N_L)} & O_{(1 \times N_L)} \\ O_{(1 \times N_L)} & z^k & O_{(1 \times N_L)} \end{bmatrix}^T$$

$$(12.5)$$

Further, the generalized translational (d_t) and rotational (d_r) displacement vectors can be combined into a single displacement vector (d) by introducing the transformation matrices (T_t and T_r), as given in Equation (12.6) where I and O indicate the identity and null matrices, respectively, while the subscript ($m \times n$) represents the size of a matrix.

$$d^k = \left(T_t + Z_r^k T_r\right) d,$$

$$d = \{u_0 \quad v_0 \quad w_0 \quad \phi_x^1 \quad \phi_x^2 \quad \dots \quad \dots \quad \phi_x^{(N_L-1)} \quad \phi_x^{N_L} \quad \phi_y^1 \quad \phi_y^2 \quad \dots \quad \dots \quad \phi_y^{(N_L-1)} \quad \phi_y^{N_L}\}^T,$$

$$T_t = \begin{bmatrix} I_{(3 \times 3)} & O_{(3 \times 2N_L)} \end{bmatrix}, \quad T_r = \begin{bmatrix} O_{(2N_L \times 3)} & I_{(2N_L \times 2N_L)} \end{bmatrix}$$

$$(12.6)$$

In the case of a plate with a small thickness compared to the in-plane dimensions, the transverse normal stress appears with a very small magnitude. Thus, it is usually omitted in the derivation of the plate's equation of motion (Reddy, 2003). Accordingly, the state of stress and the state of strain at any point in the kth layer are given in Equations (12.7a) and (12.7b), respectively. Here, the symbols σ and ε stand for the stress and strain components, respectively. The subscripts x and y indicate normal stress/strain components in the x and y directions, respectively. Further, the subscripts xy, xz, and yz signify the shear stress/strain components over the xy, xz and yz planes, respectively.

$$\sigma_b^k = \{\sigma_x^k \quad \sigma_y^k \quad \sigma_{xy}^k\}^T, \sigma_s^k = \{\sigma_{xz}^k \quad \sigma_{yz}^k\}^T \qquad (12.7a)$$

$$\varepsilon_b^k = \{\varepsilon_x^k \quad \varepsilon_y^k \quad \varepsilon_{xy}^k\}^T, \ \varepsilon_s^k = \{\varepsilon_{xz}^k \quad \varepsilon_{yz}^k\}^T \qquad (12.7b)$$

According to the displacement field (Equation (12.5)), the von Kármán nonlinear strain–displacement relations are given in Equation (12.8) where the symbol $\otimes$ represents Kronecker product.

$$\varepsilon_b^k = \left(\varepsilon_{bL} + \varepsilon_{bNL} + \mathbf{Z}_b^k \kappa_b\right), \quad \varepsilon_s^k = \left(\varepsilon_{sL} + \mathbf{Z}_s^k \kappa_s\right)$$

$$\varepsilon_{bL} = \left\{\frac{\partial u_0}{\partial x} \quad \frac{\partial v_0}{\partial y} \quad \left(\frac{\partial u_0}{\partial y} + \frac{\partial v_0}{\partial x}\right)\right\}^{\mathrm{T}},$$

$$\varepsilon_{bNL} = \frac{1}{2}\left\{\left(\frac{\partial w_0}{\partial x}\right)^2 \quad \left(\frac{\partial w_0}{\partial y}\right)^2 \quad 2\left(\frac{\partial w_0}{\partial x}\right)\left(\frac{\partial w_0}{\partial y}\right)\right\}^{\mathrm{T}},$$

$$\mathbf{k}_b = \left\{\frac{\partial \phi_x^{\mathrm{T}}}{\partial x} \quad \frac{\partial \phi_y^{\mathrm{T}}}{\partial y} \quad \left(\frac{\partial \phi_x^{\mathrm{T}}}{\partial y} + \frac{\partial \phi_y^{\mathrm{T}}}{\partial x}\right)\right\}^{\mathrm{T}},$$

$$\varepsilon_{sL} = \left\{\frac{\partial w_0}{\partial x} \quad \frac{\partial w_0}{\partial y}\right\}^{\mathrm{T}}, \quad \kappa_s = \left\{\phi_x^{\mathrm{T}} \quad \phi_y^{\mathrm{T}}\right\}^{\mathrm{T}} \tag{12.8}$$

$$\mathbf{Z}_b^k = \mathbf{I}_{(3\times 3)} \otimes \mathbf{z}^k, \quad \mathbf{Z}_s^k = \mathbf{I}_{(2\times 2)} \otimes \left(\partial \mathbf{z}^k / \partial z\right)$$

The 0–3 VEC layer is composed of a 1–3 VEC layer and two pure viscoelastic layers (Figure 12.3). So, a total of five layers ($N_L = 5$) appear in the overall layered plate (Figure 12.5). Now, for deriving the FE model, the plane of the plate is discretized using 9-node quadrilateral isoparametric elements where the edges of the rectangular graphite wafers in the same plane are followed to ensure two different elemental stacking sequences of layers, as shown in Figure 12.6. Through this FE discretization, a typical element appears with the rectangular shape having the edges in parallel to the x and y axes. Also, a typical element is comprised of isotropic layers even though there is a composite layer (1–3 VEC) in the overall layered plate (Figure 12.5). Now, the top and bottom layers of an elemental stack of layers are made of elastic isotropic materials. Also, the graphite layer possesses isotropic material properties. Therefore, according to the aforesaid state of stress and state of strain, the constitutive relations for these layers appear in the form, as given in Equation

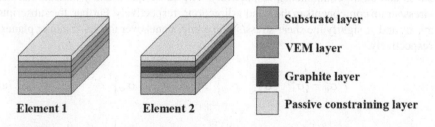

Substrate layer

VEM layer

Graphite layer

Passive constraining layer

Element 1   Element 2

FIGURE 12.6 Different elemental stacking sequence of the layers within the overall plate.

(12.9) where E^k, G^k and ν^k are the Young's modulus, shear modulus and Poisson's ratio for the kth layer, respectively.

$$\sigma_b^k = C_b^k \varepsilon_b^k, \quad \sigma_s^k = C_s^k \varepsilon_s^k$$

$$C_b^k = \frac{E^k}{1-\left(\nu^k\right)^2}\begin{bmatrix} 1 & \nu^k & 0 \\ \nu^k & 1 & 0 \\ 0 & 0 & \left(1-\nu^k\right)/2 \end{bmatrix}, \quad C_s^k = G^k \begin{bmatrix} 1 & 0 \\ 0 & 1 \end{bmatrix} \tag{12.9}$$

The viscoelastic layers in the elemental stacking sequences (Figure 12.6) are modeled using the FOD constitutive relations. Equation (12.10) illustrates the unidirectional FOD constitutive relation where E_0, E_∞, τ and α are called as relaxed elastic modulus, non-relaxed elastic modulus, relaxation time, and fractional order of time-derivative ($0<\alpha<1$), respectively (Bagley and Torvik, 1983). However, the FOD operator ($d\alpha/dt\alpha$) appearing in Equation (12.10) can be expressed using Grünwald definition, as given in Equation (12.11) (Schmidt and Gaul, 2002). In Equation (12.11), Δt denotes the incremental time-step within a specified time span; N_t is the number of terms used in the Grünwald definition and A_{j+1} is the Grünwald coefficient that can be expressed by Gamma function (Γ, Equation (12.12a)) or recurrence formulae (Equation (12.12b)) (Galucio et al., 2004).

$$\sigma(t)+\tau^\alpha \frac{d^\alpha \sigma(t)}{dt^\alpha} = E_0 \varepsilon(t)+\tau^\alpha E_\infty \frac{d^\alpha \varepsilon(t)}{dt^\alpha} \tag{12.10}$$

$$\frac{d^\alpha f(t)}{dt^\alpha} = \frac{1}{(\Delta t)^\alpha}\sum_{j=0}^{N_t} A_{j+1} f(t-j\Delta t) \tag{12.11}$$

$$A_{j+1} = \frac{\Gamma(j-\alpha)}{\Gamma(-\alpha)\Gamma(j+1)} \tag{12.12a}$$

$$A_{j+1} = \frac{j-\alpha-1}{j} A_j \tag{12.12b}$$

Now, if the unidirectional FOD constitute relation (Equation (12.10)) is applied either in the x direction or in the y direction, then one can obtain Equations 12.13a and 12.13b where the unidirectional constitutive relations are indicated by the superscript "ua". Similar constitutive relation also arises (Equations 12.13c–12.13e) for the shear strain in the $xy/xz/yz$ plane, since the shear strain in a plane appears due to the shear stress on that plane only.

$$\left(\varepsilon_x^k\right)^{ua} = \frac{1}{E_0}\sigma_x^k + \frac{\tau^\alpha}{E_0}\frac{d^\alpha \sigma_x^k}{dt^\alpha} - \frac{\tau^\alpha}{E_0}E_\infty \frac{d^\alpha \varepsilon_x^k}{dt^\alpha} \tag{12.13a}$$

$$\left(\varepsilon_y^k\right)^{ua} = \frac{1}{E_0}\sigma_y^k + \frac{\tau^\alpha}{E_0}\frac{d^\alpha \sigma_y^k}{dt^\alpha} - \frac{\tau^\alpha}{E_0}E_\infty \frac{d^\alpha \varepsilon_y^k}{dt^\alpha} \tag{12.13b}$$

$$\varepsilon_{xy}^k = \frac{1}{G_0}\sigma_{xy}^k + \frac{\tau^\alpha}{G_0}\frac{d^\alpha \sigma_{xy}^k}{dt^\alpha} - \frac{\tau^\alpha}{G_0}G_\infty \frac{d^\alpha \varepsilon_{xy}^k}{dt^\alpha} \tag{12.13c}$$

$$\varepsilon_{xz}^k = \frac{1}{G_0}\sigma_{xz}^k + \frac{\tau^\alpha}{G_0}\frac{d^\alpha \sigma_{xz}^k}{dt^\alpha} - \frac{\tau^\alpha}{G_0}G_\infty \frac{d^\alpha \varepsilon_{xz}^k}{dt^\alpha} \tag{12.13d}$$

$$\varepsilon_{yz}^k = \frac{1}{G_0}\sigma_{yz}^k + \frac{\tau^\alpha}{G_0}\frac{d^\alpha \sigma_{yz}^k}{dt^\alpha} - \frac{\tau^\alpha}{G_0}G_\infty \frac{d^\alpha \varepsilon_{yz}^k}{dt^\alpha} \tag{12.13e}$$

According to the aforesaid state of strain, the normal strains (ε_x^k and ε_y^k) at any point in a viscoelastic layer are related by Equation 12.14, where ν is the Poisson's ratio for the viscoelastic material.

$$\varepsilon_x^k = \left(\varepsilon_x^k\right)^{ua} - \nu\left(\varepsilon_y^k\right)^{ua}, \ \varepsilon_y^k = \left(\varepsilon_y^k\right)^{ua} - \nu\left(\varepsilon_x^k\right)^{ua} \tag{12.14}$$

Substituting Equations (12.13a) and (12.13b) in Equation (12.14) and then following Equations (12.13c–12.13e), the constitutive relation for a viscoelastic layer can be obtained as given in Equation (12.15):

$$\sigma_b^k + \tau^\alpha \frac{d^\alpha}{dt^\alpha}\sigma_b^k = C_b^k \varepsilon_b^k + \tau^\alpha \frac{E_\infty}{E_0}C_b^k \frac{d^\alpha}{dt^\alpha}\varepsilon_b^k, \ \sigma_s^k + \tau^\alpha \frac{d^\alpha}{dt^\alpha}\sigma_s^k = C_s^k \varepsilon_s^k + \tau^\alpha \frac{G_\infty}{G_0}C_s^k \frac{d^\alpha}{dt^\alpha}\varepsilon_s^k,$$

$$C_b^k = \frac{E_0}{1-\nu^2}\begin{bmatrix} 1 & \nu & 0 \\ \nu & 1 & 0 \\ 0 & 0 & (1-\nu)/2 \end{bmatrix}, \ C_s^k = G_0 \begin{bmatrix} 1 & 0 \\ 0 & 1 \end{bmatrix}, \ \frac{G_\infty}{G_0} = \frac{E_\infty}{E_0}$$

$$\tag{12.15}$$

Equation (12.15) can also be expressed in a simple form, as given in Equation (12.16), by introducing two anelastic stain vectors ($\bar{\varepsilon}_b^k, \bar{\varepsilon}_s^k$, Equation (12.17)) (Galucio et al., 2004). In Equation (12.16), $\bar{\varepsilon}_{bL}, \bar{\varepsilon}_{bNL}, \bar{\kappa}_b$, $\bar{\varepsilon}_{sL}$ and $\bar{\kappa}_s$ are the generalized anelastic strain vectors.

$$\sigma_b^k = \frac{E_\infty}{E_0}C_b^k\left(\varepsilon_b^k - \bar{\varepsilon}_b^k\right), \ \sigma_s^k = \frac{G_\infty}{G_0}C_s^k\left(\varepsilon_s^k - \bar{\varepsilon}_s^k\right),$$

$$\bar{\varepsilon}_b^k = \left\{\bar{\varepsilon}_x^k \quad \bar{\varepsilon}_y^k \quad \bar{\varepsilon}_{xy}^k\right\}^T \text{ or } \bar{\varepsilon}_b^k = \left(\bar{\varepsilon}_{bL} + \bar{\varepsilon}_{bNL} + Z_b^k \bar{\kappa}_b\right)$$

$$\bar{\varepsilon}_s^k = \left\{\bar{\varepsilon}_{xz}^k \quad \bar{\varepsilon}_{yz}^k\right\}^T \text{ or } \bar{\varepsilon}_s^k = \left(\bar{\varepsilon}_{sL} + Z_s^k \bar{\kappa}_s\right) \tag{12.16}$$

$$\bar{\varepsilon}_b^k + \tau^\alpha \frac{d^\alpha}{dt^\alpha} \bar{\varepsilon}_b^k = \frac{E_\infty - E_0}{E_0} \varepsilon_b^k, \quad \bar{\varepsilon}_s^k + \tau^\alpha \frac{d^\alpha}{dt^\alpha} \bar{\varepsilon}_s^k = \frac{G_\infty - G_0}{G_0} \varepsilon_s^k \quad (12.17)$$

Using the Grünwald definition for the FOD (Equations 12.11–12.12), the anelastic strain vector at the $(n + 1)^{\text{th}}$ time-step can be expressed according to Equation 12.18 by noting that $A_1 = 1$ (Schmidt and Gaul, 2002):

$$\left(\bar{\varepsilon}_b^k\right)_{n+1} = (1-c)\frac{E_\infty - E_0}{E_0}\left(\varepsilon_b^k\right)_{n+1} - c\sum_{j=1}^{N_t} A_{j+1}\left(\bar{\varepsilon}_b^k\right)_{n+1-j}$$

$$\left(\bar{\varepsilon}_s^k\right)_{n+1} = (1-c)\frac{G_\infty - G_0}{G_0}\left(\varepsilon_s^k\right)_{n+1} - c\sum_{j=1}^{N_t} A_{j+1}\left(\bar{\varepsilon}_s^k\right)_{n+1-j} \quad (12.18)$$

$$c = \frac{\tau^\alpha}{\tau^\alpha + (\Delta t)^\alpha}$$

Substituting Equation (12.18) in Equation (12.16), the constitutive relations of the viscoelastic layers at the $(n + 1)^{\text{th}}$ time-step can be obtained in the form of Equation (12.19):

$$\left(\sigma_b^k\right)_{n+1} = \left\langle 1 - c\left(\frac{E_\infty - E_0}{E_0}\right)\right\rangle C_b^k\left(\varepsilon_b^k\right)_{n+1} + \left(\frac{cE_\infty}{E_0}\right)C_b^k\sum_{j=1}^{N_t} A_{j+1}\left(\bar{\varepsilon}_b^k\right)_{n+1-j}$$

$$\left(\sigma_s^k\right)_{n+1} = \left\langle 1 - c\left(\frac{G_\infty - G_0}{G_0}\right)\right\rangle C_s^k\left(\varepsilon_s^k\right)_{n+1} + \left(\frac{cG_\infty}{G_0}\right)C_s^k\sum_{j=1}^{N_t} A_{j+1}\left(\bar{\varepsilon}_s^k\right)_{n+1-j} \quad (12.19)$$

The overall plate is considered to be subjected to a uniformly distributed transverse dynamic load ($p(\text{t})$) at the bottom surface ($z = 0$). For the corresponding vibration of the overall plate, the first variations of the total potential energy (δT_p^e) and total kinetic energy (δT_k^e) for a typical element can be obtained from Equations (12.20) and (12.21), respectively, at any instant of time (t). Here, a^e/b^e is the elemental length along the x/y direction, and the thickness coordinates over the bottom and top surfaces of the kth layer are denoted by h_k and h_{k+1}, respectively. Also, the mass density of the kth layer is represented by ρ^k:

$$\delta T_p^e = \int_0^{a^e}\int_0^{b^e}\left[\sum_{k=1}^{5}\int_{h_k}^{h_{k+1}}\left\langle\left(\delta\varepsilon_b^k\right)^{\text{T}}\sigma_b^k + \left(\delta\varepsilon_s^k\right)^{\text{T}}\sigma_s^k\right\rangle dz - \left\langle\delta w_0\, p(t)\right\rangle_{z=0}\right]dydx \quad (12.20)$$

$$\delta T_k^e = \int_0^{a^e}\int_0^{b^e}\left[\sum_{k=1}^{5}\int_{h_k}^{h_{k+1}}\left\langle\left(\delta\dot{d}^k\right)^{\text{T}}\rho^k\dot{d}^k\right\rangle dz\right]dydx \quad (12.21)$$

Using Equations (12.8 and 12.6), Equations (12.20 and 12.21) can be written in the forms, as given in Equations (12.22 and 12.23) where the matrices (N_b, M_b, N_s, M_s and $\bar{m}$) are illustrated in Equations (12.24 and 12.25):

$$\delta T_p^e = \int_0^{a^e} \int_0^{b^e} \left[\left\langle \left(\delta\varepsilon_{bL}\right)^{\mathrm{T}} + \left(\delta\varepsilon_{bNL}\right)^{\mathrm{T}} \right\rangle N_b + \left(\delta\kappa_b\right)^{\mathrm{T}} M_b \right. $$
$$\left. + \left(\delta\varepsilon_{sL}\right)^{\mathrm{T}} N_s + \left(\delta\kappa_s\right)^{\mathrm{T}} M_s + \left\langle \delta w_0 \, p(t) \right\rangle_{z=0} \right] dy dx \qquad (12.22)$$

$$\delta T_k^e = \int_0^{a^e} \int_0^{b^e} \left\langle \left(\delta\dot{d}\right)^{\mathrm{T}} \bar{m} \dot{d} \right\rangle dy dx \qquad (12.23)$$

$$N_b = \sum_{k=1}^{5} \int_{h_k}^{h_{k+1}} \sigma_b^k dz, \quad M_b = \sum_{k=1}^{5} \int_{h_k}^{h_{k+1}} \left\langle \left(\mathbf{Z}_b^k\right)^{\mathrm{T}} \sigma_b^k \right\rangle dz,$$
$$N_s = \sum_{k=1}^{5} \int_{h_k}^{h_{k+1}} \sigma_s^k dz, \quad M_s = \sum_{k=1}^{5} \int_{h_k}^{h_{k+1}} \left\langle \left(\mathbf{Z}_s^k\right)^{\mathrm{T}} \sigma_s^k \right\rangle dz \qquad (12.24)$$

$$\bar{m} = \sum_{k=1}^{5} \int_{h_k}^{h_{k+1}} \left[\left\langle \left(T_t\right)^{\mathrm{T}} + \left(T_r\right)^{\mathrm{T}} \left(\mathbf{Z}_r^k\right)^{\mathrm{T}} \right\rangle \rho^k \left(T_t + \mathbf{Z}_r^k T_r\right) \right] dz \qquad (12.25)$$

The generalized strain vectors (ε_{bL}, ε_{bNL}, κ_b, ε_{sL}, κ_s) and displacement vector (d) at any point within a typical element can be expressed in terms of the shape function matrix (N) and elemental nodal displacement vector (d^e) as given in Equation (12.26) where the operator matrices (L_{bL}, L_{bNL}, $L_b\kappa$, L_{sL}, $L_s\kappa$) can be obtained from Equation (12.8):

$$d = Nd^e, \; \varepsilon_{bL} = B_{bL}d^e, \; \varepsilon_{bNL} = B_{bNL}d^e, \; \kappa_b = B_{bк}d^e,$$
$$\varepsilon_{sL} = B_{sL}d^e, \kappa_s = B_{sк}d^e, \quad B_{bL} = L_{bL}T_tN, B_{bNL} = L_{bNL}T_tN,$$
$$B_{bк} = L_{bк}T_rN, B_{sL} = L_{sL}T_tN, \; B_{sк} = L_{sк}T_rN \qquad (12.26)$$

Using Equation (12.26) in Equations (12.22 and 12.23), the following expressions for the first variations of the elemental potential energy (δT_p^e) and kinetic energy (δT_k^e) can be obtained:

$$\delta T_p^e = \int_0^{a^e} \int_0^{b^e} \left(\delta d^e\right)^{\mathrm{T}} \left[\left(B_{bL}^{\mathrm{T}} + B_{bNL}^{\mathrm{T}}\right) N_b + B_{bк}^{\mathrm{T}} M_b + B_{sL}^{\mathrm{T}} N_s + B_{sк}^{\mathrm{T}} M_s \right. $$
$$\left. + \left\langle (\mathbf{N})^{\mathrm{T}} (T_t)^{\mathrm{T}} \{0 \quad 0 \quad 1\}^{\mathrm{T}} \right\rangle_{z=0} p(t) \right] dy dx \qquad (12.27)$$

$$\delta T_k^e = \int_0^{a^e} \int_0^{b^e} \left\langle \left(\delta \dot{d}^e \right)^{\mathrm{T}} \left(N^{\mathrm{T}} \bar{m} N \right) \dot{d}^e \right\rangle dy dx \qquad (12.28)$$

The governing equation of motion of a typical element is derived employing Hamilton's principle (Equation (12.29)). Introducing the aforesaid expressions (Equations 12.27–12.28) in Equation (12.29), the elemental governing equation of motion can be obtained in the form, as given in Equation (12.30) where the elemental mass matrix (M^e) and mechanical load vector (P_M^e) are illustrated in Equation (12.31):

$$\int_{t_1}^{t_2} \left(\delta T_k^e - \delta T_p^e \right) dt = 0 \qquad (12.29)$$

$$M^e \ddot{d}^e + \int_0^{a^e} \int_0^{b^e} \left[\left\langle B_{bL}^{\mathrm{T}} + B_{bNL}^{\mathrm{T}} \left(d^e \right) \right\rangle N_b + B_{b\kappa}^{\mathrm{T}} M_b + B_{sL}^{\mathrm{T}} N_s + B_{s\kappa}^{\mathrm{T}} M_s \right] dy dx = P_M^e p(t)$$

$$(12.30)$$

$$M^e = \int_0^{a^e} \int_0^{b^e} \left(N^{\mathrm{T}} \bar{m} N \right) dy dx, \quad P_M^e = \int_0^{a^e} \int_0^{b^e} \left[\left\langle \left(N \right)^{\mathrm{T}} \left(T_t \right)^{\mathrm{T}} \left\{ 0 \quad 0 \quad 1 \right\}^{\mathrm{T}} \right\rangle_{z=0} \right] dy dx \quad (12.31)$$

The nonlinear transient responses of the overall plate within a specified time span are evaluated using Newmark direct time integration method where the elemental governing equation (Equation (12.30)) at the $(n + 1)^{\text{th}}$ time-step can be written in a form, as given in Equation (12.32). In Equation (12.32), the matrices like N_b, M_b, N_s and M_b can be expressed in terms of the generalized strains ($\varepsilon_{bL}, \varepsilon_{bNL}, \kappa_b, \varepsilon_{sL}, \kappa_s$) and the corresponding anelastic counterparts ($\bar{\varepsilon}_{bL}, \bar{\varepsilon}_{bNL}, \bar{\kappa}_b, \bar{\varepsilon}_{sL}, \bar{\kappa}_s$), as given in Equation (12.33) by introducing Equations (12.8), (12.9) and (12.19) in Equation (12.24). The various rigidity matrices ($A_b^1, A_b^2, A_s^1, A_s^2, B_{b1}^1, B_{b1}^2, B_{s1}^1, B_{s1}^2, B_{b2}^1, B_{b2}^2, B_{s2}^1, B_{s2}^2, D_b^1, D_b^2, D_s^1, D_s^2$) appearing in Equation (12.33) are furnished in Equation (12.34), where the stiffness matrices ($C_{b1}^k, C_{b2}^k, C_{s1}^k, C_{s2}^k$) arise differently depending on the elemental stacking sequence of layers (Figure 12.6). For the two different stacking sequences over Element 1 and Element 2 (Figure 12.6), the stiffness matrices ($C_{b1}^k, C_{b2}^k, C_{s1}^k, C_{s2}^k$) are given in Equations (12.35a) and (12.35b), respectively:

$$M^e \ddot{d}_{n+1}^e + \int_0^{a^e} \int_0^{b^e} \left[\begin{array}{l} \left\langle B_{bL}^{\mathrm{T}} + B_{bNL}^{\mathrm{T}} \left(d_{n+1}^e \right) \right\rangle \left(N_b \right)_{n+1} + B_{b\kappa}^{\mathrm{T}} \left(M_b \right)_{n+1} \\ + B_{sL}^{\mathrm{T}} \left(N_s \right)_{n+1} + B_{s\kappa}^{\mathrm{T}} \left(M_s \right)_{n+1} \end{array} \right] dy dx = P_M^e p(t)_{n+1}$$

$$(12.32)$$

$$\left(N_b \right)_{n+1} = A_b^1 \left\langle \left(\varepsilon_{bL} \right)_{n+1} + \left(\varepsilon_{bNL} \right)_{n+1} \right\rangle + B_{b1}^1 \left(\kappa_b \right)_{n+1}$$

$$+ A_b^2 \sum_{j=1}^{N_t} A_{j+1} \left\langle \left(\bar{\varepsilon}_{bL} \right)_{n+1-j} + \left(\bar{\varepsilon}_{bNL} \right)_{n+1-j} \right\rangle + B_{b1}^2 \sum_{j=1}^{N_t} A_{j+1} \left(\bar{\kappa}_b \right)_{n+1-j}$$

$$
\begin{aligned}
\left(M_b\right)_{n+1} = {}& B_{b2}^1 \left\langle \left(\varepsilon_{bL}\right)_{n+1} + \left(\varepsilon_{bNL}\right)_{n+1} \right\rangle + D_b^1 \left(\kappa_b\right)_{n+1} \\
&+ B_{b2}^2 \sum_{j=1}^{N_t} A_{j+1} \left\langle \left(\bar{\varepsilon}_{bL}\right)_{n+1-j} + \left(\bar{\varepsilon}_{bNL}\right)_{n+1-j} \right\rangle + D_b^2 \sum_{j=1}^{N_t} A_{j+1} \left(\bar{\kappa}_b\right)_{n+1-j}
\end{aligned}
$$

$$
\left(N_s\right)_{n+1} = A_s^1 \left(\varepsilon_{sL}\right)_{n+1} + B_{s1}^1 \left(\kappa_s\right)_{n+1} + A_s^2 \sum_{j=1}^{N_t} A_{j+1} \left(\bar{\varepsilon}_{sL}\right)_{n+1-j} + B_{s1}^2 \sum_{j=1}^{N_t} A_{j+1} \left(\bar{\kappa}_s\right)_{n+1-j}
$$

$$
\left(M_s\right)_{n+1} = B_{s2}^1 \left(\varepsilon_{sL}\right)_{n+1} + D_s^1 \left(\kappa_s\right)_{n+1} + B_{s2}^2 \sum_{j=1}^{N_t} A_{j+1} \left(\bar{\varepsilon}_{sL}\right)_{n+1-j} + D_s^2 \sum_{j=1}^{N_t} A_{j+1} \left(\bar{\kappa}_s\right)_{n+1-j}
$$

$$\tag{12.33}$$

$$
\begin{aligned}
A_b^1 &= \sum_{k=1}^{5} \int_{h_k}^{h_{k+1}} C_{b1}^k dz, \; A_b^2 = \sum_{k=1}^{5} \int_{h_k}^{h_{k+1}} C_{b2}^k dz, \; B_{b1}^1 = \sum_{k=1}^{5} \int_{h_k}^{h_{k+1}} C_{b1}^k Z_b^k dz, \\
B_{b1}^2 &= \sum_{k=1}^{5} \int_{h_k}^{h_{k+1}} C_{b2}^k Z_b^k dz, \; B_{b2}^1 = \sum_{k=1}^{5} \int_{h_k}^{h_{k+1}} \left(Z_b^k\right)^{\mathrm{T}} C_{b1}^k dz, \\
B_{b2}^2 &= \sum_{k=1}^{5} \int_{h_k}^{h_{k+1}} \left(Z_b^k\right)^{\mathrm{T}} C_{b2}^k dz, \; D_b^1 = \sum_{k=1}^{5} \int_{h_k}^{h_{k+1}} \left(Z_b^k\right)^{\mathrm{T}} C_{b1}^k Z_b^k dz, \\
D_b^2 &= \sum_{k=1}^{5} \int_{h_k}^{h_{k+1}} \left(Z_b^k\right)^{\mathrm{T}} C_{b2}^k Z_b^k dz, \; A_s^1 = \sum_{k=1}^{5} \int_{h_k}^{h_{k+1}} C_{s1}^k dz, \; A_s^2 = \sum_{k=1}^{5} \int_{h_k}^{h_{k+1}} C_{s2}^k dz, \\
B_{s1}^1 &= \sum_{k=1}^{5} \int_{h_k}^{h_{k+1}} C_{s1}^k Z_s^k dz, \; B_{s1}^2 = \sum_{k=1}^{5} \int_{h_k}^{h_{k+1}} C_{s2}^k Z_s^k dz, \; B_{s2}^1 = \sum_{k=1}^{5} \int_{h_k}^{h_{k+1}} \left(Z_s^k\right)^{\mathrm{T}} C_{s1}^k dz, \\
B_{s2}^2 &= \sum_{k=1}^{5} \int_{h_k}^{h_{k+1}} \left(Z_s^k\right)^{\mathrm{T}} C_{s2}^k dz, \; D_s^1 = \sum_{k=1}^{5} \int_{h_k}^{h_{k+1}} \left(Z_s^k\right)^{\mathrm{T}} C_{s1}^k Z_s^k dz, \; D_s^2 = \sum_{k=1}^{5} \int_{h_k}^{h_{k+1}} \left(Z_s^k\right)^{\mathrm{T}} C_{s2}^k Z_s^k dz
\end{aligned}
$$

$$\tag{12.34}$$

$$
\left\langle C_{b1}^k = C_b^k, C_{b2}^k = 0, C_{s1}^k = C_s^k, C_{s2}^k = 0 \right\rangle_{k=1,5}
$$

$$
\left\langle
\begin{aligned}
C_{b1}^k &= \left\langle 1 - c \left(\frac{E_\infty - E_0}{E_0}\right) \right\rangle C_b^k, C_{b2}^k = \left(\frac{cE_\infty}{E_0}\right) C_b^k, \\
C_{s1}^k &= \left\langle 1 - c \left(\frac{G_\infty - G_0}{G_0}\right) \right\rangle C_s^k, C_{s2}^k = \left(\frac{cG_\infty}{G_0}\right) C_s^k
\end{aligned}
\right\rangle_{k=2,3,4}
$$

$$\tag{12.35a}$$

$$\left\langle C_{b1}^k = C_b^k, C_{b2}^k = 0, C_{s1}^k = C_s^k, C_{s2}^k = 0 \right\rangle_{k=1,3,5},$$

$$\left\langle C_{b1}^k = \left\langle 1 - c\left(\frac{E_\infty - E_0}{E_0}\right) \right\rangle C_b^k, C_{b2}^k = \left(\frac{cE_\infty}{E_0}\right) C_b^k, \right.$$

$$\left. C_{s1}^k = \left\langle 1 - c\left(\frac{G_\infty - G_0}{G_0}\right) \right\rangle C_s^k, C_{s2}^k = \left(\frac{cG_\infty}{G_0}\right) C_s^k \right\rangle_{k=2,4}$$

(12.35b)

Using Equation (12.26) in Equation (12.33) and then substituting the resulting expressions in Equations (12.32), the elemental governing equation of motion can be expressed in terms of the elemental nodal displacement vector, as given in Equation (12.36). The different stiffness matrices ($K_{bL}^e, K_{bNL}^e, K_s^e$) and the memory load vectors ($P_{vbL}^e, P_{vbN}^e, P_{vs}^e$) appearing in Equation (12.36) are illustrated in Equation (12.37). The anelastic strains ($\bar{\varepsilon}_{bL}, \bar{\varepsilon}_{bNL}, \bar{\kappa}_b, \bar{\varepsilon}_{sL}, \bar{\kappa}_s$) in Equation (12.37) at the $(n+1)^{th}$ time step can be computed following Equation (12.18):

$$M^e \ddot{d}_{n+1}^e + \left(K_{bL}^e + K_{bNL}^e \left(d_{n+1}^e \right) + K_s^e \right) d_{n+1}^e = P_M^e p(t)_{n+1} - \left\langle \left(P_{vbL}^e \right)_{n+1} + \left(P_{vbN}^e \right)_{n+1} + \left(P_{vs}^e \right)_{n+1} \right\rangle$$

(12.36)

$$K_{bL}^e = \int_0^{a^e} \int_0^{b^e} \left[\left\langle B_{bL}^T A_b^1 + B_{b\kappa}^T B_{b2}^1 \right\rangle B_{bL} + \left\langle B_{bL}^T B_{b1}^1 + B_{b\kappa}^T D_b^1 \right\rangle B_{b\kappa} \right] dy dx$$

$$K_{bNL}^e = \int_0^{a^e} \int_0^{b^e} \left[B_{bNL}^T \left\langle A_b^1 B_{bL} + B_{b1}^1 B_{b\kappa} \right\rangle + \left\langle B_{bL}^T A_b^1 + B_{bNL}^T A_b^1 + B_{b\kappa}^T B_{b2}^1 \right\rangle B_{bNL} \right] dy dx$$

$$K_s^e = \int_0^{a^e} \int_0^{b^e} \left[\left\langle B_{sL}^T A_s^1 + B_{s\kappa}^T B_{s2}^1 \right\rangle B_{sL} + \left\langle B_{sL}^T B_{s1}^1 + B_{s\kappa}^T D_s^1 \right\rangle B_{s\kappa} \right] dy dx$$

$$\left(P_{vbL}^e \right)_{n+1} = \sum_{j=1}^{N_t} A_{j+1} \int_0^{a^e} \int_0^{b^e} \left[\left\langle B_{bL}^T A_b^2 + B_{b\kappa}^T B_{b2}^2 \right\rangle \left(\bar{\varepsilon}_{bL} \right)_{n+1-j} \right.$$
$$\left. + \left\langle B_{bL}^T B_{b1}^2 + B_{b\kappa}^T D_b^2 \right\rangle \left(\bar{\kappa}_b \right)_{n+1-j} \right] dy dx$$

$$\left(P_{vbN}^e \right)_{n+1} = \sum_{j=1}^{N_t} A_{j+1} \int_0^{a^e} \int_0^{b^e} \left[B_{bNL}^T \left\langle A_b^2 \left(\bar{\varepsilon}_{bL} \right)_{n+1-j} + B_{b1}^2 \left(\bar{\kappa}_b \right)_{n+1-j} \right\rangle \\ + \left\langle B_{bL}^T A_b^2 + B_{bNL}^T A_b^2 + B_{b\kappa}^T B_{b2}^2 \right\rangle \left(\bar{\varepsilon}_{bNL} \right)_{n+1-j} \right] dx dy$$

$$\left(P_{vs}^e\right)_{n+1} = \sum_{j=1}^{N_t} A_{j+1} \int_0^{a^e}\int_0^{b^e}\left[\left\langle B_{sL}^T A_s^2 + B_{sK}^T B_{s2}^2\right\rangle\left(\bar\varepsilon_{sL}\right)_{n+1-j} + \left\langle B_{sL}^T B_{s1}^2 + B_{sK}^T D_s^2\right\rangle\left(\bar K_s\right)_{n+1-j}\right]dxdy$$

(12.37)

Assembling the elemental equations (Equation (12.36)), the following global equation of motion of the overall plate at the $(n + 1)^{th}$ time-step can be obtained,

$$M\ddot X_{n+1} + \left(K_L + K_{bNL}\left(d_{n+1}\right)\right)X_{n+1} = P_M p\left(t\right)_{n+1} - \left(P_v\right)_{n+1}$$
$$K_L = \left(K_{bL} + K_s\right); \left(P_v\right)_{n+1} = \left\langle\left(P_{vbL}\right)_{n+1} + \left(P_{vbN}\right)_{n+1} + \left(P_{vs}\right)_{n+1}\right\rangle$$

(12.38)

In Equation (12.38), M is the global mass matrix; K_L is the global linear stiffness matrix; K_{bL} and K_s are the bending and transverse shear counterparts of the global linear stiffness matrix (K_L); K_{bNL} is the global nonlinear stiffness matrix; P_M is the global load coefficient vector; P_v is the global memory load vector arising from the viscoelastic phase in the overall plate; X is the global nodal displacement vector. The solution of Equation (12.38) provides the transient response of the overall plate at the $(n + 1)^{th}$ time-step; however, it is a nonlinear equation and thus presently solved using direct iteration method.

12.4   NUMERICAL RESULTS AND DISCUSSION

The viscoelastic damping layer in the layered plate is considered to be made of either VEPC or 0–3 VEC, and the comparative damping performance of these two kinds of viscoelastic composites is presented by evaluating the nonlinear transient responses of the layered plate against a step-load. The length and width of the layered plate are considered as $a = 0.4$ m and $b = 0.4$ m, respectively. The thicknesses of the substrate plate, constrained damping layer and passive constraining layer are considered as $h = 4.5$ mm, $h_d = 1$ mm, and $h_c = 0.5$ mm, respectively. The geometrical dimensions in the construction of the 0–3 VEC layer are taken in an optimal manner, as presented in the following numerical results.

12.4.1   PROPERTIES OF THE COMPONENT MATERIALS

The substrate plate and passive constraining layer are considered to be made of aluminum (E=69 GPa, ν=0.3, ρ=2740 kg/m^3), while the material properties of graphite wafers/particles are: E=250 GPa, ν=0.3, ρ=1400 kg/m^3. The viscoelastic phase in VEPC / 0–3 VEC is made of butyl rubber. The material properties of this rubber material vary with the operating frequency and temperature (Jones, 2001). However, the layered plate is presently considered to operate at room temperature (32°C). Thus the frequency-dependent material properties of butyl rubber are taken at the same temperature from a reference (Jones, 2001). The mass density (ρ) and Poisson's ratio (ν) of the rubber material are taken as 920 kg/m^3 and 0.49, respectively. These properties of the butyl rubber are directly incorporated in the aforesaid FE model of the layered plate when the damping layer is made of 0–3 VEC. However, in the case of the use of VEPC layer, its effective properties are incorporated in the FE model. These effective material constants of VEPC can be estimated theoretically by solving two differential equations along with the relations among the material constants for an isotropic material (Gupta et al., 2020b), as given in Equations (12.39–12.41). Besides these effective properties, the effective mass density of VEPC is estimated by the rule of mixture (Equation (12.42)). In Equations (12.39–12.42), the symbols E, G, K, ν, and ρ stand for Young's modulus, shear modulus, bulk modulus, Poisson's ratio and mass density, respectively. The subscripts m and d denote the material constants for the viscoelastic matrix and inclusion of graphite particles, respectively. The superscript ($*$) indicates a complex quantity and ϕ is the volume fraction of inclusion of graphite particles in the viscoelastic matrix. Usually, the maximum volume fraction (ϕ_m) of inclusion in a particulate composite is limited to a certain value, and it is considered here as 0.63 according to the Random Closed Packing (RCP) of particles (Porfiri and Gupta, 2009). The method of solution of Equations (12.39)–(12.40)) within a frequency range is available in (Gupta et al., 2020b), and this solution provides the frequency-dependent complex material constants ($G*(\omega)$, $K*(\omega)$) of VEPC. These frequency-dependent material constants can subsequently be used to obtain the Young's modulus and Poisson's ratio of the VEPC, according to Equation (12.41):

$$\frac{dG^*(\omega)}{d\phi} = f_G^*\left(G_d, K_d, G^*(\omega), K^*(\omega)\right)\frac{G^*(\omega)}{1-\phi/\phi_m};$$

$$f_G^* = \left[\frac{5\left(G_d - G_m^*(\omega)\right)\left(3K_m^*(\omega) + 4G_m^*(\omega)\right)}{6G_d\left(K_m^*(\omega) + 2G_m^*(\omega)\right) + G_m^*(\omega)\left(9K_m^*(\omega) + 8G_m^*(\omega)\right)}\right] \tag{12.39}$$

$$\frac{dK^*(\omega)}{d\phi} = f_K^*\left(G_d, K_d, G^*(\omega), K^*(\omega)\right)\frac{K^*(\omega)}{1-\phi/\phi_m};$$

$$f_K^* = \left[\left(\frac{3K_m^*(\omega) + 4G_m^*(\omega)}{3K_m^*(\omega)}\right)\left(\frac{3K_d - 3K_m^*(\omega)}{3K_d + 4G_m^*(\omega)}\right)\right] \tag{12.40}$$

$$E = \frac{9KG}{3K+G}, \quad \nu = \frac{3K-2G}{2(3K+G)}, \tag{12.41}$$

$$\rho = \rho_d\phi + \rho_m(1-\phi) \tag{12.42}$$

The VEPC and butyl rubber phase in 0–3 VEC are modeled through the FOD viscoelastic constitutive relation, where the FOD parameters ($E_0, E_\infty, \tau, \alpha$) can be expressed in terms of the frequency-dependent storage Young's modulus ($E'(\omega)$), loss Young's modulus ($E''(\omega)$), and maximum value (η_{max}) of the material loss factor within a frequency range, as given in Equations (12.43a–12.43c) (Galucio et al., 2004). Now, the frequency-dependent properties of a viscoelastic material or composite are usually estimated experimentally or analytically in terms of the storage Young's modulus ($E'(\omega)$) and loss Young's modulus ($E''(\omega)$) where the variations of these material constants over a frequency range are commonly expressed in the form of the functional curve. These functional curves for $E'(\omega)$ and $E''(\omega)$ exhibit the maximum value (η_{max}) of the material loss factor and also appear in terms of the FOD parameters through the mathematical relations in Equations (12.43a–12.43c). So, the FOD parameters can be computed by a curve-fitting approach with reference to the functional curves for $E'(\omega)$ and $E''(\omega)$ over a frequency range. Several curve-fitting approaches are available in the open literature, which can be used to obtain the FOD parameters; however, a good result may be achieved by using the genetic algorithm (GA) (Arikoglu, 2014). In the present demonstration, the frequency-dependent material constants ($E'(\omega)$ and $E''(\omega)$) of butyl rubber are taken from (Jones, 2001) and also similar properties of VEPC for different volume fractions (ϕ) of inclusion are computed from the relations given in Equations (12.39–12.41). These material properties are further used to obtain the corresponding FOD parameters through the aforesaid curve-fitting approach in conjunction with the mathematical expressions given in Equations (12.43a–12.43c). For instance, the fitted curves in the case of butyl rubber are shown in Figures 12.7(a) and 12.7(b) for the storage Young's modulus (E') and loss Young's modulus (E''), respectively. The corresponding values of the FOD parameters are given in Table 12.1 in the row for $\phi = 0$. However, the same table (Table 12.1) also contains the data for the FOD parameters in the case of VEPC for different values of the volume fraction of inclusion (ϕ) while a frequency range is taken from 500 rad/s to 4500 rad/s:

$$E'(\omega) = \frac{E_0 + (E_\infty + E_0)(\omega\tau)^\alpha \cos(\pi\alpha/2) + E_\infty (\omega\tau)^{2\alpha}}{1 + 2(\omega\tau)^\alpha \cos(\pi\alpha/2) + (\omega\tau)^{2\alpha}} \qquad (12.43a)$$

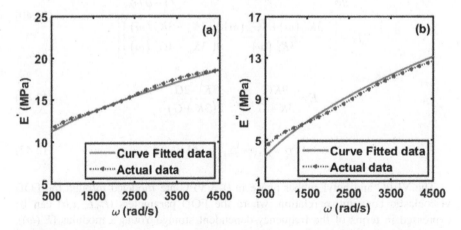

FIGURE 12.7 Fitted curves for the FOD parameters to the similar curve for the frequency-dependent material constants of Butyl rubber (a) storage Young's modulus (E') and (b) loss Young's modulus (E'').

TABLE 12.1
FOD parameters (E_0, E_∞, τ, α) for VEPC

ϕ	E_0 (MPa)	E_∞ (MPa)	α	τ (μs)
0	9.0483	194.19	0.8400	5.5500
0.1	11.0860	251.77	0.8627	5.3291
0.2	15.0000	337.70	0.8772	5.0000
0.3	23.1590	550.95	0.8946	4.5000
0.4	37.4730	934.99	0.9049	3.9709
0.5	77.9110	2163.60	0.9342	3.3700

$$E''(\omega) = \frac{(E_\infty - E_0)(\omega\tau)^\alpha \sin(\pi\alpha/2)}{1 + 2(\omega\tau)^\alpha \cos(\pi\alpha/2) + (\omega\tau)^{2\alpha}} \tag{12.43b}$$

$$\alpha = \frac{2}{\pi}\arcsin\left[\eta_{max}(E_\infty - E_0) \times \frac{2\sqrt{E_0 E_\infty} + (E_\infty + E_0)\sqrt{1 + \eta_{max}^2}}{\eta_{max}^2(E_\infty + E_0)^2 + (E_\infty - E_0)^2}\right] \tag{12.43c}$$

12.4.2 DAMPING ANALYSIS OF THE LAYERED PLATE

The damping in the layered plate is analyzed by evaluating its nonlinear transient responses, where the 0–3 VEC, VEPC, and monolithic viscoelastic (butyl rubber) layers are used separately. The edges of the layered plate are considered to be fully clamped edges to support the uniformly distributed transverse step-load (p_0) over the bottom surface ($z = 0$). However, for estimating the damping in the CLD treatment from the computed nonlinear transient responses of the plate, a performance index (I_d) is defined as given in Equation (12.44) where the attenuation of the maximum transverse displacement amplitude (W_{max}) is noted within a time span (t_s). In the present computation, a time span (t_s) is taken as 0.05 s considering the first few cycles of oscillation of the layered plate:

$$I_d = \frac{(W_{max}/h)_{t=0} - (W_{max}/h)_{t=t_s}}{(W_{max}/h)_{t=0}} \times 100 \tag{12.44}$$

Since this evaluation of damping is based on the transient responses of the layered plate, an initial verification of the present FE formulation is carried out for computation of similar responses using Newmark time integration scheme, where the nonlinear solution for every time-step is obtained by direct iteration method. For this verification, first, the layered plate is taken with the simply-supported edges and negligibly thin ($h_d \approx 0$) damping layer. The nonlinear transient responses of this plate are computed for a uniformly distributed transverse step-load (p_0) with the intensity of 100 MN/m². This result is illustrated in Figure 12.8 along with the similar reference result available in (Reddy, 2014). This comparison verifies the aforesaid FE

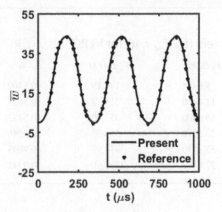

FIGURE 12.8 Verification of the present FE formulation for evaluation of the nonlinear transient response of plates. (Reference: Reddy, 2014).

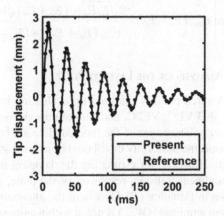

FIGURE 12.9 Verification of the present FE formulation in the modeling of constrained layer damping using FOD viscoelastic constitutive model (Reference: Galucio et al., 2004).

formulation and implementation of the solution methods in the evaluation of the nonlinear transient responses of a plate. However, for further verification of the aforesaid FE formulation in handling the FOD viscoelastic constitutive model, the reference results for a cantilever sandwich beam with the viscoelastic core is taken, as shown in Figure 12.9. Similar results obtained from the present FE formulation are also presented in the same figure, where an excellent agreement of the present results with the reference results is observed, and it confirms the accuracy of the present FE formulation in modeling CLD treatment using FOD viscoelastic constitutive model.

In the construction of the 0–3 VEC layer (Figure 12.3), the geometrical parameters appear as (a) the gap (Δ) between two consecutive rectangular graphite wafers along x/y direction, (b) number (n_g) of graphite wafers along x/y direction, and (c) thickness (h_v) of the monolithic viscoelastic layer within a given thickness (h_d) of the damping layer as it is expressed in terms of a thickness ratio ($r_v = h_v/h_d$). Here, the gap (Δ) would be small enough; however, its variation does not have much effect on the damping capability of the 0–3 VEC layer (Kumar et al., 2018). Accordingly, the gap (Δ) is considered with a value of 100 μm. Apart from this parameter (Δ), the effects of the other two parameters (n_g and r_v) on the damping capability of the 0–3 VEC layer in the CLD treatment are demonstrated in Figures 12.10–12.11. These results clarify that the damping in the layered plate is significantly dependent on the parameters n_g and r_v. Therefore, the parameters n_g and r_v would be taken in an optimal manner for effective.

Accordingly, the parameters for damping in the CLD treatment, n_g and r_v, are presently taken within the bounds as, $3 \leq n_g \leq 10$ and $0.02 \leq r_v \leq 0.22$. The corresponding two-dimensional domain of n_g and r_v is discretized to generate the grid points. At

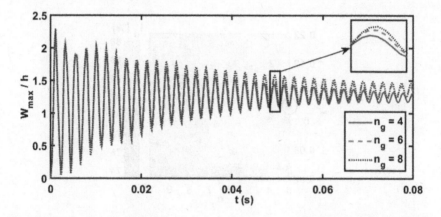

FIGURE 12.10 Nonlinear transient responses of the layered plate with the constrained 0–3 VEC layer for different numbers (n_g) of graphite wafers along the x/y direction ($r_v = 0.08$, $p_0 = 300$ kN/m^2).

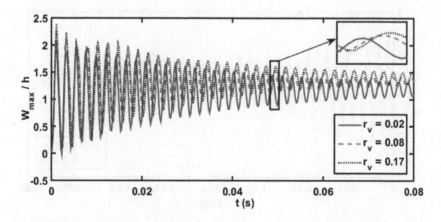

FIGURE 12.11 Nonlinear transient responses of the layered plate with the constrained 0–3 VEC layer for different values of thickness ratio (r_v) ($n_g = 6$, $p_0 = 300$ kN/m^2).

every grid point, the performance index (I_d) is evaluated for obtaining its (I_d) distribution within the same two-dimensional domain, as shown in Figure 12.12. From Figure 12.12, the maximum value of I_d is identified to obtain the optimal values of the parameters $(n_g$ and $r_v)$ as $n_g = 3$, $r_v = 0.12$ for the maximum damping in the CLD treatment of the layered plate.

With this geometrical construction of the 0–3 VEC layer, the transient response of the layered plate is evaluated, as shown in Figure 12.13. A similar response for the use of the butyl rubber layer instead of the 0–3 VEC layer is also illustrated in the same figure. These results show significantly improved damping in the CLD treatment when the 0–3 VEC layer is used in place of the monolithic viscoelastic (butyl rubber) layer. It shows the superior damping capability of the 0–3 VEC compared to that for the monolithic viscoelastic damping layer.

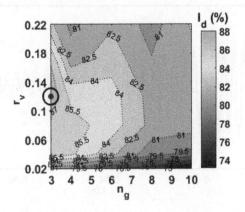

FIGURE 12.12 Contour plot of performance index (I_d) within a two-dimensional domain of n_g and r_v $(p_0 = 300$ kN/m$^2)$.

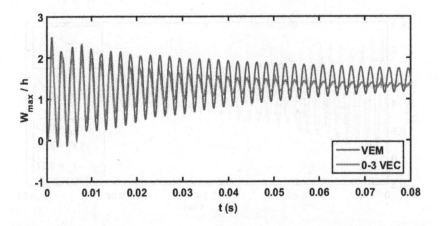

FIGURE 12.13 Nonlinear transient responses of the layered plate for the separate use of 0–3 VEC layer and monolithic viscoelastic material (VEM) layer $(p_0 = 300$ kN/m$^2)$.

Similar results are also shown in Figure 12.14 for the use of VEPC layer in place of the 0–3 VEC or monolithic viscoelastic (butyl rubber) layer. It is clear from these results that the damping in the CLD treatment varies with the volume fraction (ϕ) of graphite particles within the VEPC layer. Here, the damping in the CLD treatment increases with the increasing volume fraction (ϕ); however, it decreases after a certain value of the volume fraction (ϕ). Therefore, the VEPC would be used with an optimal value of the volume fraction (ϕ) of graphite particles. To find this optimal volume fraction of graphite particles, it is taken within a bound as $0 \leq \phi \leq 0.5$, and the variation of the performance index (I_d) is evaluated, as presented in Figure 12.15. The maximum value of I_d indicates maximum damping in the CLD treatment, and thus the optimal value of the volume fraction of inclusion (ϕ) arises as 0.3 (Figure 12.15). With this optimal volume fraction of graphite particles in the constrained VEPC layer, the transient response of the layered plate is evaluated, and this response is plotted in Figure 12.16 along with the similar responses obtained previously (Figure 12.13)

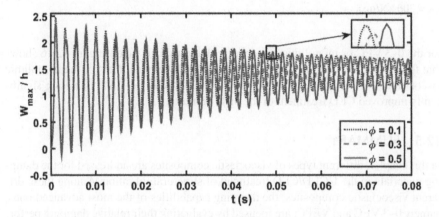

FIGURE 12.14 Nonlinear transient responses of the layered plate with the constrained VEPC layer for different values of volume fraction (ϕ) of graphite particles ($p_0 = 300$ kN/m²).

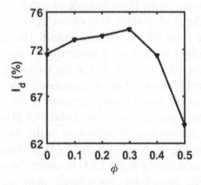

FIGURE 12.15 Variation of performance index (I_d) with the volume fraction (ϕ) of inclusion in the constrained VEPC layer ($p_0 = 300$ kN/m²).

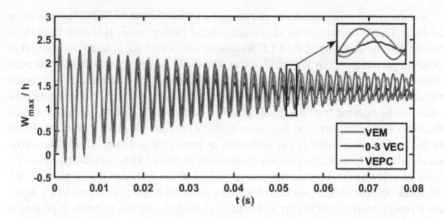

FIGURE 12.16 Nonlinear transient responses of the layered plate for the separate use of VEPC layer, 0–3 VEC layer and monolithic viscoelastic material (VEM) layer ($p_0 = 300$ kN/m^2).

for the 0–3 VEC or monolithic viscoelastic (butyl rubber) layer. Figure 12.16 shows that the damping capability of the VEPC layer is better than that for the monolithic viscoelastic (VEM) layer while the 0–3 VEC layer is the best one in having significantly improved CLD treatment of the plate vibration.

12.5 SUMMARY

In this chapter, different types of viscoelastic composites are addressed for the damping materials in the UCLD/CLD treatment of structural vibration. Among these different viscoelastic composites, the damping capabilities of the most advanced ones, namely 0–3 VEC and VEPC, are focused by evaluating their relative damping performance in the CLD treatment for passive control of plate vibration. This CLD arrangement is achieved through a layered plate configuration where the viscoelastic damping layer is constrained by a stiff constraining layer against the surface of a host plate. The constrained viscoelastic layer is made of either 0–3 VEC layer or VEPC layer, while the conventional monolithic viscoelastic layer is also considered instead of a damping composite layer for the reference result. The damping analysis in the CLD treatment is performed by deriving a geometrically nonlinear FE model of the layered plate where the special attention lies in the modeling of the 0–3 VEC layer for CLD treatment on the FOD viscoelastic constitutive model. The numerical results demonstrate the effects of the geometrical parameters in the construction of the 0–3 VEC / VEPC layer on the damping effectiveness of the CLD treatment. However, the results suggest optimal geometrical configuration of the 0–3 VEC/VEPC layer for achieving effective damping in its use for the CLD treatment. These viscoelastic composites provide better damping in the CLD treatment in comparison to the same damping treatment using monolithic viscoelastic damping materials. However, among these three kinds of damping materials, the best one is the 0–3 VEC layer for an improved CLD treatment.

REFERENCES

Alam, N., and Asnani, N.T. 1984. Vibration and damping analysis of multilayered rectangular plates with constrained viscoelastic layers. *Journal of Sound and Vibration* 97, no. 4: 597–614.

Arikoglu, A. 2014. A new fractional derivative model for linearly viscoelastic materials and parameter identification via genetic algorithms. *Rheologica Acta*, 53, no. 3: 219–233.

Arikoglu, A. 2017. Multi-objective optimal design of hybrid viscoelastic/composite sandwich beams by using the generalized differential quadrature method and the non-dominated sorting genetic algorithm II. *Structural and Multidisciplinary Optimization* 56, no. 4: 885–901.

Attipou, K., Nezamabadi, S., and Zahrouni, H. 2013. A multiscale approach for the vibration analysis of heterogeneous materials: Application to passive damping. *Journal of Sound and Vibration* 332, no. 4: 725–739.

Bagley, R.L., and Torvik, P.J. 1983. Fractional calculus – a different approach to the analysis of viscoelastically damped structures. *AIAA Journal* 21, no. 5: 741–748.

Bagley, R.L., and Torvik, P.J. 1985. Fractional calculus in the transient analysis of viscoelastically damped structures. *AIAA Journal* 23, no. 6: 918–925.

Baz, A.M. 2019. *Active and passive vibration damping*. John Wiley & Sons.

Birman, V., and Kardomateas, G.A. 2018. Review of current trends in research and applications of sandwich structures. *Composites Part B: Engineering* 142: 221–240.

Cheng, L., and Lapointe, R. 1995. Vibration attenuation of panel structures by optimally shaped viscoelastic coating with added weight considerations. *Thin-Walled Structures* 21, no. 4: 307–326.

Christensen, R. 2012. *Theory of viscoelasticity: an introduction*. Elsevier.

Cortes, F., and Elejabarrieta, M.J. 2008. Structural vibration of flexural beams with thick unconstrained layer damping. *International Journal of Solids and Structures* 45, no. 22–23: 5805–5813.

Cunha-Filho, A.G., De Lima, A.M.G., Donadon, M.V., and Leao, L.S. 2016. Flutter suppression of plates using passive constrained viscoelastic layers. *Mechanical Systems and Signal Processing* 79: 99–111.

Datta, P., and Ray, M.C. 2015. Fractional order derivative model of viscoelastic layer for active damping of geometrically nonlinear vibrations of smart composite plates. *Computers, Materials & Continua* 49, no. 1: 47–80.

de Borbon, F., Ambrosini, D., and Curadelli, O. 2014. Damping response of composites beams with carbon nanotubes. *Composites Part B: Engineering* 60: 106–110.

DeValve, C., and Pitchumani, R. 2014. Analysis of vibration damping in a rotating composite beam with embedded carbon nanotubes. *Composite Structures* 110: 289–296.

El-Sabbagh, A., and Baz, A. 2014. Topology optimization of unconstrained damping treatments for plates. *Engineering optimization* 46, no. 9: 1153–1168.

Fang, Z., Yao, L., Tian, S., and Hou, J. 2020. Microstructural topology optimization of constrained layer damping on plates for maximum modal loss factor of macrostructures. *Shock and Vibration* 2020: 8837610.

Ferreira, A.J.M., Araújo, A.L., Neves, A.M.A., Rodrigues, J.D., Carrera, E., Cinefra, M., and Soares, C.M. 2013. A finite element model using a unified formulation for the analysis of viscoelastic sandwich laminates. *Composites Part B: Engineering* 45, no. 1: 1258–1264.

Galucio, A.C., Deu, J.F., and Ohayon, R. 2004. Finite element formulation of viscoelastic sandwich beams using fractional derivative operators. *Computational Mechanics* 33, no. 4: 282–291.

Garrison, M.R., Miles, R.N., Sun, J.Q., and Bao, W. 1994. Random response of a plate partially covered by a constrained layer damper. *Journal of Sound and Vibration* 172, no. 2: 231–245.

Golla, D.F., and Hughes, P.C. 1985. Dynamics of viscoelastic structures – a time-domain, finite element formulation. *Journal of Applied Mechanics* 52, no. 4: 897–906.

Grootenhuis, P. 1970. The control of vibrations with viscoelastic materials. *Journal of Sound and Vibration* 11, no. 4: 421–433.

Gupta, A., and Panda, S. 2021. Hybrid damping treatment of a layered beam using a particle-filled viscoelastic composite layer. *Composite Structures* 262: 113623.

Gupta, A., Panda, S., and Reddy, R.S. 2020a. An actively constrained viscoelastic layer with the inclusion of dispersed graphite particles for control of plate vibration. *Journal of Vibration and Control*: 10.1177/1077546320956533.

Gupta, A., Panda, S., and Reddy, R.S. 2020b. Improved damping in sandwich beams through the inclusion of dispersed graphite particles within the viscoelastic core. *Composite Structures* 247: 112424.

Haddad, Y.M. 1995. *Viscoelasticity of engineering materials* (Vol. 378). Springer Netherlands.

Hamdaoui, M., Robin, G., Jrad, M., and Daya, E.M. 2015. Optimal design of frequency dependent three-layered rectangular composite beams for low mass and high damping. *Composite Structures* 120: 174–182.

Herrmann, A.S., Zahlen, P.C., and Zuardy, I. 2005. Sandwich structures technology in commercial aviation. *In* O.T. Thomsen, E. Bozhevolnaya, A. Lyckegaard (Eds.) *Sandwich structures 7: Advancing with sandwich structures and materials*, Springer, Dordrecht: 13–26.

Hu, H., Belouettar, S., and Potier-Ferry, M. 2008. Review and assessment of various theories for modeling sandwich composites. *Composite Structures* 84, no. 3: 282–292.

Huang, C.Y., and Tsai, J.L. 2015. Characterizing vibration damping response of composite laminates containing silica nanoparticles and rubber particles. *Journal of Composite Materials* 49, no. 5: 545–557.

Huang, P.Y., Reinhall, P.G., Shen, I.Y., and Yellin, J.M. 2001. Thickness deformation of constrained layer damping: an experimental and theoretical evaluation. *Journal of Vibration and Acoustics* 123, no. 2: 213–221.

Jones, C. J. C., and Thompson, D. J. 2000. Rolling noise generated by railway wheels with visco-elastic layers. *Journal of Sound and Vibration* 231, no. 3: 779–790.

Jones, D.I. 2001. *Handbook of viscoelastic vibration damping*. John Wiley & Sons.

Kerwin Jr, E.M. 1959. Damping of flexural waves by a constrained viscoelastic layer. *The Journal of the Acoustical society of America* 31, no. 7: 952–962.

Khalfi, B., and Ross, A. 2016. Transient and harmonic response of a sandwich with partial constrained layer damping: A parametric study. *Composites Part B: Engineering* 91: 44–55.

Khan, S.U., Li, C.Y., Siddiqui, N.A., and Kim, J.K. 2011. Vibration damping characteristics of carbon fiber-reinforced composites containing multi-walled carbon nanotubes. *Composites Science and Technology* 71, no. 12: 1486–1494.

Kumar, A., and Panda, S. 2016. Design of a 1–3 viscoelastic composite layer for improved free/constrained layer passive damping treatment of structural vibration. *Composites Part B: Engineering* 96: 204–214.

Kumar, A., Panda, S., Narsaria, V. and Kumar, A. 2018. Augmented constrained layer damping in plates through the optimal design of a 0–3 viscoelastic composite layer. *Journal of Vibration and Control* 24, no. 23: 5514–5524.

Lakes, R. and Lakes, R.S. 2009. *Viscoelastic materials*. Cambridge University Press.

Lakes, R.S. 1998. *Viscoelastic solids* (Vol. 9). CRC Press.

Lesieutre, G. A., and Bianchini, E. 1995. Time domain modeling of linear viscoelasticity using anelastic displacement fields. *Journal of Vibration and Acoustics* 117, no. 4: 424–430.

Liang, L., Huang, W., Lyu, P., Ma, M., Meng, F., and Sang, Y. 2020. Impacts of PU foam stand-off layer on the vibration damping performance of stand-off free layer damping cantilever beams. *Shock and Vibration* 2020: 8871562.

Lu, Y.P., Killian, J.W., and Everstine, G.C. 1979. Vibrations of three layered damped sandwich plate composites. *Journal of Sound and Vibration* 64, no. 1: 63–71.

Lunden, R. 1979. Optimum distribution of additive damping for vibrating beams. *Journal of Sound and Vibration* 66, no. 1: 25–37.

Madeira, J.F.A., Araújo, A.L., Soares, C.M., Soares, C.M., and Ferreira, A.J.M. 2015. Multiobjective design of viscoelastic laminated composite sandwich panels. *Composites Part B: Engineering* 77: 391–401.

Marcelin, J.L., Trompette, P., and Smati, A. 1992. Optimal constrained layer damping with partial coverage. *Finite elements in Analysis and Design* 12, no. 3–4: 273–280.

Markus, S. 1976. Damping properties of layered cylindrical shells, vibrating in axially symmetric modes. *Journal of Sound and Vibration* 48, no. 4:511–524.

Marsh, E.R., and Hale, L.C. 1998. Damping of flexural waves with imbedded viscoelastic materials. *Journal of Vibration and Acoustics* 120, no. 1: 188–193.

Masti, R.S., and Sainsbury, M.G. 2005. Vibration damping of cylindrical shells partially coated with a constrained viscoelastic treatment having a standoff layer. *Thin-walled structures* 43, no. 9: 1355–1379.

McTavish, D.J., and Hughes, P.C. 1993. Modeling of linear viscoelastic space structures. *Journal of Vibration and Acoustics* 115, no. 1: 103–110.

Mead, D.J. 2007. The measurement of the loss factors of beams and plates with constrained and unconstrained damping layers: A critical assessment. *Journal of Sound and Vibration* 300, no. 3–5: 744–762.

Mead, D.J., and Markus. S. 1969. The forced vibration of a three-layer, damped sandwich beam with arbitrary boundary conditions. *Journal of Sound and Vibration* 10, no. 2: 163–175.

Nakra, B. C. 2000. Structural dynamic modification using additive damping. *Sadhana* 25, no. 3: 277–289.

Oberst, H., and Frankenfeld, K. 1952. Damping of the bending vibrations of thin laminated metal beams connected through adherent layer. *Acustica* 2: 181–194.

Okazaki, A., Tatemichi, A., and Mirza, S. 1994. Damping properties of two-layered cylindrical shells with an unconstrained viscoelastic layer. *Journal of Sound and Vibration* 176, no. 2: 145–161.

Parthasarathy, G., Reddy, C.V.R., and Ganesan, N. 1985. Partial coverage of rectangular plates by unconstrained layer damping treatments. *Journal of Sound and Vibration* 102, no. 2: 203–216.

Porfiri, M., and Gupta, N. 2009. Effect of volume fraction and wall thickness on the elastic properties of hollow particle filled composites. *Composites Part B: Engineering* 40, no. 2: 166–173.

Pritz, T. 1996. Analysis of four-parameter fractional derivative model of real solid materials. *Journal of Sound and Vibration* 195, no. 1: 103–115.

Rajoria, H., and Jalili, N. 2005. Passive vibration damping enhancement using carbon nanotube-epoxy reinforced composites. *Composites Science and Technology* 65, no. 14: 2079–2093.

Rao, M.D. 2003. Recent applications of viscoelastic damping for noise control in automobiles and commercial airplanes. *Journal of Sound and Vibration* 262, no. 3: 457–474.

Reddy, J.N. 2003. *Mechanics of laminated composite plates and shells: theory and analysis.* CRC Press.

Reddy, J.N. 2014. *An introduction to nonlinear finite element analysis: with applications to heat transfer, fluid mechanics, and solid mechanics.* Oxford University Press, Oxford.

Ren, S., Zhao, G., and Zhang, S. 2020. A layerwise finite element formulation for vibration and damping analysis of sandwich plate with moderately thick viscoelastic core. *Mechanics of Advanced Materials and Structures* 27, no. 14: 1201–1212.

Roy, P.K., and Ganesan, N. 1996. Dynamic studies on beams with unconstrained layer damping treatment. *Journal of Sound and Vibration* 195, no. 3: 417–427.

Sahoo, S. R., and Ray M.C. 2019 Analysis of smart damping of laminated composite beams using mesh free method. *International Journal of Mechanics and Materials in Design* 14, no. 3: 359–374.

Scanlan, R. H. 1970. Linear damping models and causality in vibrations. *Journal of Sound Vibration* 13, no. 4: 499–503.

Schmidt, A., and Gaul, L. 2001. FE implementation of viscoelastic constitutive stress-strain relations involving fractional time derivatives. *Constitutive models for rubber* 2: 79–92.

Schmidt, A., and Gaul, L. 2002. Finite element formulation of viscoelastic constitutive equations using fractional time derivatives. *Nonlinear Dynamics* 29, no. 1: 37–55.

Seubert, S.L., Anderson, T.J., and Smelser, R.E. 2000. Passive damping of spinning disks. *Journal of Vibration and Control* 6, no. 5: 715–725.

Shi, J., Song, Q., Liu, Z., and Wan, Y. 2017. Formulating a numerically low-cost method of a constrained layer damper for vibration suppression in thin-walled component milling and experimental validation. *International Journal of Mechanical Sciences* 128–129: 294–311.

Sun, J., and Kari, L. 2010. *Coating methods to increase material damping of compressor blades: Measurements and modeling.* In *Turbo Expo: Power for Land, Sea, and Air* 44014:1157–1165.

Sun, W., Yan, X., and Gao, F. 2018. Analysis of frequency-domain vibration response of thin plate attached with viscoelastic free layer damping. *Mechanics Based Design of Structures and Machines* 46, no. 2: 209–224.

Swallow, W. 1939. An improved method of damping panel vibrations. *British Patent Specification* 513:171.

Tomlinson, G.R. 1990. The use of constrained layer damping in vibration control. *International Journal of Mechanical Sciences* 32, no. 3: 233–242.

Torvik, P.J., and Strickland, D.Z. 1972. Damping additions for plates using constrained viscoelastic layers. *The Journal of the Acoustical Society of America* 51, no. 3: 985–991.

Ungar, E.E., and Kerwin Jr, E.M. 1964. Plate damping due to thickness deformations in attached viscoelastic layers. *The Journal of the Acoustical Society of America* 36, no. 2: 386–392.

Wang, S., Li, Q., and Liang, S. 2020. Topology optimization of embedded and co-cured damping composite structure. *Materials Research Express* 7, no. 10: 105102.

Zarraga, O., Sarria, I., Garcia-Barruetabena, J., and Cortes, F. 2019. Dynamic analysis of plates with thick unconstrained layer damping. *Engineering Structures* 201: 109809.

Zener, C. 1948. *Elasticity and anelasticity of metals.* University of Chicago Press.

Zhai, Y., Su, J., and Liang, S. 2019. Damping properties analysis of composite sandwich doubly-curved shells. *Composites Part B: Engineering* 161: 252–262.

Zheng, H., Cai, C., Pau, G.S.H., and Liu, G.R. 2005. Minimizing vibration response of cylindrical shells through layout optimization of passive constrained layer damping treatments. *Journal of Sound and Vibration* 279, no. 3–5: 739–756.

Zheng, L., Qiu, Q., Wan, H., and Zhang, D. 2014. Damping analysis of multilayer passive constrained layer damping on cylindrical shell using transfer function method. *Journal of Vibration and Acoustics* 136, no 3. 031001.

13 Thermal Buckling and Post-Buckling Behavior of CNT-Reinforced Composite Laminated Plate

Saumya Shah

IIT (BHU), India; Meerut Institute of Engineering and Technology Meerut, India

K.K. Shukla

Motilal Nehru National Institute of Technology Prayagraj, India

Fehim Findik

Sakarya Applied Sciences University, Turkey

CONTENTS

DOI: 10.1201/9781003202233-16

13.1   INTRODUCTION

As a nanoscale graphitic structure, carbon nanotubes are fascinated in research areas because of their electronic and mechanical properties and their thermal properties. Carbon nanotubes (CNTs) are structurally different form of carbon having a cylinder type nanostructure. Nanotubes have been fabricated with length-to-diameter ratio of up to 132,000,000:1, remarkably bigger in comparison to different substances. These carbon molecules with cylindrical structures have remarkable properties, which have great value in the field of nanotechnology, electronics, optics and different fields of materials science and technology. Specifically, on account of their exceptional thermal conductivity and mechanical and electrical properties, carbon nanotubes also have utilizations as add-on to numerous structural materials.

Carbon nanotubes are hollow cylinder structures having 1 nm to 50 nm diameter range and 10 μm length. They usually comprise of only carbon atoms, and it can be made up of graphite sheet that has been rolled into a seamless cylinder.

A nanotube has a property which buckles the structures like a rubber hose upon bending rather than fracture. Carbon nanotubes, particularly single-wall carbon nanotubes (SWNTs), have acquired noteworthy electronic properties; they can be metallic or semi-metallic, based on the geometry that in what way a graphene sheet is rolled up in the form of tube. Carbon nanotubes exhibit high stiffness, strength, and resilience.

Carbon nanotubes may impart the flawless supporting material for an updated category of nanocomposite. Carbon nanotubes are generally applied as conventional carbon fiber to strengthen polymer matrix with an intention to create advanced nanocomposites, they may be also apply for the improvement of the out of plane and inter laminar properties of modern advanced composite materials. The remarkable application of carbon nanotube as strengthen material in nanocomposite has launching the requirement to investigate their exact mechanical loading and as an upcoming procedure, recognize the possibility of fracture that may appear.

The features of carbon nanotube are more complicated than that of traditional material because of influence of their mechanical properties on size and nano structure. Carbon nanotubes represent an unusual behavior of stiffness, strength, and tenacity compared with other fiber structures which generally deficient in these properties. Thermal and electrical conductivity also possess great value, and measurable to other conductive substances.

Kyung and Kim performed three-dimensional thermal buckling behavior in the case of functionally graded materials (FGM) [1]. The finite element model is implemented with an 18-node solid element to examine the fluctuation of material properties and thermal field with accuracy in the thickness direction. It has been discovered that the crucial thermal field of FGM plates is more in comparison to the fully metal plates but having lesser value when compared to the fully ceramic plates. Vodenitcharova and Zhang investigated the pure bending and bending-induced local buckling in case of a nanocomposite beam strengthens by a single-walled carbon nanotube (SWNT) [2]. In this paper, deformation of matrix has been analyzed through Airy stress-function method. It has been released that in thicker matrix layers, the SWNT buckles locally at lesser bending angles and higher flattening ratios.

Zhang and Wang reported the consequences on the mechanical behavior of multi-walled carbon nanotubes (CNTs) because of environmental temperature, through a molecular structural mechanics model whose covalent bonds are treated as dimensional Euler–Bernoulli beam [3]. An examination on the nonlinear bending of simply supported functionally graded nanocomposite plates supported by single-walled carbon nanotubes (SWCNTs) liable to a transverse consistent or waved force in thermal environment was presented by Shen [4]. The theory of FGM is employed to the nanocomposite plates strengthen by SWCNTs with less nanotube volume fractions.

Third-order shear deformation theory was represented by Aghababaei and Reddy using the nonlocal linear elastic theory of Eringen [5]. Small scale outcomes and quadratic fluctuation of shear strain and shear stress because of the plate thickness has been analyzed by third-order shear deformation theory. Shen and Zhang presented perturbation mechanism to examine the consequences of functionally graded carbon nanotubes reinforcement on the thermal buckling and post-buckling pattern of composite plates [6]. The mathematical formulation has been developed on the basis of Reddy's higher order shear deformation theory with von-Kármán type of kinematic nonlinearity.

A general third-order plate theory was presented by Reddy that describes the geometric nonlinearity and two constituent material variations through the plate thickness [7]. The mathematical formulation of third-order plate theory is dependent on power-law imbalance of the material through the thickness and the von Kármán nonlinear strains. The theory given by author in the paper helps in evolving the finite element technique and resolved the influence of the geometric non- linearity and material grading through the thickness on the bending, vibration, and buckling and post-buckling reaction of elastic plates. Zhu et al. used first-order shear deformation theory of thin-to-moderately thick composite plates supported by single-walled carbon nanotubes to show the bending and free vibration affects [8]. Authors had drawn the conclusion that, for the bending analysis, the bending deflection has affected by the CNT volume fraction, the width-to-thickness ratio, and the boundary condition. The nonlinear thermal bifurcation buckling action of carbon nanotube-reinforced composite (CNTRC) beams with surface-bonded piezoelectric layers is analyzed by Rafiee et al. [9]. The Euler–Bernoulli beam theory and von Kármán geometric nonlinearity have been applied to obtain the governing equations of piezoelectric CNTRC beam. The outcomes of the employed actuator voltage, temperature, beam geometry, boundary conditions, and volume fractions of carbon nanotubes have been explored on the buckling of piezoelectric CNTRC beams.

In other paper, the defects of thermal post-buckling behavior of functionally graded carbon nanotube-reinforced composite (FG-CNTRC) beams liable to in-plane temperature fluctuation has been investigated [10]. A parametric study for evaluating the outcomes of faulty mode, half-wave number, location and amplitude has been conducted on their thermal post-buckling performance. The outcomes indicated that the thermal post-buckling is greatly influenced to the imperfection mode, half-wave number, location, and its amplitude. Ninh concentrates on thermal torsional post-buckling of functionally graded carbon nanotube supported composite cylindrical shells with sur-bonding piezoelectric layers and immersed in an elastic medium [11]. The dispersion of strengthen through the thickness of the shells is assumed to be even and functionally graded. The fundamental equations depend on geometrically nonlinearity in von Kármán–Donnell sense have been created within the classical thin shell theory.

The thermal and mechanical buckling analysis FG-CNTRC plates has been investigated by Farzam and Hassani by the help of isogeometric analysis (IGA) according to modified couple stress theory (MCST) [12]. Buckling analysis has been done through a clarified hyperbolic shear deformation theory which fulfill the free transverse shear stress conditions on the above and below surfaces of plate in absence of shear correction factor. The influence of non-identical specifications on mechanical and thermal buckling inspection has been explored. Wang et al. investigated the thermal vibration and buckling of the FG-CNTRC quadrilateral plate with the help of meshless technique [13]. The first-order shear deformation theory (FSDT) with the involvement of the effect of thermal conditions has been used for the examining the firmness of the uni-axial and biaxial mechanical and thermal buckling.

The goal of existing analysis is to analyze the thermal pattern of CNTRC laminated plate. Micromechanics technique is aid in examining the mechanical and thermal properties of a CNTRC laminated plate. The mathematical equation depends on higher order shear deformation theory (HSDT) and von- Kármán nonlinearity. Rapid converging finite double Chebyshev series is helped in getting the logical outcomes of nonlinear governing differential equations of motion.

13.2 MATHEMATICAL MODELING

In this section, CNT is dispersed in an isotropic manner in a matrix and mixture of SWCNT has been used for the manufacturing of CNTRC layer. At the microscopic level, the composition of the carbon nanotube having major impact on universal properties of the composite.

13.2.1 Mori–Tanaka Scheme and Rule of Mixture

The Mori–Tanaka scheme and the rule of mixture are used for evaluating the fruitful material properties of CNTRC. The extended rule of mixture helps in contributing the following equations in the terms of effective Young's modulus and shear modulus,

$$E_{11} = \eta_1 V_{CNT} E_{11}^{CNT} + V_m E^m$$

$$\frac{\eta_2}{E_{22}} = \frac{V_{CNT}}{E_{22}^{CNT}} + \frac{V_m}{E^m}$$

$$\frac{\eta_3}{G_{12}} = \frac{V_{CNT}}{G_{12}^{CNT}} + \frac{V_m}{G^m}$$

(13.1)

where $E_{11}^{CNT}, E_{22}^{CNT}, G_{12}^{CNT}$ are the Young's modulus and shear modulus, respectively, of CNT and E^m and G^m are related properties for the matrix.

The relationship of carbon nanotube and matrix volume fractions is given by:

$$V_{CNT} + V_m = 1$$

(13.2)

The thermal expansion coefficients in the longitudinal and transverse directions can be manifested as:

$$\alpha_{11} = V_{CNT}\alpha_{11}^{CNT} + V_m\alpha^m$$
$$\alpha_{22} = \left(1+\upsilon_{12}^{CNT}\right)V_{CNT}\alpha_{22}^{CNT} + \left(1+\upsilon_m\right)V_m\alpha^m - \upsilon_{12}\alpha_{11} \tag{13.3}$$

where α_{11}^{CNT}, α_{22}^{CNT} and α^m are thermal expansion coefficients, and υ_{12}^{CNT} and υ_m are Poisson's ratios, respectively, of the carbon nanotube and matrix.

The assumption has been made that the material properties of nanotube and matrix is dependent on thermal condition, therefore constructive material behavior of CNTRCs, like Young's modulus, shear modulus, and thermal expansion coefficients, are dependent on temperature and position. The Poisson's ratio based on weekly on thermal condition change and position

$$\upsilon_{12} = V_{CNT}\upsilon_{12}^{CNT} + V_m\upsilon^m \tag{13.4}$$

13.2.2 MATHEMATICAL FORMULATION

This section represents the mathematical equation of real physical issue of CNTRC laminated rectangular plate liable to uni-axial loading.

1. A displacement model, presumed for the laminated composite plate, depends on HSDT. The objective is to find out the undermined values in the displacement field, (HSDT) which are the functions of x,y.
2. Strains can be calculated by von- Kármán strain-displacement relationship.
3. From the strains, stresses can be calculated by using a stress-strain relationship. Stresses and strains are related by diminished transformed stiffness coefficients, which are dependent on materials properties, E_1, E_2, v_{12}, G_{12}, G_{23}, G_{13}. These properties can be evaluated by extended rule of mixture.
4. From these stresses, the in-plane stress, moments, and transverse shear stresses can be calculated.
5. Governing equations of motions has been evaluated from Hamilton's principle, which is shown as

$$\int_{t_1}^{t_2}\delta L\,dt = 0$$

where L is Lagrangian and defined as

$$L = K - \left(U + V\right)$$

where
K = kinetic energy
U = strain energy
V = potential energy

6. Nine equations of motion are achieved with nine variables with the help of Hamilton's principle.
7. Chebyshev polynomial is used as the solution methodology.

13.2.3 DISPLACEMENT FIELD

The flawless bonding connecting the layers of composite laminated plate has been assumed which is shown in Figure 13.1. According to the HSDT, the global displacement field at any point in laminated plate is defined as:

$$u = u_0 + z\psi_x + z^2 u_1 + z^3 \phi_x$$
$$v = v_0 + z\psi_y + z^2 v_1 + z^3 \phi_y \qquad (13.5)$$
$$w = w_0$$

where the parameters u_0, v_0 and w_0 are the displacement components along the (x, y, z) coordinate directions respectively at a point on mid plane (i.e. $z = 0$). The functions ψ_x and ψ_y represented that revolves the normal to the mid plane about y and x axes, respectively. The variables u_1, v_1, ϕ_x, ϕ_y are the higher terms in Taylor's series expansion, showing higher order transverse cross-sectional deformation modes.

13.2.4 STRAIN-DISPLACEMENT RELATIONS

The real and accurate result of carbon nanotube strengthen composite laminated plate has been calculated by considering the nonlinear behavior of composite. Also, the evaluation of post-buckling pattern of laminated composite plate, geometric nonlinearity is integrated in the mathematical model. The governing differential equations get greatly nonlinear for generalized nonlinearity. So, it precludes any attempt

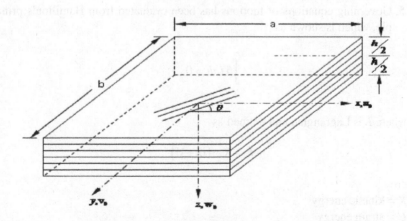

FIGURE 13.1 Geometry of laminated composite plate.

to get the solution with the help of conventional analytical methodology. Hence, in the current job, nonlinear response of laminated composite plate has been done through von- Kármán's nonlinear kinematics.

Considering Von-Kármán nonlinear kinematics, assume strains and the squares of rotations small as compared to unity and with the help of displacement field in Equation (13.6–13.10), strain-displacement relations are shown as:

$$
\begin{Bmatrix} \varepsilon_x \\ \varepsilon_y \\ \gamma_{xy} \\ \gamma_{yz} \\ \gamma_{xz} \end{Bmatrix} = \begin{Bmatrix} \varepsilon_x^0 \\ \varepsilon_y^0 \\ \gamma_{xy}^0 \\ \gamma_{yz}^0 \\ \gamma_{xz}^0 \end{Bmatrix} + z \begin{Bmatrix} \kappa_x \\ \kappa_y \\ \kappa_{xy} \\ 2v_1 \\ 2u_1 \end{Bmatrix} + z^2 \begin{Bmatrix} \varepsilon_x^1 \\ \varepsilon_y^1 \\ \gamma_{xy}^1 \\ 3\phi_y \\ 3\phi_x \end{Bmatrix} + z^3 \begin{Bmatrix} \kappa_x^1 \\ \kappa_y^1 \\ \kappa_{xy}^1 \\ 0 \\ 0 \end{Bmatrix} \tag{13.6}
$$

where

$$
\begin{Bmatrix} \varepsilon_x^0 \\ \varepsilon_y^0 \\ \gamma_{xy}^0 \\ \gamma_{yz}^0 \\ \gamma_{xz}^0 \end{Bmatrix} = \begin{Bmatrix} \dfrac{\partial u_0}{\partial x} + \dfrac{1}{2}\left(\dfrac{\partial w_0}{\partial x}\right)^2 \\[2mm] \dfrac{\partial v_0}{\partial y} + \dfrac{1}{2}\left(\dfrac{\partial w_0}{\partial y}\right)^2 \\[2mm] \dfrac{\partial u_0}{\partial y} + \dfrac{\partial v_0}{\partial x} + \left(\dfrac{\partial w_0}{\partial x}\right)\left(\dfrac{\partial w_0}{\partial y}\right) \\[2mm] \psi_y + \dfrac{\partial w_0}{\partial y} \\[2mm] \psi_x + \dfrac{\partial w_0}{\partial x} \end{Bmatrix} \tag{13.7}
$$

$$
\begin{Bmatrix} \kappa_x \\ \kappa_y \\ \kappa_{xy} \end{Bmatrix} = \begin{Bmatrix} \dfrac{\partial \psi_x}{\partial x} \\[2mm] \dfrac{\partial \psi_y}{\partial y} \\[2mm] \dfrac{\partial \psi_x}{\partial y} + \dfrac{\partial \psi_y}{\partial x} \end{Bmatrix} \tag{13.8}
$$

$$
\begin{Bmatrix} \varepsilon_x^1 \\ \varepsilon_y^1 \\ \gamma_{xy}^1 \end{Bmatrix} = \begin{Bmatrix} \dfrac{\partial u_1}{\partial x} \\[2mm] \dfrac{\partial v_1}{\partial y} \\[2mm] \dfrac{\partial u_1}{\partial y} + \dfrac{\partial v_1}{\partial x} \end{Bmatrix} \tag{13.9}
$$

$$\begin{Bmatrix} \kappa_x^1 \\ \kappa_y^1 \\ \kappa_{xy}^1 \end{Bmatrix} = \begin{Bmatrix} \dfrac{\partial \phi_x}{\partial x} \\ \dfrac{\partial \phi_y}{\partial y} \\ \dfrac{\partial \phi_x}{\partial y} + \dfrac{\partial \phi_y}{\partial x} \end{Bmatrix} \tag{13.10}$$

13.2.5 CONSTITUTIVE STRESS-STRAIN EQUATIONS

Plane stress condition in the lamina has been assumed to get the constitutive stress-strain relation for the kth layer

$$\begin{Bmatrix} \sigma_x \\ \sigma_y \\ \tau_{xy} \\ \tau_{yz} \\ \tau_{xz} \end{Bmatrix} = \begin{bmatrix} \overline{Q}_{11} & \overline{Q}_{12} & \overline{Q}_{16} & 0 & 0 \\ \overline{Q}_{12} & \overline{Q}_{22} & \overline{Q}_{26} & 0 & 0 \\ \overline{Q}_{16} & \overline{Q}_{26} & \overline{Q}_{66} & 0 & 0 \\ 0 & 0 & 0 & \overline{Q}_{44} & \overline{Q}_{45} \\ 0 & 0 & 0 & \overline{Q}_{45} & \overline{Q}_{55} \end{bmatrix} \begin{Bmatrix} \varepsilon_x - \alpha_x \Delta T \\ \varepsilon_y - \alpha_y \Delta T \\ \gamma_{xy} - \alpha_{xy} \Delta T \\ \gamma_{yz} \\ \gamma_{xz} \end{Bmatrix} \tag{13.11}$$

where $\overline{Q}_{ij}$ for $i, j = 1, 2, 3, 4, 5, 6$ are transformed reduced stiffness coefficients.

13.2.6 GOVERNING DIFFERENTIAL EQUATIONS OF MOTION

The governing equations of motion and proper boundary conditions have been evaluated through Hamilton's principle, which can be indicated as:

$$\int_{t_1}^{t_2} \delta L \, dt = 0$$

where, L is Lagrangian and defined as:

$$L = K - (U + V)$$

where
   K = Kinetic energy
   U = Strain energy
   V = Potential energy due to external loads

The equation for kinetic energy and strain energy can be written as:

$$K = \frac{1}{2}\int_{-\frac{h}{2}}^{\frac{h}{2}}\int_A \rho \left\{ \left(\frac{\partial U}{\partial t}\right)^2 + \left(\frac{\partial V}{\partial t}\right)^2 + \left(\frac{\partial W}{\partial t}\right)^2 \right\} dz dA \tag{13.12}$$

$$U = \frac{1}{2}\int_{-\frac{h}{2}}^{\frac{h}{2}}\int_A \left(\sigma_x \varepsilon_x + \sigma_y \varepsilon_y + \tau_{xy}\gamma_{xy} + \tau_{yz}\gamma_{yz} + \tau_{xz}\gamma_{xz}\right) dz dA \tag{13.13}$$

The potential energy due to transverse loads can be presented as:

$$V_q = \int_A qW\, dA \tag{13.14}$$

The governing differential equations have been evaluated using Hamilton's principle:

$$\frac{\partial N_x}{\partial x} + \frac{\partial N_{xy}}{\partial y} = 0$$
$$\frac{\partial N_y}{\partial y} + \frac{\partial N_{xy}}{\partial x} = 0 \tag{13.15}$$

$$\frac{\partial Q_x}{\partial x} + \frac{\partial Q_y}{\partial y} + N_x \frac{\partial^2 w_0}{\partial x^2} + N_y \frac{\partial^2 w_0}{\partial y^2} + 2N_{xy}\frac{\partial^2 w_0}{\partial x \partial y} = 0$$
$$\frac{\partial M_x}{\partial x} + \frac{\partial M_{xy}}{\partial y} - Q_x = 0 \tag{13.16}$$
$$\frac{\partial M_y}{\partial y} + \frac{\partial M_{xy}}{\partial x} - Q_y = 0$$

$$\frac{\partial N_x^*}{\partial x} + \frac{\partial N_{xy}^*}{\partial y} - 2S_x = 0$$
$$\frac{\partial N_y^*}{\partial y} + \frac{\partial N_{xy}^*}{\partial x} - 2S_y = 0$$
$$\frac{\partial M_x^*}{\partial x} + \frac{\partial M_{xy}^*}{\partial y} - 3Q_x^* = 0 \tag{13.17}$$
$$\frac{\partial M_y^*}{\partial y} + \frac{\partial M_{xy}^*}{\partial x} - 3Q_y^* = 0$$

The equation of motion represented in terms of displacement as:

$$A_{11}\frac{\partial^2 u_0}{\partial x^2} + A_{66}\frac{\partial^2 u_0}{\partial y^2} + 2A_{16}\frac{\partial^2 u_0}{\partial x \partial y} + A_{16}\frac{\partial^2 v_0}{\partial x^2} + A_{26}\frac{\partial^2 v_0}{\partial y^2} + \left(A_{12} + A_{66}\right)\frac{\partial^2 v_0}{\partial x \partial y}$$

$$+ B_{11}\frac{\partial^2 \psi_x}{\partial x^2} + B_{66}\frac{\partial^2 \psi_x}{\partial y^2} + 2B_{16}\frac{\partial^2 \psi_x}{\partial x \partial y} + B_{16}\frac{\partial^2 \psi_y}{\partial x^2} + B_{26}\frac{\partial^2 \psi_y}{\partial y^2} + \left(B_{12} + B_{66}\right)\frac{\partial^2 \psi_y}{\partial x \partial y}$$

$$+ D_{11}\frac{\partial^2 u_1}{\partial x^2} + D_{66}\frac{\partial^2 u_1}{\partial y^2} + 2D_{16}\frac{\partial^2 u_1}{\partial x \partial y} + D_{16}\frac{\partial^2 v_1}{\partial x^2} + D_{26}\frac{\partial^2 v_1}{\partial y^2} + \left(D_{12} + D_{66}\right)\frac{\partial^2 v_1}{\partial x \partial y}$$

$$+ E_{11}\frac{\partial^2 \phi_x}{\partial x^2} + E_{66}\frac{\partial^2 \phi_x}{\partial y^2} + 2E_{16}\frac{\partial^2 \phi_x}{\partial x \partial y} + E_{16}\frac{\partial^2 \phi_y}{\partial x^2} + E_{26}\frac{\partial^2 \phi_y}{\partial y^2} + \left(E_{12} + E_{66}\right)\frac{\partial^2 \phi_y}{\partial x \partial y}$$

$$+ \left(A_{11}\frac{\partial^2 w_0}{\partial x^2} + A_{66}\frac{\partial^2 w_0}{\partial y^2} + 2A_{16}\frac{\partial^2 w_0}{\partial x \partial y}\right)\left(\frac{\partial w_0}{\partial x}\right) + \left\{A_{16}\frac{\partial^2 w_0}{\partial x^2} + A_{26}\frac{\partial^2 w_0}{\partial y^2} + \left(A_{12} + A_{66}\right)\frac{\partial^2 w_0}{\partial x \partial y}\right\}\left(\frac{\partial w_0}{\partial x}\right)$$

$$- \frac{\partial}{\partial x}\left[\Delta T \sum_{K=1}^{N} \int_{Z_{K-1}}^{Z_K}\left(\overline{Q_{11}}\alpha_x + \overline{Q_{12}}\alpha_y + \overline{Q_{16}}\alpha_{xy}\right)dz\right] - \frac{\partial}{\partial y}\left[\Delta T \sum_{K=1}^{N} \int_{Z_{K-1}}^{Z_K}\left(\overline{Q_{16}}\alpha_x + \overline{Q_{26}}\alpha_y + \overline{Q_{66}}\alpha_{xy}\right)dz\right] = 0$$

$$(13.18)$$

Similarly, other equations have been written.

13.2.7 TRANSFORMATION OF GOVERNING DIFFERENTIAL EQUATIONS INTO NON-DIMENSIONAL FORM

Then the equation of motion presented in non-dimensional form as:

$$A_1\frac{\partial^2 U_0}{\partial X^2} + A_2\frac{\partial^2 U_0}{\partial Y^2} + A_3\frac{\partial^2 U_0}{\partial X \partial Y} + A_4\frac{\partial^2 V_0}{\partial X^2} + A_5\frac{\partial^2 V_0}{\partial Y^2} + A_6\frac{\partial^2 V_0}{\partial X \partial Y}$$

$$+ A_7\frac{\partial^2 \psi_X}{\partial X^2} + A_8\frac{\partial^2 \psi_X}{\partial Y^2} + A_9\frac{\partial^2 \psi_X}{\partial X \partial Y} + A_{10}\frac{\partial^2 \psi_Y}{\partial X^2} + A_{11}\frac{\partial^2 \psi_Y}{\partial Y^2} + A_{12}\frac{\partial^2 \psi_Y}{\partial X \partial Y}$$

$$+ \frac{\partial W_0}{\partial X}\left[A_{13}\frac{\partial^2 W_0}{\partial X^2} + A_{14}\frac{\partial^2 W_0}{\partial Y^2} + A_{15}\frac{\partial^2 W_0}{\partial X \partial Y}\right] + \left[AD_{16}\frac{\partial^2 W_0}{\partial X^2} + A_{17}\frac{\partial^2 W_0}{\partial Y^2} + A_{18}\frac{\partial^2 W_0}{\partial X \partial Y}\right]\frac{\partial W_0}{\partial Y}$$

$$+ A_{19}\frac{\partial^2 U_1}{\partial X^2} + A_{20}\frac{\partial^2 U_1}{\partial Y^2} + A_{21}\frac{\partial^2 U_1}{\partial X \partial Y} + AD_{22}\frac{\partial^2 V_1}{\partial X^2} + A_{23}\frac{\partial^2 V_1}{\partial Y^2} + A_{24}\frac{\partial^2 V_1}{\partial X \partial Y}$$

$$+ A_{25}\frac{\partial^2 \phi_X}{\partial X^2} + AD_{26}\frac{\partial^2 \phi_X}{\partial Y^2} + A_{27}\frac{\partial^2 \phi_X}{\partial X \partial Y} + A_{28}\frac{\partial^2 \phi_Y}{\partial X^2} + A_{29}\frac{\partial^2 \phi_Y}{\partial Y^2} + A_{30}\frac{\partial^2 \phi_Y}{\partial X \partial Y} = 0$$

$$(13.19)$$

All other equations have been written in similar way.

13.2.8 BOUNDARY CONDITIONS

The related boundary conditions are given as:

(a) Simply supported immovable edges:

$$u_0 = v_0 = w_0 = \psi_y = u_1 = v_1 = \phi_y = M_x = M_x^* = 0 \text{ at } X = \pm 1 \qquad (13.20)$$

$$u_0 = v_0 = w_0 = \psi_x = u_1 = v_1 = \phi_x = M_y = M_y^* = 0 \text{ at } Y = \pm 1 \qquad (13.21)$$

(b) Clamped immovable edges:

$$u_0 = v_0 = w_0 = \psi_x = \psi_y = u_1 = v_1 = \phi_x = \phi_y = 0 \text{ at } X = \pm 1, Y = \pm 1 \qquad (13.22)$$

13.3 METHODOLOGY OF SOLUTION

An analytical approach on the basis of a finite double Chebyshev series for evaluating the governing nonlinear equations of motion of the laminated composite plate liable to variable loading conditions accompanying the suitable boundary conditions and initial conditions is presented in this section. The rapidly converging finite double Chebyshev series is applied for the spatial discretization of the differential equations along with appropriate domain boundary conditions. It results in transformation of a mathematical model from a system of coupled partial differential equations to a system of algebraic equations. The governing nonlinear equations of motion are linearized by quadratic extrapolation method and resolved, insistently.

13.3.1 SPATIAL DISCRETIZATION TECHNIQUE

In structural mechanics, the problems are formulated either regarding partial differential equations or by means of an equivalent variational principle. The differential equations governing the moderate to large deformation pattern of plates are nonlinear and are difficult to manageable to exact results. As a necessity the approximate numerical technique mainly the finite element technique has been used by various investigators. The analytical solutions based on Fourier series that are limited to classical (Levy-type) boundary conditions only have been used for the solution despite their poor convergence properties. Other numerical techniques like Finite Difference, Differential Quadrature and Rayleigh-Ritz etc. have also been used by many researchers. The widespread availability of the computing facility suggests examination, assessment, and correction of existing methodology and creation of latest techniques to resolve the complex nonlinear problems of plates analytically, particularly due to the limited availability of the experimental results. In this thesis, an analytical technique regarding Chebyshev polynomials has been employed for the spatial discretization of the governing equations of motion.

Fox and parker have done the detailed study of the effects of Chebyshev polynomials [14]. The appropriate prime effects and repetition relations have been included in this section for the finalization. A Chebyshev series can converge infinite orders even when the functions are nonperiodic and ends up with discontinuities. The spectral methods (Chebyshev series technique etc.) use global basis function in which each basis function is a polynomial of high degree, which is non-zero, excluding segregated points over the whole computational domain. When rapid continual matrix solvers are implemented, the spectral methods can be much more beneficent in comparison of finite element method for several classes of problems. The major limitations of the spectral techniques are that they are generally much crucial to program and irregular domain inflicts great privation in getting accurate and efficient results.

The Chebyshev polynomial closer to a function $F(x)$ in the range $-1 \leq x \leq 1$ is minimaximal, i.e., if the unrevealed consistent values in the Chebyshev series are examined by assembling the zeros of the Chebyshev polynomials as an assembled points then the outcome series will offer a lower of the higher complete error. Moreover, if a specified function is represented as a aggregate of Chebyshev polynomials, with the help of property of orthogonal functions, it has been observed that out of all the ultra-spherical polynomials, the Chebyshev polynomials converge the solutions in a speedy way.

The ith term in a Chebyshev polynomial is shown by:**

$$T_i(x) = \cos(i\theta); \quad \cos\theta = x; \quad -1 \leq x \leq 1 \tag{13.23}$$

$$T_0(x) = \cos(0) = 1 \tag{13.24}$$

$$T_1(x) = \cos(\theta) = x \tag{13.25}$$

$$T_2(x) = \cos(2\theta) = 2\cos^2\theta - 1 = 2x^2 - 1 \tag{13.26}$$

$$x^2 = \frac{T_0(x) + T_2(x)}{2} \tag{13.27}$$

From trigonometric identity,

$$2\cos n\theta \cos\theta = \cos(n+1)\theta + \cos(n-1)\theta \tag{13.28}$$

Using Equations (13.24–13.28), the recurrence relation can be written as:

$$T_{n+1}(x) + T_{n-1}(x) = 2xT_n(x) \tag{13.29}$$

The displacement functions $\eta(x, y, z)$ and the loadings are estimated in space domain and represented as:

$$(\overline{u_0}, \overline{v_0}, \overline{w_0}, \overline{\psi_x}, \overline{\psi_y}, \overline{u_1}, \overline{v_1}, \overline{\phi_x}, \overline{\phi_y})$$

$$= \sum_{i=0}^{M} \sum_{j=0}^{N} \overline{\delta_{ij}} (\overline{u_{ij}}, \overline{v_{ij}}, \overline{w_{ij}}, \overline{\psi_{xij}}, \overline{\psi_{yij}}, \overline{u_{1ij}}, \overline{v_{1ij}}, \overline{\phi_{xij}}, \overline{\phi_{yij}}) T_i(x) T_j(y); \quad -1 \leq X, Y \leq 1 \tag{13.30}$$

where, M and N are the number of parameters in finite degree double Chebyshev series. The spatial derivatives of the function are expressed as:

$$\overline{u_{ij}^{rs}} = \sum_{i=0}^{M-r}\sum_{j=0}^{N-s}\delta_{ij}\left(u_{ij}\right)^{rs}T_i\left(x\right)T_j\left(y\right); -1 \le x, y \le 1 \tag{13.31}$$

where, r and s are the orders of derivatives with respect to x and y, respectively. The term δ_{ij} defined in Equations (13.27–13.28) takes the following values:

$$\delta = 0.25 \text{ if } i = 0 \ \& \ j = 0$$

$$\delta = 0.5 \ \text{ if } i = 0 \ \& \ j \ne 0 \text{ or } i \ne 0 \ \& \ j = 0$$

$$\delta = 1.0 \text{ otherwise}$$

The derivative function $\overline{u_{ij}^{rs}}$ is calculated with the help of recurrence relations:

$$\left(\overline{u}_{i-1,j}\right)^{rs} = \left(\overline{u}_{i+1,j}\right)^{rs} + 2i\left(\overline{u}_{i,j}\right)^{(r-1)s}$$

$$\left(\overline{u}_{i,j-1}\right)^{rs} = \left(\overline{u}_{i,j+1}\right)^{rs} + 2j\left(\overline{u}_{i,j}\right)^{r(s-1)}$$

Nonlinearity in the governing equations emerges because of the product of the dependent values. The nonlinear terms are linearized at all levels of marching variables. A typical nonlinear function G at step J as product of two dependent values u_{ij}^{rs} and v_{ij}^{rs} is represented as:

$$G_J = \left[\sum_{i=0}^{M-r}\sum_{j=0}^{N}\delta_{ij}\overline{u}_{ij}T_i\left(x\right)T_j\left(y\right)\right]_J\left[\sum_{k=0}^{M}\sum_{l=0}^{N-s}\delta_{ij}\overline{v}_{ij}T_k\left(x\right)T_l\left(y\right)\right]_J \tag{13.32}$$

The variables appearing in the Equation (13.31) have been evaluated through quadratic extrapolation method at every stage J and represented as:

$$\left(\overline{u}_{ij}, \overline{v}_{ij}\right)_J = \eta_1\left(\overline{u}_{ij}, \overline{v}_{ij}\right)_{J-1} + \eta_2\left(\overline{u}_{ij}, \overline{v}_{ij}\right)_{J-2} + \eta_3\left(\overline{u}_{ij}, \overline{v}_{ij}\right)_{J-3} \tag{13.33}$$

The product of two Chebyshev polynomials showing in Equation (13.31) is represented as:

$$T_i\left(x\right)T_j\left(y\right)T_k\left(x\right)T_l\left(y\right)$$
$$= \left[T_{(i+k)}\left(x\right)T_{(j+l)}\left(y\right) + T_{(i+k)}\left(x\right)T_{(j-l)}\left(y\right) + T_{(i-k)}\left(x\right)T_{(j+l)}\left(y\right) + T_{(i-k)}\left(x\right)T_{(j-l)}\left(y\right)\right]/4 \tag{13.34}$$

The displacement functions in the nonlinear variables in the governing equations have been evaluated at every stage of the marching variables with the help of Equations (13.31–13.33). The method explained above has been employed in discretizing and linearizing the governing nonlinear equilibrium equations in space state. The loads (marching variables) are incremented in little stages and nonlinear variables have been calculated at every stage of the marching values and shifted to right hand side. The governing equations ultimately reduced to a set of linear concurrent equations and represented as:

$$\sum_{i=0}^{M-2}\sum_{j=0}^{N-2}F_k\left(u_{ij},v_{ij},w_{ij},\psi_{xij},\psi_{yij},u_{1ij},v_{1ij},\varphi_{xij},\varphi_{yij}\right)T_i\left(X\right)T_j\left(y\right)=0;\left(k=1\,to\,9\right) \quad (13.35)$$

The suitable sets of boundary conditions have been discrete in the same way and showed in the terms of linear concurrent equations.

The set of producing equations offers $9(M-1)*(N-1)$ algebraic equations. Boundary conditions with all edges clamped (CCCC)), all edges simply supported (SSSS) have been given to $(18M+18N+36)$ and $(18M+18N+28)$ algebraic equations, respectively. $9(M+1)$ $(N+1)$ $+36$ and $9(M+1)$ $(N+1)$ $+28$ number of equations have been obtained for above cases, respectively. The aggregate quantity of unspecified coefficients obtained $9(M+1)$ $(N+1)$. This clearly indicates that the final quantity of equations is higher in comparison of overall sum of undefined terms. This thing is also obtained for further boundary conditions. Distinct and appropriate result can be obtained by multiple regression technique depends on least–square error. The set of linear equations have been represented in the matrix term as:

$$Aa = Q \quad (13.36)$$

where A is $(p \times q)$coefficient matrix, a is $(q \times 1)$ displacement coefficient vector, Q is $(p \times 1)$ load vector. Multiple regression technique produces:

$$a = \left(A^T A\right)^{-1} A^T Q \quad (13.37)$$

$$a = BQ \quad (13.38)$$

The methodology of the solution presented herein is coded in Fortran-77 [15]. In order to confirm the precision and to validate the solving technique, convergence studies and comparison of the outcomes are carried out and are shown in subsequent sections.

13.4   RESULTS AND DISCUSSION

The outcomes obtained through the buckling of carbon nanotube strengthen composite laminated plate have been discussed in this section. The buckling temperatures for different conditions are obtained and discussed for harmonic cross-ply laminated plate liable to in-plane loading.

13.4.1 CONVERGENCE STUDY

Convergence analysis has been accomplished for obtaining the stability and accuracy in solution. The convergence test has been performed on all edges simply supported immovable, harmonic cross-ply CNTRC laminated plate. The following material properties have been used:

For matrix (PMMA):

$$E^m = 2.5 \text{GPa}, V_m = 0.88, G^m = 0.932 \text{GPa}, \upsilon^m = 0.34, \alpha^m = 45.0 * 10^{-6}/\text{K}$$

For SWNT:

$$E^{CNT} = 5.6466 * 10^3 \text{ GPa}, V_{CNT} = 0.12, G_{12}{}^{CNT} = 1944.5 \text{GPa}, \upsilon_{12}{}^{CNT} = 0.175,$$

$$\alpha_{11}{}^{CNT} = 3.4584 * 10^{-6} / \text{K}, \alpha_{22}{}^{CNT} = 5.1682 * 10^{-6}/\text{K}$$

Convergence of the buckling thermal value of symmetric [0/90/90/0], cross-ply, simply supported square plate thick plate (a/h=10) liable to uni-axial compression is studied and presented in Figure 13.2. The results specified that solution is become convergent when total terms in the Chebyshev series is more than 8.

13.4.2 VALIDATION STUDY

For the validation of results, values of buckling temperatures are match up with those given in the paper Shen [6].

Table 13.2 shows buckling temperatures of CNTRC laminated plate when dissimilar variables of beta and nanotube volume fraction have been taken. Outcomes are compared with the values given in paper [6].

TABLE 13.1

Convergence of non-dimensional buckling temperature for all edges simply supported immovable CNTRC laminated plate [0/90/90/0] liable to uni-axial compression. (h = 2mm, beta = 10, VCNT = 0.12)

S. No.	M,N	Buckling Temperature (^{0}K)
1	5	450
2	6	422
3	7	414
4	8	398
5	9	385
6	10	385

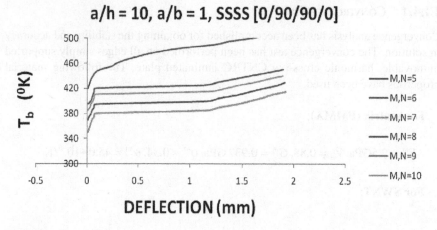

FIGURE 13.2 Convergence of the buckling temperature of simply supported, square CNTRC laminated plate.

TABLE 13.2

Comparison of buckling thermal values (in °K) for CNTRC laminated plate [0/90/0/90] under stable temperature increment (all four edges of plate are simply supported immovable)

BETA	VCNT=0.12		VCNT=0.17		VCNT=0.28	
	Shen (2010)	Current Value	Shen (2010)	Current Value	Shen (2010)	Current Value
10	388.19	385	399.44	395	391.62	390
20	339.42	325	343.00	340	344.08	342

13.4.3 Results and Discussions

Figures 13.3 and 13.4 shows the consequences of discrete values of nanotube volume fractions on the buckling temperature (T_b) for simply supported and clamped boundary conditions liable to a consistent temperature increment. It shows that buckling temperature is highest for the volume fraction 0.17. Also, the buckling thermal value for all volume fractions (=0.12, 0.17 and 0.28) is high regarding all edges clamped fixed in comparison of all edges simply supported immovable.

Figure 13.5 and 13.6 represents the consequences of nanotube volume fraction for the case of a/h =20 and two dissimilar boundary conditions, i.e., all edges simply supported immovable and all edges clamped immovable respectively. In contrast for the case of a/h =10, buckling temperature is highest for the nanotube volume fraction 0.28. Here also, buckling temperature is more for the boundary condition all edges clamped immovable in comparison of all edges simply supported immovable.

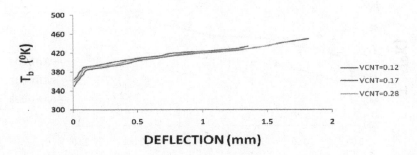

FIGURE 13.3   Effect of nanotube volume fraction on thermal buckling behavior (all edges simply supported immovable).

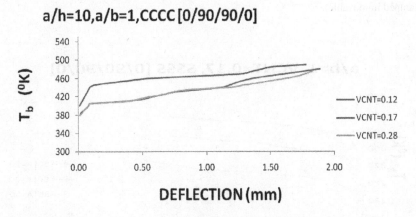

FIGURE 13.4   Effect of nanotube volume fraction on thermal buckling property (all edges clamped immovable).

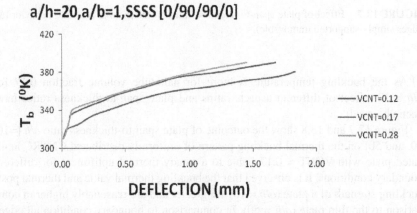

FIGURE 13.5   Effect of nanotube volume fraction on thermal buckling behavior (all edges simply supported immovable).

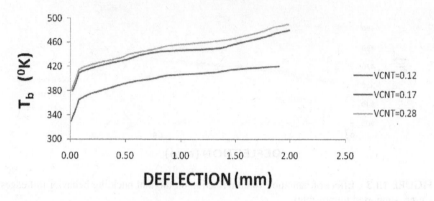

FIGURE 13.6 Effect of nanotube volume fraction on thermal buckling pattern (all edges clamped immovable).

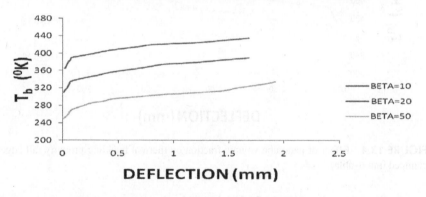

FIGURE 13.7 Effect of plate span-to-thickness ratio on the thermal buckling behavior (all edges simply supported immovable).

As the buckling temperature is more for nanotube volume fraction 0.17 for $a/h = 10$, effect of different aspects ratios and plate span-to-thickness ratios have been observed.

Figure 13.7 and 13.8 show the outcome of plate span-to-thickness ratio a/h (=10, 20, and 50) on the thermal buckling pattern of uniformly distributed CNTRC laminated plate with VCNT = 0.17 liable to a steady thermal upliftment for different boundary conditions. It is observed that the buckling thermal value and thermal postbuckling strength of a plate ($a/h = 10$) are greater and are reasonably higher in comparison to the thin plate ($a/h = 50$). In comparison to boundary condition all edges

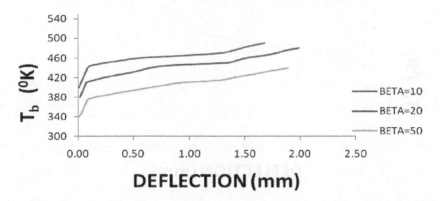

FIGURE 13.8 Effect of plate span-to-thickness ratio on the thermal buckling behavior (all edges clamped immovable).

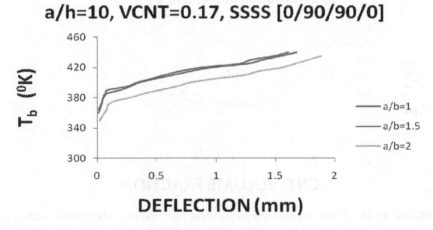

FIGURE 13.9 Effect of plate aspect ratio on the thermal buckling behavior (all edges simply supported immovable).

clamped immovable, in case of all edges simply supported immovable, there is large variation in the buckling thermal value for three variables of plate span-to-thickness ratio.

Figure 13.9 and 13.10 represents the consequences of plate aspect ratio (a/b = 1, 1.5, and 2) on the thermal buckling pattern of uniformly distributed CNTRC laminated plate with nanotube volume fraction 0.17 liable to steady thermal increment. It is anticipated that these outcomes represented that the buckling thermal value is less for high value of plate aspect ratio.

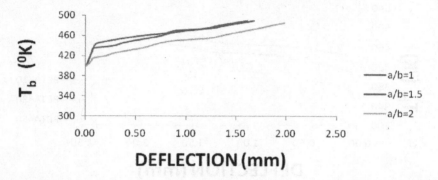

FIGURE 13.10 Effect of plate aspect ratio on the thermal buckling behavior (all edges clamped immovable).

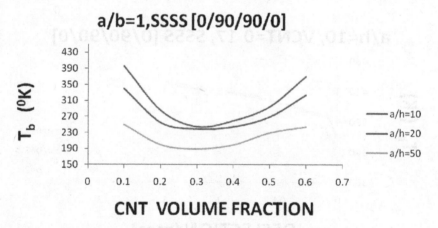

FIGURE 13.11 Effect of plate span-to-thickness ratio and nanotube volume fraction on thermal buckling pattern (all edges simply supported and immovable).

Figure 13.11 and 13.12 shows the buckling temperatures for discrete variables of nanotube volume fractions at different plate span-to-thickness ratio. It has been noticed that that buckling thermal value is highest for the minimum variable of plate thickness ratio (a/h = 10). The value of plate thickness ratio is higher when there is decrement in buckling temperature for both boundary conditions. Also, it is observed that as the value of nanotube volume fraction decreases value of buckling temperature decreases up to VCNT = 0.3, after that buckling temperature increases with nanotube volume fraction.

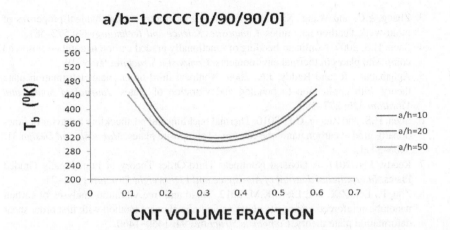

FIGURE 13.12   Effect of plate span-to-thickness ratio and nanotube volume fraction on thermal buckling pattern (all edges clamped immovable).

13.5   CONCLUSION

In the current work, thermal behaviors of the CNT strengthen composite laminated plate is evaluated liable to uniform temperature fluctuation. The mathematical equation pattern of selected physical issue is done by using HSDT and von- Kármán's nonlinearity. The Chebyshev polynomial applied for the spatial discretization of the field variables is found to be efficient in resolving the boundary value problem.

The main conclusions drawn from work are given below:

- Its outcomes represented that buckling temperature of CNTRC laminated plate is lower for high value of nanotube volume fraction up to value 0.3 and again increases for further values of volume fraction.
- Value of buckling temperature for different nanotube volume fraction based on the plate span-to-thickness ratio.
- Buckling temperature is low for square plate in comparison to rectangular plate.
- The outcomes reveal that buckling temperature is greater in case of boundary condition all edges clamped immovable as compared to that for all edges simply supported immovable.

REFERENCES

1. Kyung, S.N. and Kim, J.H. 2004. Three-dimensional thermal buckling analysis of functionally graded materials. *Composites* 35: 429–437.
2. Vodenitcharova, T. and Zhang, L.C. 2006. Bending and local buckling of a nanocomposite beam reinforced by a single-walled carbon nanotube. *International Journal of Solids and Structures* 43: 3006–3024.

3. Zhang, Y.C. and Wang, X. 2008. Effects of temperature on mechanical properties of multi-walled carbon nanotubes. *Composites Science and Technology* 68: 572–581.

4 Shen, H.S. 2009. Nonlinear bending of functionally graded carbon nanotube-reinforced composite plates in thermal environments. *Composite Structures* 91: 9–19.

5. Aghababaei, R. and Reddy, J.N. 2009. Nonlocal third-order shear deformation plate theory with application to bending and vibration of plates. *Journal of Sound and Vibration* 326: 277–289.

6. Shen, H.S. and Zhang, C.L. 2010. Thermal buckling and postbuckling behavior of functionally graded carbon nanotube-reinforced composite plates. *Materials and Design* 31: 3403–3411.

7. Reddy, J.N. 2011. A General Nonlinear Third-Order Theory of Functionally Graded Plates. *International Journal of Aerospace and Lightweight Structures* 1: 1–21.

8. Zhu, P., Lei, Z.X. and Liew, K.M. 2012. Static and free vibration analyses of carbon nanotube-reinforced composite plates using finite element method with first order shear deformation plate theory. *Composite Structures* 94: 1450–1460.

9. Rafiee, M., Yang, J. and Kitipornchai, Siritiwat 2013. Thermal bifurcation buckling of piezoelectric carbon nanotube reinforced composite beams. *Computer and Mathematics with Applications* 7: 1147–1160.

10. Wu, H., Kitipornchai, Sritawat and Yang, J. 2017. Imperfection sensitivity of thermal post-buckling behaviour of functionally graded carbon nanotube-reinforced composite beams. *Applied Mathematical Modelling* 42: 735–752.

11. Ninh, Dinh Gia. 2018. Nonlinear thermal torsional post-buckling of carbon nanotube-reinforced composite cylindrical shell with piezoelectric actuator layers surrounded by elastic medium. *Thin-Walled Structures* 123: 528–538.

12. Farzam, A. and Hassani, B. 2018. Thermal and mechanical buckling analysis of FG carbon nanotube reinforced composite plates using modified couple stress theory and isogeometric approach. *Composite Structures* 206: 774–790.

13. Wang, J. F., Cao, S.H. and Zhang, W. 2021. Thermal vibration and buckling analysis of functionally graded carbon nanotube reinforced composite quadrilateral plate. *European Journal of Mechanics – A/Solids* 85: 104–105.

14. Fox, L. and Parker, J.B. 1996. *Chebyshev Polynomials in Numerical Analysis*. Oxford University Press, London.

15. Rajaraman, V. *Computer Programming in Fortran 77 (with an introduction to Fortran 90)*, Fourth edition. PHI Learning Pvt. Ltd., Delhi.

16. Carbon Nanotechnologies Incorporated. July 3, 2002 http://www.cnanotech.com/.

17. Gibson, R. F. 1994. *Principles of Composite Material Mechanics*. McGraw- Hill Publications, New York.

18. Reddy, J.N. n.d. *Mechanics of Laminated Composite Plates and Shells*. Second edition. CRC Press, Taylor & Francis, Boca Raton, FL..

14 Mesoscale Analysis of Polymer-CNT Composites for Evaluation of Elasto-Plastic and Thermo-Elastic Properties

Gaurav Arora

Composites Design and Manufacturing Research Group,
Indian Institute of Technology Mandi, India

Himanshu Pathak

Composites Design and Manufacturing Research Group,
Indian Institute of Technology Mandi, India

CONTENTS

DOI: 10.1201/9781003202233-17

14.1   INTRODUCTION

The small units or molecules called monomers form a large macromolecule termed polymer when repeated by chemical bonding. There are unlimited blends of monomers possible to form a polymer. The simple form of the polymer is named homopolymer, has one sort of monomer unit. The complexity is conceivable in the structure of polymers to achieve an extensive network by cross-linking. The cross-linking intensely changes the properties of the polymers. Some examples of cross-linking polymers acting as single units are high-density polyethylene and rubber bands. Copolymers have various structural monomer units. The material's physical and chemical nature is highly dependent on the quantities of the monomer units present. Polymers are either organic or inorganic. The primary backbone of the molecular chain has carbon atoms, then the class of polymers is called organic polymers, otherwise inorganic polymers. Organic polymers are subcategorized into natural and synthetic polymers. Some popular natural and synthetic polymers are as follows:

a) Natural polymers: cellulose, glycogen, starch, seaweed, and vegetable gums are polysaccharides. albumin, collagen, DNA, globulin, insulin, and RNA are polypeptides. Rubber and polyisoprene are hydrocarbons.
b) Synthetic polymers: These are of two types, i.e., thermoplastics and thermosets.
   i) Thermoplastics: These polymers are stable at a specific temperature and soften at a specific elevated temperature. Upon cooling, below the glass transition temperature is again stable. Acrylate resin, fluorocarbon resin, linear polyethylene, polystyrene, poly-lactic acid, and polyglycolic acid are examples of thermoplastic polymers.
   ii) Thermosets: These polymers mostly form covalent bonds, unlike thermoplastics, upon heating. The physical properties can't be altered after curing these polymers. Alkyds, cross-linked polyethylene, polyesters, and phenolics are examples of thermoset polymers.

Polymers are used in almost every working condition. Polymers are used in aerospace, medical, automotive, electrical, and electronics applications. Few advantages and applications of polymers for their successful use in these applications are as follows:

• Reasonable cost, lightweight, and density, resistance to corrosion, absorbing radar waves, and high elasticity are the advantages of fabricating the aircraft's windows and coverings (Lim, 2012).
• Compatible with imaging systems, easy molding, corrosion resistance, good thermal and electrical insulators are the advantages of polymers for manufacturing implantable medical devices (Maddock et al., 2012).
• High mechanical strength and quantum efficiency are the advantages of thermosetting polymers for fabricating light-emitting diodes (LEDs) (Roitman et al., 1998).
• Low weight, electrical conductivity, low thermal conductivity, low thermal expansion, and flammability are the properties required for using polymers in automotive and space applications (Lim, 2012).

Apart from these advantages, there are a few disadvantages of polymers that should be taken care of before using them in structural applications. High viscosity, poor creep resistance, low impact resistance, low fracture toughness, low tensile strength, low modulus, high processing time, temperature, and pressures are the disadvantages of thermoset and thermoplastic polymers.

Therefore, to eradicate these disadvantages, polymers are reinforced with several inclusions. The inclusion of various shapes and sizes is used as reinforcing materials to tailor the polymers' (electrical, mechanical, and thermal) properties. Thus, the tailored material is known as polymer composite. The reinforcement of nano-level or nano-size or nano-dimension, when used to enhance the properties, the material produced or formed, is termed as polymer nanocomposites (PNCs). After Sumio Iijima's discovery (Iijima, 1991) in 1991, the carbon nanotubes (CNTs) have revolutionized the research field for their use in fabricating PNCs. An aspect ratio greater than 1000, Young's modulus 1 TPa, tensile strength 50–100 GPa, and interfacial adhesion between CNTs and polymers attracted their use for structural applications (Arash et al., 2014; Iijima, 1991; Pandey, 2015; Pantano, 2018; Silvestre et al., 2016). CNTs have single, double, and multi-walls, thus called single-walled CNTs or SWCNTs, double-walled CNTs or DWCNTs, and multi-walled CNTSs or MWCNTs. MWCNTs are heterogeneous compared to SWCNTs. Also, MWCNTs have inferior electrical, mechanical, and thermal properties compared to SWCNTs. But the advantages of easy and cheap manufacturing of MWCNTs attracted their use in PNCs.

Literature has shown that CNTs are the best reinforcement for polymers (Arora et al., n.d., Arora and Pathak, 2020, 2021; Clifton et al., 2020; Lu and Tsai, 2013; Moore and Shi, 2014; Parandoush and Lin, 2017; Srivastava et al., 2017; Sturm et al., 2010; Yadav et al., 2020). The modeling of polymer-CNT composites is not new, but accurate and efficient mechanical properties' prediction is still ongoing. Mesoscale models predict the polymer-CNT composites' effective mechanical properties explicitly to use at the macro scale (Arora and Pathak, 2019c, 2020; Bhatnagar et al., 2021; Grabowski et al., 2017; Padmanabhan et al., 2020). A concept of representative volume element (RVE) has been used to model the polymer-CNT composites (Arora and Pathak, 2019a, 2019b, 2019c, 2020; Bhatnagar et al., 2021; Grabowski et al., 2017; Padmanabhan et al., 2020; Weidt and Figiel, 2014). The effect of CNT characteristics like aspect ratio, orientation, volume fraction, and interphase was studied using the RVE method (Grabowski et al., 2017; Sun et al., 2021; Weidt and Figiel, 2014) (Ansari et al., 2017; Kumar and Srinivas, 2014; Mohammadi and Yas, 2016). Elastic, elasto-plastic, visco-elastic, electro-mechanical, etc., responses of the composites were investigated using the RVE method (Grabowski et al., 2017; Pierard et al., 2004; Weidt and Figiel, 2014). The homogenization and finite element methods (FEM) predicted the mesoscale's effective properties accurately and efficiently (Chwał and Muc, 2015; Gómez-Pachón et al., 2013; Pakseresht et al., 2021; Savvas et al., 2016; Zhai et al., 2018). Therefore, the chapter will focus on the CNT characteristics studied with homogenization and FEM using RVE to evaluate the effective mechanical and thermal properties.

14.2 HOMOGENIZATION AND FEM TECHNIQUES

Homogenization schemes aim to replace the real heterogeneous material with a fictitious homogenous material whose behavior is the same at the macro-level. Every material is heterogeneous at a certain level. The continuum micromechanics approach defines the RVE from the mechanical viewpoint to analyze the micro-level heterogeneous material. The main aim of mean-field homogenization (MFH) schemes is to calculate approximate volume averages of the strains and stresses at the constituent and RVE level.

14.2.1 MORI–TANAKA METHOD (MTM) FOR THERMO-ELASTIC COMPOSITES

Consider the thermo-elastic composites subjected to isothermal conditions. Each phase, i.e., polymer (denoted by 'pr') and nanofiller (denoted by 'nr') being homogenous (individually), follow the following constitutive equations:

$$\sigma_{pr}(x) = C_{pr}\,(\varepsilon_{pr}(x) - \alpha_{pr}\Delta T) \quad \text{or}$$
$$= C_{pr}\,\varepsilon_{pr}(x) + \beta_{pr}\Delta T \tag{14.1}$$

$$\sigma_{nr}(x) = C_{nr}\,(\varepsilon_{nr}(x) - \alpha_{nr}\Delta T) \quad \text{or}$$
$$= C_{nr}\,\varepsilon_{nr}(x) + \beta_{nr}\Delta T \tag{14.2}$$

C and α represent the elastic stiffness and thermal expansion coefficient in Equations 14.1 and 14.2 of the respective constituent where $\beta = -C : \alpha$. The RVE is subjected to linear displacement along the x-axis corresponding to macroscopic strain and a uniform change in temperature. For a two-phase model, the Mori–Tanaka model for an isothermal case can be defined based on their strain concentration tensor $S\varepsilon$. The relation among mean strain (overall reinforcements) and macroscopic strain in thermo-elasticity are as follows:

$$\langle \varepsilon \rangle_{V_{nr}} = S^\varepsilon : E + a^\varepsilon\, \Delta T \tag{14.3}$$

where

$$a^\varepsilon \equiv (S^\varepsilon - I) : (C_{nr} - C_{px})^{-1} : (\beta_{nr} - \beta_{px}),$$
$$E = c_{pr} \langle \varepsilon \rangle_{V_{pr}} + c_{nr} \langle \varepsilon \rangle_{V_{nr}} \tag{14.4}$$

c_i and c_{ii} represent the volume fraction of the matrix and nanofiller. Therefore, the scheme predicts the macro response of the composite as follows:

$$\langle \sigma \rangle = \bar{C} : E + \bar{\beta}\, \Delta T \tag{14.5}$$

where,

$$\overline{\mathbf{C}} = \left[c_{nr}\,\mathbf{C}_{nr} : \mathbf{S}^{\varepsilon} + (1 - c_{nr})\mathbf{C}_{pr} \right] : \left[c_{nr}\,\mathbf{S}^{\varepsilon} + (1 - c_{nr})\mathbf{I} \right]^{-1},$$
and
$$\overline{\beta} = c_{nr}\,\mathbf{C}_{nr} + c_{pr}\,\mathbf{C}_{pr} - c_{nr}\,(\mathbf{C}_{pr} - \mathbf{C}_{nr}) : \mathbf{a}^{\varepsilon}, \tag{14.6}$$
and
$$\overline{\alpha} = -\overline{\mathbf{C}}^{-1} : \overline{\beta}$$

The macro thermal expansion coefficient is a function of each phase's stiffness, and macro stiffness is represented in Equation 14.6.

14.2.2 Mori–Tanaka Method (MTM) for Elastic Composites

In a micro-macro analysis, the macro strain $\overline{\varepsilon}$ is known, and the macro stress $\overline{\sigma}$ is computed. An RVE of domain Ω and boundary $\partial\Omega$ when subjected to linear boundary conditions (LBC) thus the macro-stresses and strains can be related to average stresses and strains.

The strain averages per phase are related to $\mathbf{B}^{\varepsilon}$, strain concentration tensor by the following relation:

$$\langle \varepsilon \rangle_{\Omega_{nr}} = \mathbf{B}^{\varepsilon} : \langle \varepsilon \rangle_{\Omega_{pr}} \tag{14.7}$$

The per-phase average strains are related to the macro strains $\langle \varepsilon \rangle$ as:

$$\langle \varepsilon \rangle_{\Omega_{pr}} = \left[c_{nr}\,\mathbf{B}^{\varepsilon} + c_{pr}\mathbf{I} \right]^{-1} : \langle \varepsilon \rangle$$
$$\langle \varepsilon \rangle_{\Omega_{nr}} = \mathbf{B}^{\varepsilon} : \left[c_{nr}\,\mathbf{B}^{\varepsilon} + c_{pr}\mathbf{I} \right]^{-1} : \langle \varepsilon \rangle \tag{14.8}$$

where '**I**' is the fourth-order symmetric identity tensor in Equation (14.8).

The homogenization models are based on Eshelby's fundamental solution. The solution is based on a single ellipsoidal inclusion (I) of stiffness $\mathbf{C}_{nr}$ reinforced in an infinite matrix of stiffness $\mathbf{C}_{pr}$. The far strain $\overline{\varepsilon}$, i.e., linear displacement, when applied to RVE represents that the strain inside the ellipsoidal inclusion is uniform, and their relationship is given as:

$$\varepsilon(x) = \mathbf{H}^{\varepsilon}\left(\mathrm{I}, \mathbf{C}_{pr}, \mathbf{C}_{nr} \right) : \overline{\varepsilon} \tag{14.9}$$

where '$\mathbf{H}^{\varepsilon}$' is the strain concentration of single inclusion.

$$\mathbf{H}^{\varepsilon}\left(\mathrm{I}, \mathbf{C}_{pr}, \mathbf{C}_{nr} \right) = \left[\mathbf{I} + \xi : \left[(\mathbf{C}_{pr})^{-1} : \mathbf{C}_{nr} - \mathbf{I} \right] \right]^{-1} \tag{14.10}$$

where 'ξ' is Eshelby's tensor, which depends on the geometry of the inclusion and matrix's stiffness.

Considering the inclusions of the same size, same orientation, and stiffness C_{ii} and strain concentration tensor (represented in Equation (14.7)) for any homogenization model, the macro stiffness $\bar{C}$ is represented as:

$$\bar{C} = \left[c_{nr} \, C_{nr} \, : \, B^\varepsilon \, + \, c_{pr} C_{pr} \right] : \left[c_{nr} \, B^\varepsilon \, + \, c_{pr} \, I \right]^{-1} \qquad (14.11)$$

14.2.3 Mesoscale Finite Element Method

The mesoscale finite element method utilizes the 3-D RVE to evaluate the composites' effective elastic properties. The composites are assumed orthotropic initially and subjected to small displacement (Zhang et al., 2018). The generalized Hooke's law for any composite is represented as:

$$\overline{\sigma}_{ij} = C_{ijkl} \, \overline{\varepsilon}_{kl} \qquad (14.12)$$

where C_{ijkl} is the effective stiffness tensor. In the unit-volume, the average stresses, $\overline{\sigma}_{ij}$ and average strains, $\overline{\varepsilon}_{kl}$ are defined as follows (Zhang et al., 2018):

$$\overline{\sigma}_{ij} = \frac{1}{V} \int_V \sigma_{ij} \, dV \quad \text{and} \quad \overline{\varepsilon}_{ij} = \frac{1}{V} \int_V \varepsilon_{ij} \, dV \qquad (14.13)$$

where V is the volume of the unit cell. The periodic boundary conditions are employed on the 3-D RVE represented in Table 14.1.

The small displacement is known, and the forces on the surfaces can be calculated or $\overline{\varepsilon}_{kl}$ are known and $\overline{\sigma}_{ij}$ can be determined by the following relation (Zhang et al., 2018):

$$\overline{\sigma}_{ij} = \frac{\left(F_i \right)_j}{S_j} \qquad (14.14)$$

TABLE 14.1
Periodic boundary conditions

Loading Direction	Effective Elastic Moduli	Periodic Boundary Conditions
11	E_{11}	$\varepsilon_{11} = 0.1$, $\varepsilon_{22} = 0$, $\varepsilon_{33} = 0$, $\gamma_{12} = 0$, $\gamma_{13} = 0$, $\gamma_{23} = 0$
22	E_{22}	$\varepsilon_{11} = 0$, $\varepsilon_{22} = 0.1$, $\varepsilon_{33} = 0$, $\gamma_{12} = 0$, $\gamma_{13} = 0$, $\gamma_{23} = 0$
33	E_{33}	$\varepsilon_{11} = 0$, $\varepsilon_{22} = 0$, $\varepsilon_{33} = 0.1$, $\gamma_{12} = 0$, $\gamma_{13} = 0$, $\gamma_{23} = 0$
12	G_{12}	$\varepsilon_{11} = 0$, $\varepsilon_{22} = 0$, $\varepsilon_{33} = 0.1$, $\gamma_{12} = 0$, $\gamma_{13} = 0$, $\gamma_{23} = 0$
13	G_{13}	$\varepsilon_{11} = 0$, $\varepsilon_{22} = 0$, $\varepsilon_{33} = 0$, $\gamma_{12} = 0$, $\gamma_{13} = 0.1$, $\gamma_{23} = 0$
23	G_{23}	$\varepsilon_{11} = 0$, $\varepsilon_{22} = 0$, $\varepsilon_{33} = 0$, $\gamma_{12} = 0$, $\gamma_{13} = 0$, $\gamma_{23} = 0.1$

where $(\mathbf{F}_i)_j$ is the ith resultant forces on the jth surface and $\mathbf{S}j$ is the jth surface area.

The computed stress-strain curves are then used to define the various properties of the composites. Also, the linear range of the stress-strain curves can be used to determine the elastic moduli of the composites using the following relations (Zhang et al., 2018):

$$\mathbf{E}_{ii} = \frac{\bar{\sigma}_{ii}}{\bar{\varepsilon}_{ii}}, \quad v_{ij} = \frac{-\bar{\varepsilon}_{jj}}{\bar{\varepsilon}_{ii}}, \quad \mathbf{G}_{ij} = \frac{\bar{\sigma}_{ij}}{\bar{\varepsilon}_{ij}} \quad (i,j = 1,2,3) \tag{14.15}$$

14.3 APPLICATION OF HOMOGENIZATION AND FEM TECHNIQUES

The mesoscale model uses the constituents' properties mentioned in Tables 14.2 and 14.3 to predict the isothermal polymer-CNT composites' mechanical responses. Tables 14.2 and 14.3 represent the isotropic properties of CNTs and elastoplastic properties of the polymer, respectively. CNTs are assumed to be linear, homogenous, and isotropic. An elastic, homogenous, and linear coating of Young's modulus 3.2 GPa and Poisson ratio's 0.3 has been considered over the CNTs. All cases have perfect bonding assumptions between the constituents. Mori–Tanaka and FEM are the two techniques employed to predict polymer-CNT composites' effective properties. Two different RVEs of aligned CNT and random CNTs in the polymer matrix are shown in Figure 14.1.

TABLE 14.2
Isotropic properties of CNTs (Arora and Pathak, 2019c)

Constituent	Young's Modulus (E)	Poisson's Ratio (ν)	Volume Fraction (%)	Density (kg/m³)
CNT	1 TPa	0.3	2.75	2200

TABLE 14.3
Elastoplastic properties of the polymer (Arora and Pathak, 2019c)

Constituent	Polymer
Young's modulus (E)	3.2 GPa
Poisson's ratio (ν)	0.3
Volume fraction	97.25 %
Density	950 kg/m³
Plasticity model	J_2
Hardening law	Exponential model
Hardening modulus	22.2 MPa
Hardening exponent	101.39

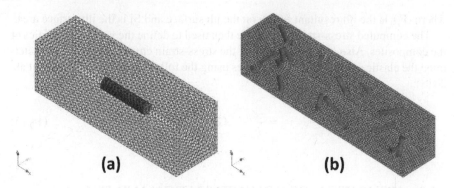

FIGURE 14.1   RVE of (a) aligned CNT and (b) random CNTs.

14.3.1   APPLICATION OF MORI–TANAKA TO POLYMER-CNT COMPOSITES

The simple tool which evaluates any effective property of the composites can be provided by mean-field (MF) micromechanics. Mori–Tanaka scheme has been applied to the composite, considering the constituents' linear elastic, homogeneous, and isotropic behavior. In this study, the first-order homogenization scheme has been employed. The multilevel method with incremental linearization, to relate the average strains in the CNT and matrix, has been selected. M^T scheme calculates the effective properties based on the fourth-order tensor. It produces the result either in the form of elastic moduli of the composite or stiffness or compliance matrix. The CNT and polymer's individual properties used for MF simulation are presented in Tables 14.2 and 14.3.

14.3.1.1   Effect of Volume Fraction and Orientation of CNTs on the Elastic Properties

Mori–Tanaka analysis of polymer-CNT composites has been discussed in this section. Effect of volume fraction (V_f) and aspect ratio (AR) of the aligned and random CNTs on the elastic moduli are the main focus of this section. Tables 14.4 and 14.5 represent the computational results of elastic moduli variation with the volume fraction of aligned and random CNTs of aspect ratio 20. It has been observed that axial Young's modulus has doubled as the volume fraction of CNTs increased from 1% to 4% in the case of aligned CNTs. Transverse Young's modulus has also increased, but a significant improvement was not noticed.

Similarly, a minor improvement of 4.8% and 6.2% in the in-plane and transverse shear modulus of aligned CNTs composites has been observed, respectively. In the random distribution of CNTs in the polymer matrix, Young's modulus, shear modulus, and bulk modulus have increased, but Poisson's ratio is decreased, represented in Table 14.5. The stress-strain curves of the aligned and random cases are shown in Figures 14.2 and 14.3, respectively. It is evident that as the volume fraction increases, the composite strength increases. A non-linear elastic range has been observed in both cases.

TABLE 14.4
Variation of elastic moduli with the volume fraction of aligned CNTs of aspect ratio 20

Volume Fraction of CNTs, V_f (%)	Axial Young's Modulus, E_x (GPa)	Transverse Young's Modulus, E_y (GPa)	In-plane Poisson's Ratio, ν_{yz}	Transverse Poisson's Ratio, ν_{xy}	In-plane Shear Modulus, G_{yz} (GPa)	Transverse Shear Modulus, G_{xy} (GPa)
1	4.949	3.352	0.3406	0.3004	1.250	1.256
2	6.722	3.456	0.3609	0.3008	1.270	1.281
3	8.528	3.542	0.3729	0.3012	1.290	1.307
4	10.365	3.619	0.3808	0.3016	1.310	1.334

TABLE 14.5
Variation of elastic moduli with the volume fraction of random CNTs of aspect ratio 20

Volume Fraction of CNTs, V_f (%)	Young's Modulus, E (GPa)	Poisson's Ratio, ν	Shear Modulus, G (GPa)	Bulk Modulus, K (GPa)
1	3.539	0.2962	1.365	2.895
2	3.884	02930	1.502	3.127
3	4.235	0.2901	1.641	3.363
4	4.591	0.2876	1.783	3.603

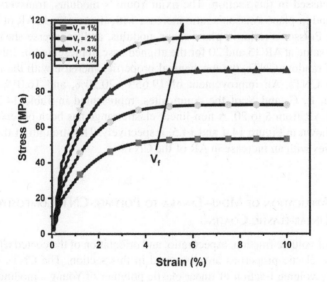

FIGURE 14.2 Stress-strain curve of aligned CNTs composites of different volume fractions (V_f 1% to 4%).

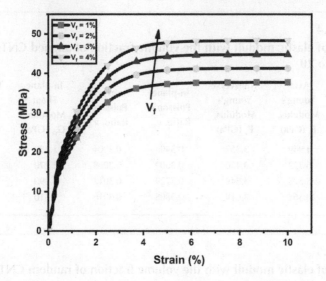

FIGURE 14.3 Stress-strain curve of random CNTs composites of different volume fractions (V$_f$ 1% to 4%).

14.3.1.2   Effect of Aspect Ratio and Orientation of CNTs on the Elastic Properties

The effect of aspect ratio on the aligned and random CNTs (V$_f$ 2.75%) in the polymer is discussed in this section. The axial Young's modulus, transverse Young's modulus, and in-plane Poisson's ratio increase with an increase in AR of the CNTs. Transverse Poisson's ratio, in-plane shear modulus, and transverse shear modulus remain the same at AR 15 and 20 for the aligned case, represented in Table 14.6. In the case of random CNTs, the mechanical properties increase with the increase in AR of the CNTs. An improvement of 19.65%, 20.30%, and 16.01% has been observed in E, G, and K of the composites, represented in Table 14.7, with an increase in AR from 5 to 20. A non-linear elastic range has been observed in both the case, shown in Figure 14.4 and 14.5, respectively. The strength of the composites increases with an increase in AR of the CNTs.

14.3.2   APPLICATION OF MORI–TANAKA TO POLYMER-CNT COMPOSITES WITH A LINEAR-ELASTIC COATING

The effect of volume fraction, aspect ratio, and orientation of the coated CNTs on the composites' elastic properties are discussed in this section. The CNTs are coated with a 0.1% volume fraction of linear-elastic polymer of Young's modulus 3.2 GPa and Poisson's ratio 0.3. The polymer and CNTs properties are the same as represented in Tables 14.2 and 14.3.

TABLE 14.6
Variation of elastic moduli with the aspect ratio of aligned CNTs of 2.75% volume fraction

Aspect Ratio	Axial Young's Modulus, E_x (GPa)	Transverse Young's Modulus, E_y (GPa)	In-plane Poisson's Ratio, ν_{yz}	Transverse Poisson's Ratio, ν_{xy}	In-plane Shear Modulus, G_{yz} (GPa)	Transverse Shear Modulus, G_{xy} (GPa)
5	3.979	3.388	0.3168	0.3006	1.287	1.304
10	5.116	3.443	0.3398	0.3010	1.285	1.302
15	6.523	3.489	0.3576	0.3011	1.284	1.301
20	8.074	3.522	0.3704	0.3011	1.284	1.301

TABLE 14.7
Variation of elastic moduli with the aspect ratio of random CNTs of 2.75% volume fraction

Aspect Ratio	Young's Modulus, E (GPa)	Poisson's Ratio, ν	Shear Modulus, G (GPa)	Bulk Modulus, K (GPa)
5	3.465	0.2973	1.335	2.848
10	3.652	0.2954	1.410	2.975
15	3.887	0.2931	1.503	3.131
20	4.146	0.2908	1.606	3.304

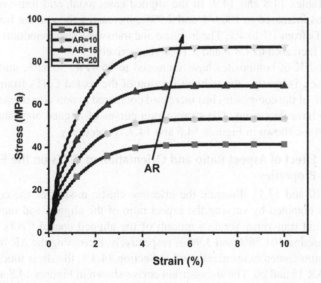

FIGURE 14.4 Stress-strain curve of aligned CNTs composites of different aspect ratios (AR 5 to 20).

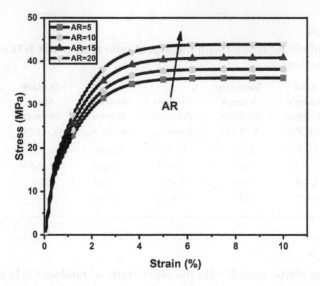

FIGURE 14.5 Stress-strain curve of random CNTs composites of different aspect ratios (AR 5 to 20).

14.3.2.1 Effect of Volume Fraction and Orientation of CNTs on the Elastic Properties

In comparison to the discussion in Section 14.3.1, it is observed that the moduli increase with the increase in the volume fraction of the coated CNTs. A remarkable increase in the moduli has been observed for both aligned and random cases represented in Tables 14.8 and 14.9. In the aligned case, axial, and transverse Young's modulus has increased to 108.9% and 7.99%, increasing the volume fraction of the coated CNTs from 1% to 4%. The in-plane and transverse shear modulus of the composites has increased to 4.8% and 6.14%, respectively.

E, G, and K of composites have increased to 26.41%, 30.62%, and 19.32% in random cases, increasing the volume fraction of the coated CNTs from 1% to 4%. The strength of the composites has increased compared to non-coated cases, whereas the moduli have decreased. The stress-strain curves of aligned and random coated CNTs cases are shown in Figures 14.6 and 14.7, respectively.

14.3.2.2 Effect of Aspect Ratio and Orientation of CNTs on the Elastic Properties

Tables 14.10 and 14.11 illustrate the effective elastic moduli of the coated CNTs composites obtained by varying the aspect ratio of the aligned and random CNTs. The axial and transverse Young's moduli of the aligned coated CNTs composites have increased to 101.86% and 3.90%, respectively, increasing the AR from 5 to 20. Similar to non-coated cases mentioned in Section 14.3.1, the shear moduli show no change at AR 15 and 20. The stress-strain curves shown in Figures 14.8 and 14.9, for the aligned and random cases, have shown an increase in the composites' strength with an increase in AR.

TABLE 14.8
Variation of elastic moduli with the volume fraction of aligned coated CNTs of AR 20

Volume Fraction of CNTs, V_f (%)	Axial Young's Modulus, E_x (GPa)	Transverse Young's Modulus, E_y (GPa)	In-plane Poisson's Ratio, ν_{yz}	Transverse Poisson's Ratio, ν_{xy}	In-plane Shear Modulus, G_{yz} (GPa)	Transverse Shear Modulus, G_{xy} (GPa)
1	4.931	3.350	0.3402	0.3007	1.250	1.255
2	6.691	3.454	0.3603	0.3014	1.269	1.280
3	8.481	3.539	0.3723	0.3021	1.289	1.306
4	10.301	3.615	0.3801	0.3028	1.310	1.332

TABLE 14.9
Variation of elastic moduli with the volume fraction of random coated CNTs of AR 20

Volume Fraction of CNTs, V_f (%)	Young's Modulus, E (GPa)	Poisson's Ratio, ν	Shear Modulus, G (GPa)	Bulk Modulus, K (GPa)
1	3.491	0.2946	1.365	2.837
2	3.797	0.2900	1.502	3.014
3	4.103	0.2861	1.642	3.197
4	4.413	0.2827	1.783	3.385

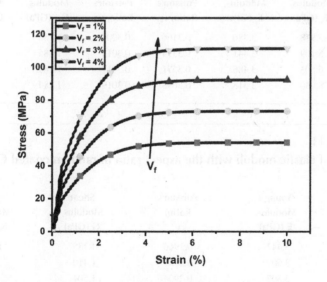

FIGURE 14.6 Stress-strain curve of aligned coated CNTs composites of different volume fractions (V_f 1% to 4%).

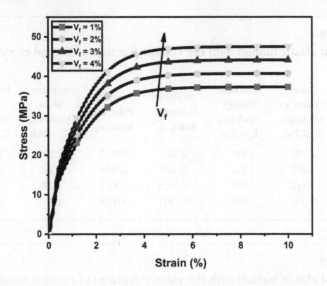

FIGURE 14.7 Stress-strain curve of random coated CNTs composites of different volume fractions (V_f 1% to 4%).

TABLE 14.10
Variation of elastic moduli with the aspect ratio of aligned coated CNTs of 2.75% V_f

Aspect Ratio	Axial Young's Modulus, E_x (GPa)	Transverse Young's Modulus, E_y (GPa)	In-plane Poisson's Ratio, ν_{yz}	Transverse Poisson's Ratio, ν_{xy}	In-plane Shear Modulus, G_{yz} (GPa)	Transverse Shear Modulus, G_{xy} (GPa)
5	3.978	3.386	0.3166	0.3007	1.286	1.303
10	5.109	3.441	0.3394	0.3013	1.285	1.300
15	6.503	3.486	0.3571	0.3017	1.284	1.299
20	8.030	3.518	0.3698	0.3019	1.284	1.299

TABLE 14.11
Variation of elastic moduli with the aspect ratio of random coated CNTs of 2.75% V_f

Aspect Ratio	Young's Modulus, E (GPa)	Poisson's Ratio, ν	Shear Modulus, G (GPa)	Bulk Modulus, K (GPa)
5	3.447	0.2966	1.335	2.825
10	3.607	0.2937	1.410	2.914
15	3.807	0.2904	1.504	3.026
20	4.026	0.2871	1.607	3.151

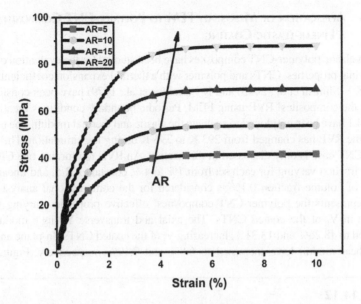

FIGURE 14.8 Stress-strain curve of aligned coated CNTs composites of different AR (5 to 20).

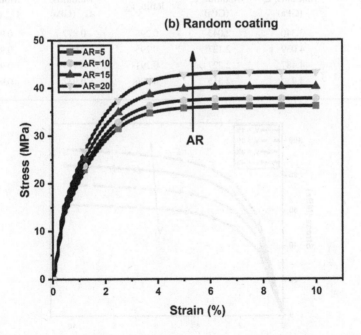

FIGURE 14.9 Stress-strain curve of random coated CNTs composites of different AR (5 to 20).

14.3.3 Application of Mesoscale FEM to Polymer-CNT Composites with a Linear-elastic Coating

Thermo-elastic polymer-CNT composites have been studied for the evaluation of elastic and thermal properties. CNTs and polymer with a thermal expansion coefficient of $-1.2 \times 10^{-5} K^{-1}$ (Shirasu et al., 2015) and K^{-1} (Kardos et al., 1979) have been considered to analyze the composites' RVE using FEM. Periodic boundary conditions mentioned in Table 14.1 have been employed to evaluate the elastic and thermal moduli. The temperature of the RVE has changed from 293 K to 393 K during the simulation. The aligned coated CNT case is only discussed in this section. An RVE reinforcing one CNT with volume fraction varying for each set from 1% to 4%, constant AR 5, and linear-elastic coating of volume fraction 0.1% is considered for the computational analysis. Table 14.12 represents the polymer-CNT composites' effective properties varying with an increase in V_f of the coated CNTs. The axial and transverse Young's modulus has increased to 48.26% and 13.31%, increasing V_f of the coated CNTs. In-plane and transverse shear moduli have increased to 6.8% and 7.19%, respectively. Figure 14.10

TABLE 14.12

Variation of elastic moduli with the aspect ratio of aligned coated CNTs of different V_f

Volume Fraction of CNTs, V_f (%)	Axial Young's Modulus, E_x (GPa)	Transverse Young's Modulus, E_y (GPa)	Transverse Poisson's Ratio, ν_{xy}	In-plane Shear Modulus, G_{yz} (GPa)	Transverse Shear Modulus, G_{xy} (GPa)
1	3.340	2.043	0.298	0.877	0.904
2	4.039	2.127	0.295	0.900	0.924
3	4.487	2.227	0.291	0.919	0.948
4	4.952	2.315	0.286	0.937	0.969

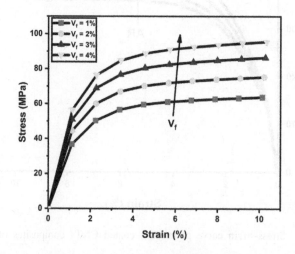

FIGURE 14.10   Stress-strain curve of aligned coated CNTs composites of different V_f (1% to 4%).

epresents the stress-strain curves at different V_f of the coated CNTs. The von-Mises contour plots are represented in Figure 14.11. It is evident from the figure that maximum stress has been at the ends of CNT, and the linea-elastic coating has also taken the stress.

Table 14.13 represents the thermo-elastic moduli of the coated CNTs composites obtained by RVE-FEM analysis. In comparison to isothermal composites (Table 14.12), the elastic moduli of these composites have decreased. The decrease in axial Young's modulus has reduced from 5% to 2% as the V_f of coated CNTs has increased from 1% to 4%. The axial and transverse coefficient of thermal expansion has also reduced as the V_f of the coated CNTs increases. Figure 14.12 represents the stress-strain curve of the composites along the longitudinal direction. The % elongation has increased compared to isothermal composites (Figure 14.10) as the composites' plastic range increased with an increase in temperature. The linear elastic range of the composites has reduced. The transversely isotropic nature of the composites has been obtained as represented in Table 14.13. The von-Mises contours of coated CNT thermo-elastic composites are shown in Figure 14.13.

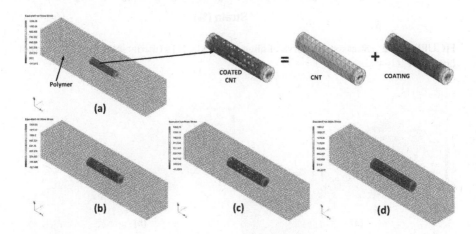

FIGURE 14.11   von-Mises contours of aligned coated CNT composites of different V_f, (a to d) – (1–4%).

TABLE 14.13

Variation of elastic moduli with the aspect ratio of aligned coated CNTs (thermo-elastic composites) of different V_f

CNT, V_f (%)	E_x (GPa)	E_y (GPa)	ν_{xy}	G_{yz} (GPa)	G_{xy} (GPa)	α_{xx} (10^{-5} K^{-1})	α_{yy} (10^{-5} K^{-1})
1	3.170	1.786	0.299	0.878	0.892	2.25	11.43
2	3.836	1.871	0.297	0.902	0.922	2.21	11.39
3	4.388	1.955	0.294	0.920	0.946	2.18	11.34
4	4.848	2.029	0.289	0.937	0.966	2.15	11.29

Note: α_{xx} and α_{yy} represent the longitudinal and transverse thermal expansion coefficient.

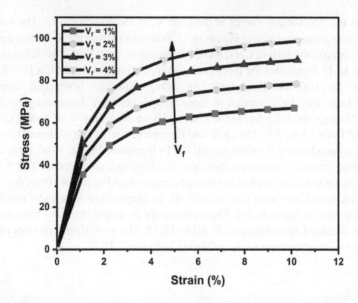

FIGURE 14.12   Stress-strain curve of aligned coated CNTs (thermo-elastic composites) of different V_f (1–4%).

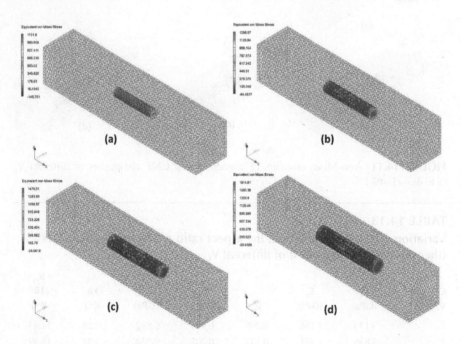

FIGURE 14.13   von-Mises contours of aligned coated CNT composites: thermo-elastic composites of different V_f, (a to d) – (1–4%).

14.4 SUMMARY

The chapter includes the Mori–Tanaka and FEM scheme's mathematical background for the elastic and thermo-elastic analysis of polymer-CNT composites. Aligned and random non-coated and coated CNTs composites have been analyzed for the practical evaluation of the elastic and thermal properties using the schemes. The effect of volume fraction, aspect ratio, orientation, and coating are the study's main variables. The schemes have revealed the transversely isotropic nature of the composites. The stress-strain curves for each case have proved that the composites' ultimate stress increases with an increase in the volume fraction or aspect ratio of the CNTs. Thermoelastic composites' analysis revealed that the moduli (elastic and thermal) have reduced under the temperature effect.

REFERENCES

Ansari, R., Rouhi, S., & Eghbalian, M. (2017). On the elastic properties of curved carbon nanotubes/polymer nanocomposites: A modified rule of mixture. *Journal of Reinforced Plastics and Composites*, *36*(14), 991–1008. doi: 10.1177/0731684417699075

Arash, B., Wang, Q., & Varadan, V. K. (2014). Mechanical properties of carbon nanotube/polymer composites. *Scientific Reports*, *4*(1), 1–8. doi: 10.1038/srep06479

Arora, G., & Pathak, H. (2019a). Multi-scale fracture analysis of fibre-reinforced composites. *Materials Today: Proceedings*, *18*, 687–695. doi: 10.1016/j.matpr.2019.06.469

Arora, G., & Pathak, H. (2019b). Numerical study on the thermal behavior of polymer nanocomposites. *Journal of Physics: Conference Series*, *1240*(1). doi: 10.1088/1742-6596/1240/1/012050

Arora, G., & Pathak, H. (2019c). Modeling of transversely isotropic properties of CNT-polymer composites using meso-scale FEM approach. *Composites Part B: Engineering*, *166*, 588–597. doi: 10.1016/J.COMPOSITESB.2019.02.061

Arora, G., & Pathak, H. (2020). Experimental and numerical approach to study mechanical and fracture properties of high-density polyethylene carbon nanotubes composite. *Materials Today Communications*, *22*(December 2019), 100829. doi: 10.1016/j.mtcomm.2019.100829

Arora, G., & Pathak, H. (2021). Nanoindentation characterization of polymer nanocomposites for elastic and viscoelastic properties: Experimental and mathematical approach. *Composites Part C: Open Access*, *4*(December 2020b), 100103. doi: 10.1016/j.jcomc.2020.100103

Arora, G., Pathak, H., & Zafar, S. (n.d.). *Fabrication and characterization of microwave cured high-density polyethylene/carbon nanotube and polypropylene/carbon nanotube composites*. doi: 10.1177/0021998318822705

Bhatnagar, A. S., Gupta, A., Arora, G., Padmanabhan, S., & Burela, R. G. (2021). Mean-field homogenization coupled low-velocity impact analysis of nano fibre reinforced composites. *Materials Today Communications, December 2020*, 102089. doi: 10.1016/j.mtcomm.2021.102089

Chwał, M., & Muc, A. (2015). Transversely isotropic properties of carbon nanotube/polymer composites. *Composites Part B: Engineering*, *88*. doi: 10.1016/j.compositesb.2015.11.009

Clifton, S., Thimmappa, B. H. S., Selvam, R., & Shivamurthy, B. (2020). Polymer nanocomposites for high-velocity impact applications – A review. *Composites Communications*, 17, pp. 72–86). Elsevier Ltd. doi: 10.1016/j.coco.2019.11.013

Gómez-Pachón, E. Y., Sánchez-Arévalo, F. M., Sabina, F. J., Maciel-Cerda, A., Campos, R. M., Batina, N., Morales-Reyes, I., & Vera-Graziano, R. (2013). Characterisation and modelling of the elastic properties of poly(lactic acid) nanofibre scaffolds. *Journal of Materials Science*, *48*(23), 8308–8319. doi: 10.1007/s10853-013-7644-7

Grabowski, K., Zbyrad, P., Uhl, T., Staszewski, W. J., & Packo, P. (2017). Multiscale electro-mechanical modeling of carbon nanotube composites. *Computational Materials Science*, *135*, 169–180. doi: 10.1016/j.commatsci.2017.04.019

Iijima, S. (1991). Helical microtubules of graphitic carbon. *Nature*, *354*(6348), 56–58. doi: 10.1038/354056a0

Kardos, J. L., Raisoni, J., Piccarolo, S., & Halpin, J. C. (1979). Prediction and measurement of the thermal expansion coefficient of crystalline polymers. *Polymer Engineering & Science*, *19*(14), 1000–1009. doi: 10.1002/pen.760191407

Kumar, P., & Srinivas, J. (2014). Numerical evaluation of effective elastic properties of CNT-reinforced polymers for interphase effects. *Computational Materials Science*, *88*, 139–144. doi: 10.1016/j.commatsci.2014.03.002

Lim, J. (2012). Introduction to aerospace materials. *Journal of the Korean Medical Association*, *55*(7). doi: 10.5124/jkma.2012.55.7.649

Lu, T.-C., & Tsai, J.-L. (2013). Characterizing load transfer efficiency in double-walled carbon nanotubes using multiscale finite element modeling. *Composites Part B: Engineering*, *44*(1), 394–402. doi: 10.1016/J.COMPOSITESB.2012.04.059

Maddock, N. A., James, N. L., McKenzie, D. R., & Patrick, J. F. (2012). Technological advances for polymers in active implantable medical devices. In J. Paulo Davim (Ed.), *The Design and Manufacture of Medical Devices*. Elsevier Masson SAS (pp. 239–272). doi: 10.1533/9781908818188.239

Mohammadi, S., & Yas, M. H. (2016). Modeling of elastic behavior of carbon nanotube-reinforced polymers by accounting the interfacial debonding. *Journal of Reinforced Plastics and Composites*, *35*(20), 1477–1489. doi: 10.1177/0731684416655608

Moore, A. L., & Shi, L. (2014). Emerging challenges and materials for thermal management of electronics. *Materials Today*, *17*(4), 163–174. doi: 10.1016/J.MATTOD.2014.04.003

Padmanabhan, S., Gupta, A., Arora, G., Pathak, H., Burela, R. G., & Bhatnagar, A. S. (2020). Meso–macro-scale computational analysis of boron nitride nanotube-reinforced aluminium and epoxy nanocomposites: A case study on crack propagation. *Proceedings of the Institution of Mechanical Engineers, Part L: Journal of Materials: Design and Applications*, *0*(0), 1–16. doi: 10.1177/1464420720961426

Pakseresht, M., Ansari, R., & Hassanzadeh-Aghdam, M. K. (2021). An efficient homogenization scheme for analyzing the elastic properties of hybrid nanocomposites filled with multiscale particles. *Journal of the Brazilian Society of Mechanical Sciences and Engineering*, *43*(1), 3. doi: 10.1007/s40430-020-02709-4

Pandey, J. K. (2015). Processing, performance and application. In Pandey, J. K., Takagi, H., Nakagaito, A. N., Kim, H. -J. (Eds.), *Handbook of Polymer Nanocomposites*. (pp. 13–40). Springer-Verlag Berlin Heidelberg. doi: 10.1007/978-3-642-45232-1

Pantano, A. (2018). Mechanical properties of CNT/polymer. In Roham Rafiee (Ed.), *Carbon Nanotube-Reinforced Polymers: From Nanoscale to Macroscale* (pp. 201–232). Elsevier Inc. doi: 10.1016/B978-0-323-48221-9.00009-1

Parandoush, P., & Lin, D. (2017). A review on additive manufacturing of polymer-fiber composites. *Composite Structures*, *182*, 36–53. doi: 10.1016/j.compstruct.2017.08.088

Pierard, O., Friebel, C., & Doghri, I. (2004). Mean-field homogenization of multi-phase thermo-elastic composites: A general framework and its validation. *Composites Science and Technology*, *64*(10–11), 1587–1603. doi: 10.1016/j.compscitech.2003.11.009

Roitman, D. B., Antoniadis, H., Hueschen, M., Moon, R., & Sheats, J. R. (1998). Polymer thermosetting organic light-emitting devices. *IEEE Journal on Selected Topics in Quantum Electronics*, *4*(1), 58–65. doi: 10.1109/2944.669467

Savvas, D., Stefanou, G., Papadopoulos, V., & Papadrakakis, M. (2016). Effect of waviness and orientation of carbon nanotubes on random apparent material properties and RVE size of CNT reinforced composites. *Composite Structures*, *152*, 870–882. doi: 10.1016/j.compstruct.2016.06.009

Shirasu, K., Yamamoto, G., Tamaki, I., & Ogasawara, T. (2015). Negative axial thermal expansion coef fi cient of carbon nanotubes: Experimental determination based on measurements of coef fi cient of thermal expansion for aligned carbon nanotube reinforced epoxy composites. *Carbon*, *95*, 904–909. doi: 10.1016/j.carbon.2015.09.026

Silvestre, J., Silvestre, N., & De Brito, J. (2016). Polymer nanocomposites for structural applications: Recent trends and new perspectives Polymer nanocomposites for structural applications: Recent trends and new perspectives. *Mechanics of Advanced Materials and Structures*, 23, 1263–1277. doi: 10.1080/15376494.2015.1068406

Srivastava, V. K., Gries, T., Veit, D., Quadflieg, T., Mohr, B., & Kolloch, M. (2017). Effect of nanomaterial on mode I and mode II interlaminar fracture toughness of woven carbon fabric reinforced polymer composites. *Engineering Fracture Mechanics*, *180*, 73–86. doi: 10.1016/j.engfracmech.2017.05.030

Sturm, S., Zhou, S., Mai, Y. W., & Li, Q. (2010). On stiffness of scaffolds for bone tissue engineering – A numerical study. *Journal of Biomechanics*, *43*(9), 1738–1744. doi: 10.1016/j.jbiomech.2010.02.020

Sun, Y., Hu, Y., & Liu, M. (2021). Elasto-plastic behavior of graphene reinforced nanocomposites with hard/soft interface effects. *Materials and Design*, *199*, 109421. doi: 10.1016/j.matdes.2020.109421

Weidt, D., & Figiel, Ł. (2014). Finite strain compressive behaviour of CNT/epoxy nanocomposites: 2D versus 3D RVE-based modelling. *Computational Materials Science*, *82*, 298–309. doi: 10.1016/j.commatsci.2013.10.001

Yadav, R., Tirumali, M., Wang, X., Naebe, M., & Kandasubramanian, B. (2020). Polymer composite for antistatic application in aerospace. In *Defence Technology* (Vol. 16, Issue 1, pp. 107–118). China Ordnance Society. doi: 10.1016/j.dt.2019.04.008

Zhai, S., Zhang, P., Xian, Y., Zeng, J., & Shi, B. (2018). Effective thermal conductivity of polymer composites: Theoretical models and simulation models. *International Journal of Heat and Mass Transfer*, *117*, 358–374. doi: 10.1016/j.ijheatmasstransfer.2017.09.067

Zhang, C., Curiel-Sosa, J. L., & Bui, T. Q. (2018). Meso-scale finite element analysis of mechanical behavior of 3d braided composites subjected to biaxial tension loadings. *Applied Composite Materials*, 1–19. doi: 10.1007/s10443-018-9686-0

15 Analysis of Magnetic Abrasive Flow Machining (MAFM) Process Parameters for Internal Finishing of Al/SiC/Al$_2$O$_3$/ REOs Composites Using Box–Behnken Design

Vipin Kumar Sharma
Meerut Institute of Engineering & Technology, India

Mayur Sharma
Daikin Air-Conditioning India Private Limited, India

Janardhan Gorti
National Institute of Technology, Rourkela, India

Vinod Kumar and Ravinder Singh Joshi
Thapar Institute of Engineering and Technology Patiala, India

Yogesh Kumar Singla
Case Western Reserve University, USA

CONTENTS

DOI: 10.1201/9781003202233-18

15.1   INTRODUCTION

In this era of modernization and globalization, the means of transportation – especially aviation – demands a material that will help passengers to reach their destination safely without any loss of life by applying a zero damage policy to the aircraft. Considering the aircraft safety, the need for reliable aluminum housings for aircraft actuators and O-rings arises. The flight control is totally dependent upon a sophisticated assembly called an actuator. The O-rings used in the aircraft require a seat on housing of high finish, otherwise the chance of leakage may occur. Therefore, surface quality of the finished engineering components is a very vital factor in improving the functional properties like corrosion resistance, hydrophobicity, wear resistance, and fatigue properties etc. (Nagdeve et al., 2018; Singh et al., 2018). A small burr or scratch may cause big loss to the industries such as energy loss in engine parts, fatigue failure of aerospace devices and malfunctioning of the components etc. So, in order to make the components free from any burr or scratch, industries are willing to spend increasing amounts of money. Over the past four decades, industries are using the traditional types of finishing processes like lapping, honing to get the required surface finish on the machine components. However due to their limitation to be used for particular geometries and materials, these methods cannot be used for the finishing of complex and intricate shapes to get the required degree of surface finish. In order to nullify the effect of traditional methods, advanced finishing processes have been developed, which are used to machine difficult parts internally as well as externally in all types of materials including oxides and non-oxide-based ceramics, MMCs, polymers, refractory materials by (Pandey and Pandey, 2019).

The life of product is very much affected by their surface textures that include its lay, roughness and waviness (Kenton, 2009). The thrust on various finishing processes increased continuously due to the rise in demand of miniaturize products with micro to nano level finish. In addition, the expansion of new kind of materials also leads to demand of advanced finishing processes. Out of the existing finishing processes, magnetic abrasive flow machining processes are gaining much importance due to the better surface finish capability with least damage to the surface. Another reason is getting attention by researchers toward magnetic abrasive-based finishing processes due to its ability of controlling the finishing forces during process of finishing (Jain, 2009). Magnetic abrasive finishing process is one of the advanced finishing processes which use the magnetic abrasive particles to perform the task of finishing. In this process magnetic particles magnetically join with each other in a way that it lies between two magnetic poles. The magnetic particles in the direction along magnetic lines of forces form a flexible magnetic abrasive brush (FMAB). Relative motion between FMAB and the work piece generates abrasion between them result in smooth surface formation on work piece. The magnetic abrasive finishing method (MAF) is used for the burrs removal on the surface of small parts made from ferromagnetic or nonmagnetic materials (Ko, 2007;

Yamaguchi et al., 2007; Kim and Kwak, 2010; Zou et al., 2010; Kala and Pandey, 2015;Barman and Das, 2017).

The present study is aimed to develop a magnetic abrasive finish technique to remove the scratch and burrs from the internal surface of the rare earth-based Al-6061 composite housing by optimizing process parameters using the Box–Behnken design approach. This approach aspires to produce parts within very close tolerance limits and ultimately helps in reducing the cost.

15.2 EXPERIMENTAL DETAILS

15.2.1 Preparation of Hybrid Composites and MAFM Setup

The aluminum housing made of Al-6061 alloy-based composites reinforced with SiC, Al_2O_3, and CeO_2 were prepared by means of stir casting route. These components were used for making samples for MAFM process. The detailed reinforcement characteristics like average size of particles, its purity and alloying elements present in Al-6061 alloy as suggested by the author previously (Sharma et al., 2020). A hydraulically powered experimental setup for MAFM process has been designed and fabricated. This setup has been designed keeping in view the fundamental mechanism of the process and basic functional requirements of different parts (Tzeng et al., 2007). In MAFM process, magnetic abrasive medium carrying abrasives is extruded through the work piece passage to be finished utilized two opposed cast iron cylinders under the presence of external magnetic field (Liu et al., 2013). The shearing of the magnetic abrasive medium near the workpiece surface contributes to material removal rate and surface finish (Lin et al., 2007). Extrusion of the magnetic medium through the passage formed in the work piece fixture is accomplished by driving two opposed pistons in medium containing cylinders using hydraulic actuators in desired manner using designed hydraulic circuit (Singh et al., 2006; Piexto, 1987). The workpiece fixture designed along the motion of the working fluid in aluminum housing is shown in the Figures 15.1(a–c). The developed MAFM setup with different components such as: double acting hydraulic cylinder (EN- 8 steel capable of withstanding up to 10 MPa pressure) with piston arrangements, workpiece holding tooling arrangement, circular fixture of aluminum with appropriate design, two coil type magnets for generating effect (0–2 Tesla), abrasive media and hydraulic fluid control unit comprises of extrusion pressure gauge, pressure variation system with hydraulic power pack (two horse power motor, with two directional control valves, and two pressure gauges with 3000 psi) and flange support structure is shown in Figure 15.2.

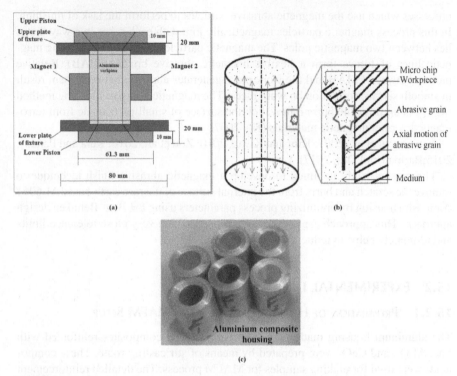

FIGURE 15.1 (a) Fixture drawing along aluminum workpiece, (b–c) medium reciprocation motion while internal finishing of aluminum housing.

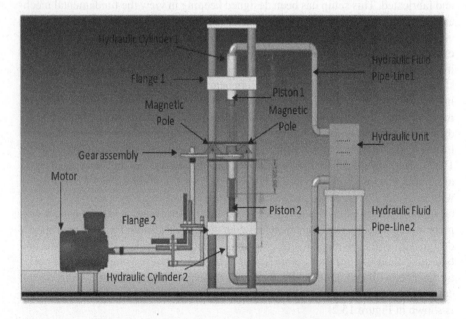

FIGURE 15.2 Magnetic abrasive finishing process setup along with the composite housing.

15.2.2 PLANNING FOR EXPERIMENTS

Two ways the MAF process setup (shown in Figure 15.2) is used for fine finishing of Al-6061 composite internally by utilizing the pressure of magnetic abrasive particles. The hollow parts of the composites are prepared using turning, drilling, and reaming. The specification of each prepared specimen is: 12 mm internal diameter, 19 mm external diameter, and 38 mm length. The talysurf surface tester is used for measuring surface roughness of each part by measuring the average roughness (R_a). The various input parameters used for MAF process are given in Table 15.1. Seventeen experiments were performed on the basis of Box–Behnken design taking three control factors at three levels with change in surface roughness values as shown in Table 15.2. Finally, in order to observe the better resolution of the finished

TABLE 15.1

Process parameters and their levels

				Levels		
S. No	Process Parameters	Notations	Units	1	2	3
1	Magnetic flux density (MFD)	A	Tesla	0.25	0.5	0.75
2	Number of cycles	B	Stroke	80	160	240
3	Extrusion pressure	C	MPa	1.4	2.8	4.2

MAP Grit Size: 200 microns
Working gap: 1.00 mm

TABLE 15.2

Experimental conditions for finishing Al-6061 hybrid composite specimen using the MAFM process

Run	Magnetic Flux Density (Tesla)	Number of Cycles (stroke)	Extrusion Pressure (MPa)	Initial Roughness(R_1)	Final Roughness (R_2)	ΔR_a (µm)	% improvement
1	0.5	160	2.8	1.38	0.5	0.88	63.7
2	0.5	160	2.8	1.34	0.45	0.89	66.4
3	0.5	240	4.2	1.39	0.45	0.94	67.6
4	0.75	240	2.8	1.43	0.45	0.98	68.53
5	0.5	80	4.2	1.35	0.43	0.92	68.1
6	0.5	240	1.4	1.20	0.36	0.84	70
7	0.5	160	2.8	1.21	0.34	0.87	71.9
8	0.25	240	2.8	2.11	1.37	0.74	35.07
9	0.75	160	4.2	1.51	0.48	1.03	68.2
10	0.75	80	2.8	1.41	0.5	0.91	64.5
11	0.5	80	1.4	1.13	0.38	0.75	66.37
12	0.5	160	2.8	1.20	0.35	0.85	70.8
13	0.25	160	1.4	1.26	0.59	0.67	53.17
14	0.25	160	4.2	1.35	0.61	0.74	54.81
15	0.25	80	2.8	1.24	0.56	0.68	54.84
16	0.75	160	1.4	1.21	0.37	0.84	69.4
17	0.5	160	2.8	1.19	0.32	0.87	73.1

surface of the composites, the samples were investigated via atomic force microscopy analysis.

15.3 EXPERIMENTAL RESULTS AND DISCUSSIONS

15.3.1 RESULTS OF MAFM

15.3.1.1 Analysis of Variance and Mathematical Model for Surface Roughness

The results and analysis for output response, i.e., surface roughness from ANOVA,
are discussed. The adequacy of the model for the output response (surface roughness) is depicted in Table 15.3. Quadratic model using backward elimination
approach for surface roughness is recommended by design expert 6.0.7® software.
The backward elimination method removes insignificant terms to adjust the fitted
quadratic models, and in the present work, the backward elimination process is used
to eliminate the insignificant terms.

TABLE 15.3
Adequacy checking of model of surface roughness

Sequential Model Sum of Squares

Source	Sum of Squares	Degrees of Freedom	Mean Square	F-Value	Prob. > F (p-value)	Remarks
Mean	12.20	1	12.20			
Linear	0.15	3	0.050	45.49	< 0.0001	
2FI	0.00485	3	0.00162	1.71	0.2286	
Quadratic	0.00852	3	0.00284	20.82	0.0007	Suggested
Cubic	0.000075	3	0.0000250	0.11	0.9476	
residual	0.000880	4	0.000220			
Total	12.36	17	0.73			

Lack of Fit Tests

Source	Sum of Squares	Degrees of Freedom	Mean Square	F-Value	Prob. > F (p-value)	Remarks
Linear	0.013	9	0.00149	6.79	0.0404	
2FI	0.00859	6	0.00143	6.51	0.0455	
Quadratic	0.0000750	3	0.0000250	0.11	0.9476	Suggested
Cubic	0.000	0				
Pure error	0.000880	4	0.000220			

Model Summary Statistics

Source	Standard Deviation	R-Squared	Adjusted R-Squared	Predicted R-Squared	Press	Remarks
Linear	0.033	0.9130	0.8930	0.8413	0.026	
2FI	0.031	0.9425	0.9080	0.8015	0.033	
Quadratic	0.012	0.9942	0.9868	0.9844	0.002575	Suggested
Cubic	0.015	0.9947	0.9786			

In addition to this normal plot of residuals and residual versus predicted plots has been shown in Figure 15.3(a–b). The data of surface roughness is normally distributed. Most of residuals are falling on the straight line, implying that errors are normally distributed. The results of surface roughness measurement after ANOVA test for composite samples finished by MAFM process are discussed in following sections. Adequacy of the model is checked using ANOVA at 95% confidence level as discussed in above section and presented in Table 15.4. It shows that F-value of the model is 210.69 and corresponding p-value is 0.0001. Thus, quadratic model is significant at 95% confidence level. The p-value of lack of fit is greater than 0.05. Thus, lack of fit is insignificant. Further, lack of fit value of 0.90 implies that it is not significant relative to pure error. Moreover, R^2 value is 0.98. This confirms that accuracy and general ability of polynomial model is good, and it suggests that the quadratic model can be used to navigate in the design space. The regression equations for change in surface roughness in coded and actual form as follows:

Final equation in terms of coded factors:

$$(\mathbf{R})_a = +0.87 + 0.12 \times A + 0.030 \times B + 0.066 \times C - 0.044 \times A^2 + 0.030$$
$$\times A \times C - 0.018 \times B \times C \qquad (15.1)$$

Final equation in terms of actual factors:

$$\Delta(\mathbf{R})_a = +0.31667 + 0.92944 \times \text{Magnetic Flux Density} + 8.12500 \times 10^{-4}$$
$$\times \text{Number of cycles} + 0.029464 \times \text{Extrusion Pressure} - 0.70444$$
$$\times (\text{Magnetic Flux Density})^2 (\text{Nagdeve et al.,2018}) + 0.085714$$
$$\times \text{Magnetic Flux Density} \times \text{Extrusion Pressure} - 1.56250 \times 10^{-4}$$
$$\times \text{Number of cycles} \times \text{Extrusion Pressure} \qquad (15.2)$$

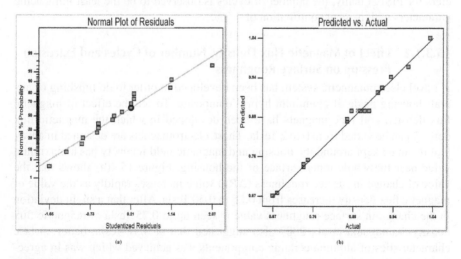

FIGURE 15.3 (a) Normal probability plots of residuals for surface roughness. (b) Plot of actual versus predicted of surface roughness.

TABLE 15.4
Results of ANOVA for effect of parameters (main and interaction) on surface roughness

Source	Sum of Squares	DOF	Mean Square	F- Value	Prob > F	Remarks	Contribution (%)
Model	0.16	6	0.027	210.69	<0.0001	Significant	
A	0.11	1	0.11	836.10	<0.0001	Significant	68.75
B	0.00720	1	0.00720	55.68	<0.0001	Significant	4.5
C	0.035	1	0.035	271.55	<0.0001	Significant	21.87
A^2	0.008210	1	0.00821	63.49	<0.0001	Significant	5.13
AC	0.00360	1	0.00360	27.84	0.0004	Significant	2.25
BC	0.00123	1	0.00123	9.47	0.0117	Significant	0.769
Residual	0.00129	10	0.00129				
Lack of Fit	0.000413	6	0.0000688	0.31	0.9009	Not significant	
Cor Total	0.16	16					

Standard Deviation 0.011 R^2 0.9922
Mean 0.85 Adjusted R^2 0.9874
Coefficient of Variation 1.34 Predicted R^2 0.9821
PRESS 0.00296 Adequate Precision 50.022
Legend: A – Magnetic Flux Density, B – Number of Cycles, C – Extrusion Pressure

Table 15.4 clearly shows the percentage contribution of process parameters on the percentage improvement in surface roughness (PISF) as observed from the ANOVA results. It is observed that the percent contribution of magnetic flux density with a value 68.75% suggested to be the most influencing factors on PISF. After that the extrusion pressure with percent contribution of 21.87% is second influencing parameters for PISF. Finally, the number of cycles is observed to be the least influencing parameters with percent contribution of 4.5%.

15.3.1.2   Effect of Magnetic Flux Density, Number of Cycles and Extrusion Pressure on Surface Roughness

A novel electromagnetic system has been developed in order to do finishing of aircraft housing made of aluminum hybrid composite. To see the effect of magnetic flux density, coil type magnets have been developed in which the magnetic flux density can be varied from 0 to 2 Tesla. These electromagnets are designed in a way that it can be kept around the housing and magnetic field intensity has a maximum value near the whole inner surface of the housing. Figure 15.4(a) shows that the value of change in surface roughness (ΔRa) value increases rapidly as the value of magnetic flux density increases from 0.25 to 0.50 Tesla. After that a slight deviation in the change in surface roughness value is seen up to 0.75 Tesla of magnetic flux density. Yamaguchi et al., 2007 also suggested that at 0.37 Tesla, better surface characteristics of alumina ceramic components was achieved which was in agreement with the present surface roughness results. The improvement in the surface roughness was also suggested at low value of magnetic flux density (0.50 Tesla)

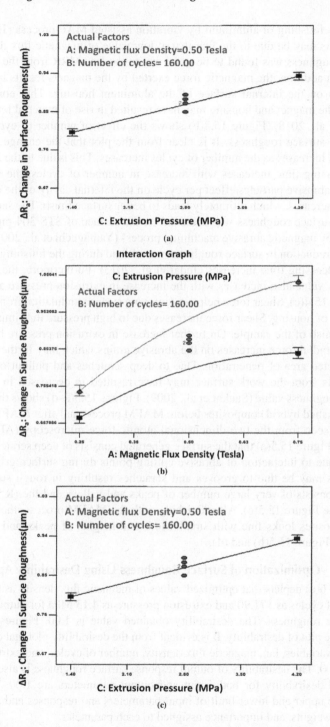

FIGURE 15.4 One factor plot to analyze the effect of machining parameters on the change in surface roughness.

during the finishing of aluminum by vibration assisted MAF process (Judal et al., 2013). This may be due to the fact that at low value of magnetic flux density, the surface roughness was found to be good due to the less MRR from the surface of sample. In addition, the magnetic force exerted by the magnets results in a normal force acts on the internal surface of the aluminum housing. This normal force between the magnet and housing may have resulted in rise of material removal rate (Zhang et al., 2018). Figure 15.4(b) shows the effect of number of cycles on the change in surface roughness. It is clear from the plot that the change in surface roughness increases as the number of cycles increases. This is due to the fact that as the processing time increases with increase in number of cycles, the number of magnetic abrasive particles effect per cycle on the internal cavity of the cylindrical housing increases, which ultimately leads to better surface finish. The same kind of result of surface roughness was achieved on the surface of STS 304 pipes during finishing by magnetic abrasive machining process (Yamaguchi et al., 2007). Similar results of reduction in surface roughness were found during the finishing of sample as the processing time increases (Judal et al., 2013). Furthermore, the change in roughness value also increases with the increase in extrusion pressure as depicted by Figure 15.4(c). Shear force helps in removing surface undulation from the internal cavity of housing. Shear force increases due to high pressure thus improving the surface finish of the sample. On further increase in extrusion pressure beyond 2.8 MPa, the radial force increases on the abrasive grains which may further increases the projected area of penetration. Due to deep scratches and pullout of the reinforcements from the work surface may have resulted in decreased in change in surface roughness value (Sankar et al., 2009). Figures 15.5(a–d) shows the topography of finished hybrid composites before MAFM process and after MAFM process. It can be seen from the two-dimensional atomic force microscopy (AFM) image shown in Figure 15.5(a) that the surface generated consists of deep scratches having grooves due to interaction of abrasive cutting points during surface grinding process. This may be due to grooves and scratches resulting in rough surface. The surface consists of very large number of peaks and valleys and the (R_a) values as seen in the Figure 15.5(c). After finishing with the MAFM process, the surface of the composites looks fine with small width and heights of peaks and valleys as shown in Figure 15.5(b) and (d).

15.3.1.3 Optimization of Surface Roughness Using Desirability Approach

Figure 15.6(a) depicts that optimized values of magnetic flux density as 0.75 Tesla, number of cycles as 171.90 and extrusion pressure as 4.13 MPa for output response as surface roughness. The desirability obtained value is 1.00. Figures 15.6(b–c) depicts the plot of desirability. It is evident from the desirability plot that desirability for input variables, i.e., magnetic flux density, number of cycles, and extrusion pressure is 1.00. The desirability of output response surface roughness is also 1.00. The combined desirability for both input and output parameters are 1.00. Table 15.5 shows the upper and lower limit of input parameters and responses and also shows the goal, weights, and importance assigned to each parameter.

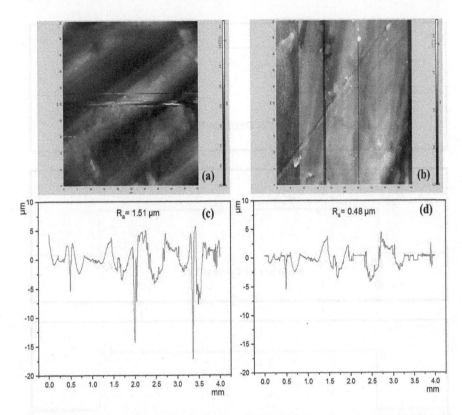

FIGURE 15.5 AFM image of the cross-section: (a) before MAFM; (b) after MAFM; (c) surface profile before MAFM; (d) surface profile after MAFM.

15.3.1.4 Confirmatory Experiments

Confirmation experiment was conducted to verify the improvement in performance characteristics. The confirmatory experiment was conducted to check the percentage of model prediction error between experimental result and predicted result as shown in Table 15.6. The error is lies within ±5% that confirms the excellent repeatability of the results. The error was calculated using the formula as shown in Equation 15.3. The error can be calculated by comparing the experimental results with the predicted results.

$$\%\textbf{Prediction error} = \frac{Experimental\ result - Predicted\ result}{Experimental\ result} \times 100 \qquad (15.3)$$

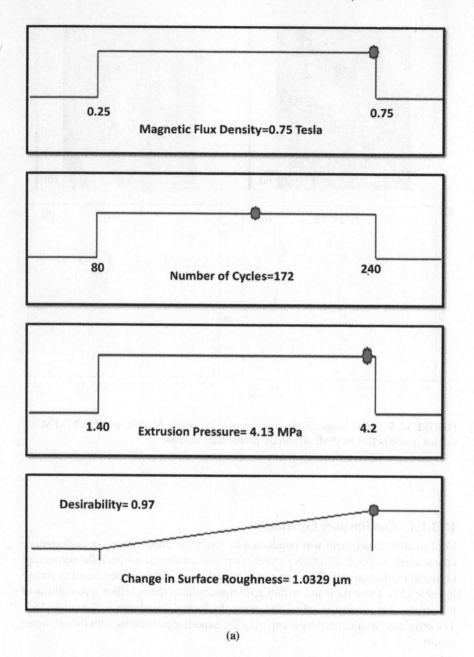

(a)

FIGURE 15.6 (a) Response optimization using desirability plot. *(Continued)*

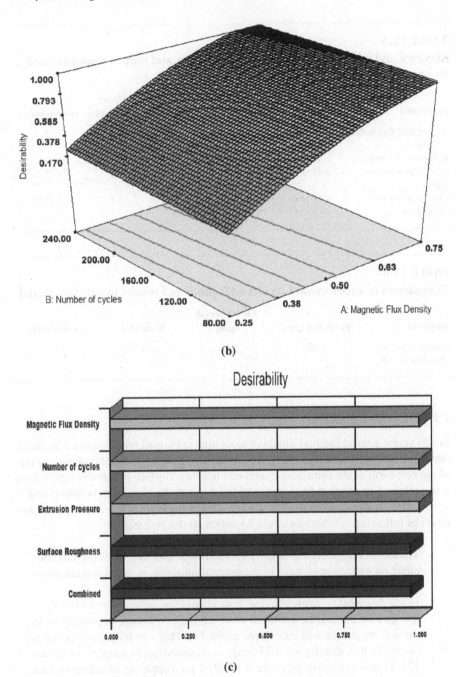

(b)

(c)

FIGURE 15.6 (CONTINUED) (b) 3-D Surface desirability plot (c) Desirability analysis for hybrid composites using bar graph.

TABLE 15.5
Maximal and minimal limits of input parameters and output responses and its importance

Parameters	Goal	Lower Limit	Upper Limit	Lower Weight	Upper Weight	Importance
A: Magnetic flux density (Tesla)	is in range	0.25	0.75	1	1	3
B: Number of cycles	is in range	80	240	1	1	3
C: Extrusion pressure (MPa)	is in range	1.4	4.2	1	1	3
Change in surface roughness	maximize	0.67	1.03	1	1	3

TABLE 15.6
Comparison of experimental results with predicted results to validate model

Response	Predicted (µm)	Experimental (µm)	Error (%)	Desirability
Change in Surface Roughness (ΔR_a)	1.08	1.032	4.65	0.97

15.4 CONCLUSIONS

In this study, a novel internal finishing technique using coil type magnet was developed and the experimental results have proved its capability to finish internal portion of the rare earth oxide aluminum composite housing. Surface roughness mechanism was explained in detail. After successful finishing of the aluminum housing, finally atomic force microscopy is used to see the 3D morphology of the finished composites. The following conclusions could be drawn on this research.

- The use of magnetic flux density and extrusion pressure has a predominant effect on surface roughness in the internal finishing of the aluminum housing as compared to the number of cycles.
- Using desirability approach, the best change in surface roughness R_a as low as 1.03 µm can be achieved at which the percentage improvement in surface roughness was calculated as 68.2%. This condition may occur at magnetic flux density of 0.75 Tesla corresponding to number of cycles 171.90 and extrusion pressure 4.13 MPa for preparing ultrafine surface finish.
- The addition of magnetic abrasive particles in the coil type magnetically driven finishing process can greatly increase the material removal rate and produce a relatively good surface finish on aluminum housing.

- A significant improvement of 69.4% in surface roughness is achieved when magnetic abrasive particles are used in the internal portion of housing with 1 mm gap between coil type magnet and workpiece at magnetic flux density of 0.75 Tesla, number of cycles as 171.90 and extrusion pressure as 4.13 MPa.
- Good agreements between the experimental and predicted values were obtained using developed model.

REFERENCES

Barman A, Das M. (2017). Design and fabrication of a novel polishing tool for finishing free form surfaces in magnetic field assisted finishing (MFAF) process. *Precis Eng.* 49: 61–68.

Jain VK. (2009). Magnetic field assisted abrasive based micro-nano-finishing. *J Mater Process Technol.* 209 (20):6022–6038.

Judal KB, Yadava V, Pathak D. (2013). Experimental investigation of vibration assisted cylindrical- magnetic abrasive finishing of aluminium work piece. *Mater Manuf Process.* 28:1196–1202.

Kala P, Pandey P. (2015).Comparison of finishing characteristics of two paramagnetic materials using double disc magnetic abrasive finishing. *J Manuf Process.* 17: 63–77.

Kenton T. (2009). The future of mechanical surface finishing. *Met Finish,* 107(5):22–24.

Kim TW, Kwak JS. (2010). A study on deburring of magnesium alloy plate by magnetic abrasive polishing. *Int J Precis Eng Manuf.* 11(2): 189–194.

Ko SL. (2007). Micro deburring for precision parts using magnetic abrasive finishing method. *J Mater Process Technol.* 187: 19–25.

Lin CT, Yang LD, Chow HM. (2007). Study of magnetic abrasive finishing in free-form surface operations using the Taguchi method. *Int J Adv Manuf Tech.* 34(1–2):122–130.

Liu ZQ, Chen Y, Li Y, Zhang X. (2013). Comprehensive performance evaluation of the magnetic abrasive particle. *Int J Adv Manuf Tech.* 68(1–4): 631–640.

Nagdeve L, Sidpara A, Jain VK, Ramkumar J. (2018). On the effect of relative size of magnetic particles and abrasive particles in MR fluid-based finishing process. *Mach Sci Technol.* 22(3):493–506.

Pandey K, Pandey PM. (2019). Surface roughness modeling in chemically etched polishing of Si (100) using double disk magnetic abrasive finishing *Mach Sci Technol.* 23(5):824–846.

Piexto JL. (1987).Hierarchical model selection in polynomial regression models. *The American Statistician.* 41(4):311–313.

Sankar MR, Ramkumar J, Jain VK. (2009). Experimental investigation and mechanism of Material Removal in Nano finishing of MMCs using Abrasive Flow Finishing (AFF) process. *Wear.* 266(7):688.

Sharma VK, Kumar V, Joshi RS, Sharma D. (2020). Experimental analysis and characterization of SiC and RE oxides reinforced Al-6063 alloy based hybrid composites. *Int J Adv Manuf Tech.* https://doi.org/10.1007/s00170-020-05228-7.

Singh DK, Jain VK, Raghuram V. (2006). Experimental investigations into forces acting during a magnetic abrasive finishing process. *Int J Adv Manuf Tech.* 30(7–8):652–662.

Singh S, Sankar MR, Jain V. (2018). Simulation and experimental investigations into abrasive flow nano-finishing of surgical stainless steel tubes. *Mach Sci Technol.* 22(3):454–475.

Tzeng HJ, Yan BH, Hsu RT, Lin YC. (2007). Self-modulating abrasive medium and its application to abrasive flow machining for finishing micro channel surfaces. *Int J Adv Manuf Tech.* 32: 1163–1169.

Yamaguchi H, Shinmura T. Ikeda R. (2007). Study of internal finishing of austenitic stainless steel capillary tubes by magnetic abrasive finishing. *J Manuf Sci Eng.* 129(5):885–892.

Zhang J, Hu J, Wang H, Kumar AS, Chaudhari A. (2018) A novel magnetically driven polishing technique for internal surface finishing. *Precis Eng.* 54:222–232.

Zou YH, Jiao AY, Aizawa T. (2010). Study on plane magnetic abrasive finishing process-experimental and theoretical analysis on polishing trajectory. *Adv Mat Res.* 2126: 1023–1028.

Index